The Foodservice and Purchasing Specification Manual

STUDENT EDITION

The Foodservice and Purchasing Specification Manual

STUDENT EDITION

LEWIS REED

John Wiley & Sons, Inc.

This book is printed on acid-free paper. ♾

Published by John Wiley & Sons, Inc., Hoboken, New Jersey
Published simultaneously in Canada

Library of Congress Cataloging-in-Publication Data:

Reed, Lewis.
Specs : the foodservice and purchasing specification manual / Lewis Reed. — Student ed.
p. cm.
Includes index.
ISBN-13: 978-0-471-69695-7 (cloth)
ISBN-10: 0-471-69695-1 (cloth)
1. Food service purchasing. 2. Food service management. I. Title.
TX911.3.P8R443 2006
647.95'068'7—dc22

2005005781

Printed in the United States of America

10 9 8 7 6 5 4 3 2 1

Contents

Introduction

Foodservice purchasing has become as highly developed and complex as the foodservice industry itself. Thus, instead of having the person who prepares the food do the ordering, the logical step was to have a purchasing agent do the ordering as a separate function. SPECS is intended to assist all parts of the growing hospitality industry in three areas: substance, practicality, and theoretical knowledge.

This book contains information on purchasing policies, foods, quality controls, and storage and handling procedures, for students, entrepreneurs, and corporate foodservice personnel. It includes a detailed chapter on developing a purchasing system, complete with sample forms for bids, ordering, receiving, and issuing.

Many food items are presented, with descriptions, detailed purchasing specifications, and quality, size, and packaging information in ten chapters covering food products. Quality controls and federal regulations are covered in detail.

This book also discusses topics such as decisions pertaining to cutting your own steaks or buying them fabricated, the use of convenience foods, and processed and prepared products.

SPECS has been revised to make it more practical for daily use. The book has been organized to provide information ranging from the very general to the specific. Readers can determine how much information they need at a particular moment, such as from the varieties of lemons or ordering specifications of size, grade, and pack. The information in all areas has been updated to reflect current practices.

The theory behind specifications is also discussed. Why do we need specifications? Who develops them? Who uses them? Most important, can you use SPECS to save time and money? The answer is yes. SPECS provides information on purchasing and how to incorporate this function into the total foodservice industry.

Internet Resources

www.hospitality-industry.com

Hospitality directory for professionals, which includes the related areas of associations, investment, beverage, lodging, consultants, design, restaurants and bars, sales and marketing, food, technology, travel, and human resources.

www.hospitalitynet.org

Current information in the news, including topics such as industry news, financial reports, market reports, job opportunities, hotel schools, industry links, and events.

Acknowledgments

This student edition of SPECS offers an overview of an important aspect of the hospitality industry—food specifications and purchasing. Success in any field is dependent on many factors, and the ability to write specifications and to know the product you are buying, storing, and serving to your customers is critical. I hope this book helps on your journey to success.

As Raymond Peddersen, the original author of this book, said in his acknowledgments, "No work of this scope can be done alone." I have developed great respect for the amount of work and detail that made the first edition an industry-wide reference. This made my job of updating and revising the material a realistic vision, rather than a tedious chore. I owe a great debt to him for beginning this immense project and doing it well.

Within the industry, I have been a waiter, a purchasing agent, a cook, a chef, a manager, and an owner. All these experiences have helped broaden my knowledge and appreciation of this industry. I owe much to all the chefs, managers, and owners who trained and supported me in this field.

I have also studied and taught hotel and restaurant management. My thanks to Santa Barbara City College, California Polytechnic State University at Pomona, University of Las Vegas, Nevada, and Washington State University for various stages of my education and the opportunity to teach. Certain instructors always influence students beyond educating them. To John Dunn, John Stephanelli, and Charles Levinson, I am greatly indebted.

In addition, I thank Andy Divine of the University of Denver and Marshall Shafkowitz of Culinard: The Culinary Institute of Virginia College, for reviewing this book.

My wife, Debbie, helped me to complete the first book and encouraged me to keep writing. Only a true friend would put up with my long hours working on this project and in front of the computer—Thanks. My sons, Christopher and Jeffrey, were extremely patient and understanding. Chris has started a fresh chapter in his life with his new wife, Linnea, and Jeff is a sophomore in college. My mom and dad have always supported me in everything, and this project has been no exception. To *all* my family, I owe many thanks.

Special thanks to Nigar Hale of John Wiley & Sons, who directed this version for instructional purposes. Thanks for your support and e-mails.

This book has been more pleasure than work for me, and I hope that it will help to make your job the same.

Lewis S. Reed Jr.

CHAPTER

1

Purchasing Policies

PURCHASING

Purchasing can be defined as the procurement of goods or services for an operation. However, the reality is that purchasing is not just a clerical function, but a vital and dynamic part of any business system. Cost control begins with purchasing; profits are developed on the basis of buying the right items, in the right quantity, at the right price, for the right purpose. Purchasing and the hospitality industry are tightly interrelated, as the quality and quantity of products purchased can be both the raw materials and the end products. Successful purchasing is but one key to quality control: It is not possible to produce high-quality meals from low-quality ingredients, but it is possible to produce low-quality meals from high-quality ingredients.

WHAT ARE SPECIFICATIONS?

Specifications can be defined as a detailed list of requirements or particulars in specific terms. Bolts and screws that are purchased for industrial use carry exacting specifications. There are definite measures that can be applied—percentages of different metals in the alloy, physical size, type and depth of threads, and so on. Equally definite measures or specifications have been applied to food. This book was developed to promote the use of exacting standards in foodservice purchasing. It will help the reader to differentiate between the poorly defined terms "high" and "low" quality, and to take advantage of all the updated knowledge available at this time on the measures of quality in foods.

Various agencies and departments of the federal government have established detailed specifications for virtually all types of raw products available in today's market. The accumulations of these specifications cover thousands of printed pages, and an individual product specification may average about eight pages. These standards and controls were established by the government to ensure disease-free and unadulterated products to the public and to protect consumers from hazardous practices of food processors. Foodservice operators cannot possibly know all the details about every product, but it is essential that they know the most important ones in order to do the best job in purchasing. Thus, the most pertinent details on a broad range of food and food products available in the United States are included here.

The function of this book is similar to that of a dictionary. SPECS is a ready reference to the standards by which you measure the foods you specify for purchase and inspect upon delivery. These specifications are the means to ensure that a foodservice operation is getting the value it is paying for.

Caveat emptor is the Latin phrase for "let the buyer beware." The foodservice buyer must beware of many things, and specifications help to narrow a product to a set of requirements that can be checked. Products should also be purchased for their intended use. The smart food buyer will buy food that is of no higher or lower quality than the needs call for. Purchasing U.S. No. 1 tomatoes (perfect color, firmness, etc.) for use in a sandwich or sauce means using a costlier product where U.S. No. 2 tomatoes (slight blemishes) would serve equally well and be cost-effective. Conversely, only a U.S. No. 1 tomato should be used where tomato wedges or slices are open to sight and are part of the merchandising, such as on a salad at a fine dining establishment.

Another example of improper buying is to specify USDA Grade A canned cling peaches (score of 90 points, this is used to compare points with a grade B product, or better) for broiling or service on salad plates. The quality level for this item should be USDA Grade B (score of 84 points or better), because the allowable defects in the B grade are few and not very visible, whereas the price difference is usually substantial between the two grades.

HOW TO USE SPECIFICATIONS

The purveyor and the operator must be able to communicate in the same language. It is the buyer's job to know what he or she is buying, and the purveyor's job to know what he or she is selling. The buyer who fails to specify exactly what is requested may get a product other than what was wanted and has cheated his employer. The purveyor who delivers a product inferior to what was ordered has cheated the buyer and the company.

The purveyors generally operate as middlemen, purchasing from a source and selling to the operator. In order to purchase products to sell, the purveyors must know specifications. It is best to avoid buying from purveyors who do not know federal specifications or who do not provide evidence of grade stamps or grade certificates.

It is not practical for the buyer to provide a purveyor with full oral or written specifications every time a quotation is taken or an order is placed. It is essential that both the buyer and the seller know they are contracting for the same product. Food purchasing can be done in person or by phone, fax, or computer, and it is important that the operator provide detailed specifications for the operation. There are three ways in which this can be accomplished:

Method 1: Provide purveyors with a detailed list of the food items that you may purchase from them.

Method 2: Develop your specifications for products and provide these to purveyors. The specifications in this book are presented in such a manner as to facilitate their use as models.

Method 3: Provide purveyors with a list of the products you use, and they will work with you to help develop specifications to best suit your needs with the products they carry.

THE PURCHASING FUNCTION: WHAT IS IT?

Purchasing food and supplies is somewhat unique in that it goes beyond specifying products and placing orders. In foodservice, the functions related to purchasing (forecasting, supplier selection, ordering, and receiving) are generally done by the same person. In most large-volume operations, a division of these responsibilities will occur, although most foodservice operations do not have a sufficiently large volume. This chapter discusses these related functions.

Forecasting

If you do not know how many customers you are going to serve, you cannot know how much to buy. Therefore, you must predict how much will be served based on past sales, reservations, and other relevant factors. Some foodservice operators have been making good predictions for a long time, which results in good controls, fewer headaches, and controllable operating expenses. Bad predictions lead to operators having to spend time figuring out what to serve when they run out, or how to reuse leftovers. Bad forecasting is bad business, which affects profit margins.

A spreadsheet that gives a history of food items served can be used as a record to indicate the percentage of sales on one item as compared with other menu items. This is the operation's menu mix. In almost every operation the consumption pattern of one item of a class (entrées, desserts, etc.), as opposed to any number of competing menu items of that class, will be fairly consistent. Moreover, in most operations there is a pattern as to the total number of customers served on a regular basis. In a restaurant this may be a function of the day of the week, or the season of the year, or both. In an institution, it may reflect many factors, but this information is vital in planning for proper purchasing.

By keeping careful sales records with customer count patterns, an operator is usually able to estimate closely (forecast) the number of customers expected on any given day. When this figure is multiplied by the usual consumption percentage of any given menu item, the result is an estimate of the number of portions needed.

Here is an example of this application: Assume a standardized recipe for 100 portions of roast beef that calls for 30 lb of IMPS No. 167-Knuckle. If the forecast is for 357 portions, then the buyer will know to purchase 3.57 × 30 lb, or 107 lb, of IMPS No. 167-Knuckle.

A more sophisticated system is to use these records to calculate a moving average. A moving average is simply the average number of portions served over a certain period of time. The simplest moving average to calculate is one based on ten serving times. To calculate:

- **a.** Add the number of portions served the last ten consecutive times the item has been served.
- **b.** Divide the total by 10.
- **c.** Next time the item is served, subtract from the total (a) the number of portions served 11 times ago, and add the number of portions served this time.
- **d.** Divide the new total (c) by 10.
- **e.** Repeat (c) and (d) each time the item is served.

The moving average may then be used as your forecast.

HOW TO WRITE A PURCHASING MANUAL

A purchasing manual is not part of a bureaucratic process; it is essential to better management of the purchasing function. The *Guide to Purchasing*, of the National Association of Purchasing Management, declares that no purchasing manager has an excuse for not having at least an informal statement of purchasing policy and a collection of purchasing procedures. A good manual can provide continuity and consistency, which are otherwise hard to maintain. It can help to train and integrate new employees quickly. If purchasing is being centralized for the first time, a manual serves to clarify the value and authority of the function and its relations to other departments.

Much has been written about constructing purchasing manuals. Most of the advice deals with massive and elaborate manuals requiring much time and effort. In fact, many instructions on writing purchasing manuals are longer than the manual itself needs to be. This section is a practical guide to the purposes, contents, and preparation of purchasing manuals. It emphasizes the "how" and then backs it up with some of the "why." The approach is based on the classic management formula: Plan—Do—Control.

Planning a purchasing manual means, essentially, deciding on its contents. These fall into two basic categories: policies and procedures.

Policies are broad statements of the purchasing department's main objectives and relationships. It provides information for administration, department heads, suppliers, and for purchasing itself. Procedures, in contrast, are rather detailed outlines of the methods and routines that purchasing uses to achieve its objectives. They almost entirely concern internal operations. This distinction is not always perfectly clear in manuals. Because there are two different audiences, there should be, ideally, two purchasing manuals, allowing for flexibility in preparation and circulation. The most realistic way to attack the problem is to consider policies and procedures as two separate parts of one manual.

The Policy Section

The policy section should contain essentially a mission statement from top administration, expressing support for the authority of the purchasing function, stating the general quality level required, and outlining as many specific policies as are appropriate.

Endorsement by administration lends an element crucial to the success of any program: management commitment. The endorsement is evidence that the policies in the manual are not being forced for purchasing convenience, but are meant to further the overall policies and objectives of the company. By adding the administration's authority, the endorsement makes observance of the policies and general acceptance of the purchasing role far more likely.

The objectives, scope, and responsibility of purchasing can be covered in many different ways. None can be totally understood without some reference to the other aspects of the purchasing role. Two other helpful, although not essential, items frequently appear in policy manuals. When used, they should be placed near the beginning.

One is an organization chart. It should show at least the reporting relationship of the purchasing manager to the administration and the basic organization. Larger segments of the total organization should be included only if they help to clarify the purchasing role.

The second popular option is a code of ethics. Ethical questions arise frequently in purchasing, and few manuals ignore them completely. Many deal with such questions as they relate to specific policies such as acceptance of gifts, conflicts of interest, and so on.

To determine the specific policies a manual should cover, keep in mind what a policy is supposed to do. It represents a solution or a decision involving a problem or question that is so common as to require action repeatedly. By framing a policy, you provide a generalized solution and avoid a long process every time the problem comes up. Therefore, identifying such problems is the first step in formulating policy.

Every purchasing function has its own special problems. However, some are so common that they should be addressed in every manual. Most of these involve basic questions related to three areas: authority, supplier relations, and ethics. There are also a few miscellaneous questions. Perhaps the best way to deal with these questions is to list some topics that ought to be resolved by specific policy statements, if the general statement of policy does not settle them. For example:

Which signature is required on a requisition?

What are the purchasing department's responsibilities and authority to challenge need or specifications?

Who may initiate or maintain contacts with suppliers or potential suppliers?

What are the criteria of supplier selection?

When is competitive bidding required?

Who takes part in negotiation?

Who may commit funds for supplies and equipment?

What limits, if any, are there to this authority?

What authority or responsibilities does purchasing share with other departments?

What is the policy on cooperative purchasing arrangements?

Are there limits on sales calls?

Who handles requests for samples or literature?

What are the limits on accepting gifts or entertainment from suppliers?

What rules govern potential conflicts of interest?

Will the purchasing department make personal purchases for employees and staff members?

On what committees or organizations, inside or outside, is purchasing represented?

This list is by no means complete, but it indicates the general types of problems that may be easiest to settle by policy.

It is reasonable, but not essential, to formulate policy before procedures. In some cases, however, a newly hired purchasing manager may find it necessary to straighten out systems and procedures first and be content for a while with just the sketchiest statement of policy. Treating the two areas as separate, although related, permits this flexibility.

The Procedures Section

The procedures section of a manual can also be prepared according to the Plan—Do—Control formula. Ideally, it should enable a competent employee to perform the routines necessary to keep the supply cycle going.

Some purchasing managers begin the procedures section of a manual with job descriptions. If systems and procedures are described accurately and completely, the roles of all purchasing employees should be clear enough to allow job descriptions to become backup documents.

One way to begin may be to record a calendar of the typical workweek. It should highlight recurring events that are assigned specific time slots: supplier calls, meetings of committees, supply distribution days, inventory report deadlines, and so on. Such a calendar is easier both to prepare and to grasp than a job description. It simplifies employment and ensures that no important task will be overlooked.

The heart of the procedures section is the detailed descriptions of systems and routines. A good way to plan this section is to think through the procurement cycle, jotting down each major action. Then a description is written, in sequence, of subevents. Many of these actions can be coordinated with forms or records; a copy of each should be included. Events in the procurement cycle, and the forms involved, may include:

Determination of need: Requisition or other request to purchase; reorder point on inventory records

Supplier selection: Supplier, product, or past buying records; invitation-to-bid form

Ordering: Purchase order; list or file of blanket orders in force; special forms for lease, rental, trial, or other nonpurchase procurement; inventory file; outstanding order file

Delivery: Receiving report; inventory records

In cataloging these events, be conscious of the questions who, what, when, where, and how. Play down the "why" aspects unless they emphasize or clarify something that may otherwise be overlooked.

Instructions for purchase orders should specify:

Who actually prepares them?

Who signs them? (Do some require more than one signature?)

What data are required, and where are they to be found?

When are they prepared—under what circumstances, at what time of day or week, and about how long before expected need?

Where—to what offices or files—do copies go?

Some purchases are exceptional. They may be onetime, small-value orders, paid from petty cash; rush orders requiring special attention; emergency orders placed outside normal office hours; or orders for capital items. Special procedures for handling these situations should be included.

The arrangement of this material can follow any of several patterns. Perhaps the best is first to cover fully the most standard procedure. Then, for exceptions, give instructions only to cover differences in procedure. For steps that are the same for all types of purchases, refer the user to the standard procedure by page and numbered paragraph. Here is a sample statement of purchasing objectives, scope, and responsibility that might be found in a purchasing manual.

THE ROLE OF PURCHASING

1. Objectives. The primary objective of the purchasing department is to ensure availability of supplies and equipment in quantity and quality consistent with set standards, at the most favorable prices consistent with those standards. The secondary objective is to ensure that the supply cycle is accomplished at minimum administrative cost.
2. Scope. The policies and procedures established by the purchasing department for procurement shall be standard.
3. Authority and limitations. Consistent with item 2, only purchasing has the authority to commit the organization to spending funds for supplies and equipment. In the case of high cost or capital items, this authority is subject to approval by the manager. Purchasing has the authority to question any and all aspects of a requisition, including need and specifications.
4. Responsibility. Purchasing shall maintain systems, procedures, and records adequate to meet its objectives. It shall evaluate these, and their overall performance, periodically, in the interests of improvement. It shall prepare such reports and other data as the administration directs.
5. Relationships with other departments. Purchasing shall cooperate fully in filling the procurement needs of all departments. To this end, it shall seek all information relevant to good procurement and share it as dictated by the best interests of the organization. It shall exercise its authority in those interests and shall not infringe on the authority of other departments.
6. Relations with suppliers. In dealing with suppliers, purchasing shall pursue the good of the organization vigorously, within the bounds of fairness, courtesy, and good commercial practice.

WHAT TO CHARGE?

Sales Analysis—Percentage Weighting

The job of setting menu prices is not the responsibility of the purchasing agent. It is his or her responsibility to inform management of fluctuating prices in commodities so that adjustment to menu pricing can be made. The price of lettuce in the wintertime can be double the summer season price. Salmon, halibut, and even beef, fluctuate with availability and demand. Evaluating sales by menu item is essential to determine not only what to charge, but also what should—and should not—be on the menu. One way to make a sales analysis is to tabulate physically the number of portions served for each item over a specific period of time. The period of time chosen should be no less than a week but need be no longer than a month. To illustrate, let us use a one-week period and a menu of 100 items.

Step 1. Tabulate the number of sales of each item for one week.

Step 2. Multiply the total of each item by its selling price (1,700 coffees at $1.50 each = $2,550).

Step 3. Add the totals for each of the 100 menu items from Step 2 together. This grand total should equal the total sales dollar for the week.

Step 4. Divide each of the totals in Step 2 by the grand total from Step 3. The quotient of this division is the percentage of the total sales dollar represented by each item. (Example: If coffee sales are $2,550 and grand total sales are $25,500, then coffee sales represent 10 percent of total sales.)

Step 5. Multiply the number of each item sold (from Step 1) by the item's food cost (1,700 coffees at $0.21 food cost = $357.00).

Step 6. Add the totals of the food costs from Step 5 together. This figure will be less than your total food cost (from inventory) for the week because of food items used that are not part of the menu item food cost (cream and sugar for coffee, condiments, etc.). Divide this total rough food cost by the actual total food cost (from inventory).

Example: Rough food cost—$728.75
Actual food cost—$765.00 = 95.3 percent

The difference of 4.7 percent (100.0 percent – 95.3 percent) should be added to the food cost of each menu item. This is done by multiplying each Step 5 figure by 1.047. This grand total should equal the actual food cost (coffee: $0.021 × 1.047 = $0.022).

Step 7. Divide the food cost of each item by its selling price. The result of this division is the food cost percentage of that item (coffee: $0.22/$1.50 = 0.1466 or 14.7 percent).

Step 8. Divide the total food cost from Step 6 by the grand total sales from Step 3. The result of this division is the average food cost percentage for all items.

Example: $765 (total food cost)/$2,550 = 0.30 or 30 percent.

DETERMINING PRICING

Break-Even Operations

In institutions where the objective is to break even on foodservice operating costs, but not make a profit, the selling price must absorb labor and other overhead expenses. If, in our example operation, total operating costs are 10 percent greater than current sales, then the operator has two choices:

(a) Raise the selling price of each item by 10 percent. This is often possible; for instance, by charging an increased price of $1.65 for a cup of coffee.

(b) Raise the selling price of only enough items to generate a 10 percent gain in total dollar sales. In the example operation, the total sales were $25,500 and the total operating costs $28,050. The 10 percent difference ($2,550) may be obtained by adding $0.05 or $0.10 to the price of items with above-average sales (Step 2, above) to result in additional revenue of $2,550. Which items are best for this?

It is likely that the items with the highest food cost percentage will be the entrées and that entrées will account for 30 to 50 percent of the total number of items sold. Because meat is the major component of most entrées, and because meat prices generally rise faster than those of staple items, it is advisable to raise entrée prices first and then, if necessary, resort to raising the prices of other items.

Another point in favor of raising entrée prices is that most customers order entrées, whereas fewer customers eat both entrées and several other items. If the price of a beverage, appetizer, or dessert item goes up, the customer may eliminate those but rarely will eliminate the entrée.

Profit-Making Operations

If the institution's objective is to make a profit, then the sales objective must include the total costs of operation plus the profit. If the objective is a 10 percent profit and the total sales are $25,500, then selling prices must be similarly altered to bring in total sales of $28,050. Presuming that purchasing procedures, specifications, and recipes are at the most economical level, you can use the data derived from Steps 1–8 (given in an earlier section) to lower the food cost. Isolate those items that represent an insignificant number of sales, or sales dollars, or a food cost percentage well above the average. These are the candidates to be taken off the menu. Replace them with more popular items or items having a lower-than-average food cost percentage.

Projecting Food Cost and Sales Prices for a New Operation

Step 1. Write the menu for the operation.

Step 2. Develop the standardized recipe card for each item.

Step 3. Obtain current market prices for all recipe items; cost out each recipe.

Step 4. Forecast the number of menu items that will be sold for a one-week period.

Step 5. Using the methods explained in Steps 5 and 6 in the preceding section "Sales Analysis," determine the total food cost dollars.

Step 6. Make a forecast of the total nonfood operating costs and add to this the total food cost dollars. This figure is the total operating cost.

Step 7. Use the procedures explained earlier in "Break-even Operations" and/or "Profit-making Operations" to establish the selling price of each menu item. Add another 10 percent to cover the cost of items not in the recipes (cream and sugar for coffee, condiments, etc.) and 1 to $1\frac{1}{2}$ percent for each month until opening day to cover raw food price increases.

(Straight percentage markup is one method of setting prices.)

Ethics—Morality—Money

Food purveyors operate on a small profit margin. They are selling merchandise that is perishable; if the stock is not sold while it is fresh, it will either deteriorate and have to be sold at a discount, or will spoil and have to be discarded. Such conditions tend to make food purveyors fiercely competitive. Food buyers are purchasing merchandise that can vary considerably in quality and therefore in price. The price difference between grades of a product is often as great as 20 percent.

The avenue to dishonesty in food purchasing and purveying is paved with temptation. It is not difficult for a purveyor to arrange kickbacks to food buyers in exchange for an operation's exclusive business. The food buyer who accepts such an arrangement has no choice but to accept whatever is shipped to the operation, without complaint. What is more serious is that he or she has committed a felony by accepting a bribe. There is no excuse for the purveyor who offers a bribe; such a person should be reported to his or her trade association and to the attorney general of the state. Likewise, there is no excuse for the food buyer who solicits a bribe; that individual should be reported to his or her employer and dismissed from the foodservice industry. This is a direct form of theft.

There are other, more subtle forms of theft. Consider these possibilities:

1. All the major food purveyors of any given type of product (such as fresh fruits and vegetables) may agree to set a floor, or minimum, price on certain products or on all products. Such price fixing is illegal, except for floor prices set by manufacturers of certain products in states having Fair Trade Practice laws.
2. The purveyors of a given type of product establish territories. Each company gets a section that the other companies agree to respect. Such arrangements may or may not be illegal, but they restrict the food buyer's ability to buy the best product at the best possible price.

Bribes can take forms other than money. For instance:

1. The premium offer. The buyer selects a gift from a sales incentive catalog in exchange for points earned. This is a perfectly honest marketing technique, and it is ethical to participate in such a program provided that the price or the price plus the value of the premium is the best available on the specified product. Moreover, the premium is to be owned and used exclusively by the operation that is paying for the food purchased. The problem with premium programs is that the catalogs usually contain 90 percent or more consumer products that seem to tempt the buyer to use his or her position to obtain premiums for personal use. The cost of these premium programs, of course, is reflected in the price of the food products purchased.
2. An offer allowing a buyer to purchase foods for personal use at discount prices. If all purveyors of a class extend this privilege, then there is nothing wrong with such an offer. If the buyer's ability to make such purchases is limited to one or two purveyors of a class, then the buyer may be tempted to "play favorites."
3. Christmas and/or birthday presents. In some areas the giving of gifts by purveyors to food buyers is a common practice. Such presents usually include liquor, perfume, clocks, and other inexpensive items. The acceptance of such gifts may not influence the buyer, but the cost is inevitably passed on. Acceptance of such gifts is bad policy.
4. The donation. Restaurants and institutions that ask purveyors to donate food or money to employee parties, annual picnics, favorite charities, or to the institutions themselves are, in essence, making such contributions a condition of doing business with them. The cost of these donations is inevitably passed on in purchase prices. A request for such a contribution is a form of asking for a kickback and should be discouraged.

PURCHASING SYSTEMS

The relationship between the buyer and the purveyor must be one of trust and confidence. There are shady practices to be avoided on both sides. Some purveyors deliver supplies that are inferior to those ordered. This will continue until the buyer stops the practice by rejecting shipments that do not meet the specifications. To be sure, purveyors who attempt this should be condemned for failing to keep faith with the buyer. However, the buyer should be condemned for failure to verify that the product received is the product that was ordered.

Specifications are of no use if they are applied only to the buying and no further. The person(s) responsible for receiving the food must be as thoroughly familiar with the buying specifications as the buyer must. The receiving person(s) must have reference materials readily available in the receiving area. These materials should include the *Meat Buyers Guide* (National Association of Meat Purveyors), this book, and any other references that contain clear and concise descriptions and pictures of foods.

Single-Source Buying

Single-source buying occurs frequently: The buyer simply orders what he or she needs from one purveyor and is billed at a percentage rate over cost. Most small operators cannot purchase from several sources and are limited to "one-stop shopping." Large corporations can negotiate lower costs with a large purveyor if they agree to buy all their goods from that purveyor.

Competitive Buying

Competitive buying means taking bids from two or more purveyors for any given item. The taking of bids may be done by telephone, mail, fax, computer, or in person. All these methods require copies of the specifications to be in the hands of the purveyor. See Figures 1.1, 1.2, and 1.3.

Article	Qty on hand	Qty needed	Quotation			Article	Qty on hand	Qty needed	Quotation			Article	Qty on hand	Qty needed	Quotation		
			#1	#2	#3				#1	#2	#3				#1	#2	#3
VEGETABLES						VEGETABLES (Cont.)						FRUIT (Cont.)					
Artichokes						Tomatoes						Strawberries					
Asparagus																	
Asparagus Tip												Tangerines					
Beans, Green						Turnips, White						Watermelon					
Beans. Lima						Turnips, Yellow											
Beans, Wax																	
Beets						Watercress											
Beet Tops																	
Broccoli																	
Brussels Sprouts																	
Cabbage Green																	
Cabbage Red						FRUIT											
Kale						Apples											
Pumpkin						Bananas											
Radish																	
Rhubarb						Cherries											
Romaine																	
Sage						Figs											
Sauerkraut																	
Shallots						Grapes, Red											
Sorrel						Grapes, Green											
Spinach																	
Squash						Lemons											
						Limes											
Tarragon																	
Thyme						Pears											

Figure 1.1 Inventory and Quote List for Produce

SELECTING SOURCES

Telephone, fax, computer, or mail bids may be compiled on a form. The most straightforward method is to assign the order for each product to the purveyor who has submitted the lowest bid on the item. This is possible when there are enough orders going to each purveyor. Note, however, that it costs the purveyor money to make a stop and deliver. The delivery expenses (drivers, trucks, fuel, etc.) divided by the purveyor's number of deliveries typically reveals a cost-per-stop of at least $50, and sometimes as much as $100. It is unfair to expect a purveyor to deliver orders of a dollar amount less than enough to make fair profit. So in the interest of good business relations, the buyer should attempt to consolidate the orders.

Attached please find quotation sheets which must be returned to:

P.D.Q. Cafes
Attn: Lewis Reed, Buyer
123 Suite Street
Los Angeles, CA 90667

no later than 12:00 noon on March 24, 20XX

SUBMITTED BY: ______________________

DATE: ____________ PRICES WILL REMAIN IN EFFECT UNTIL: ____________

This bid represents a four-week supply of staple supplies which will be scheduled for a single shipment between March 31 through April 4, 20XX. We have listed the estimated order quantity. Do not submit quotations on those items which you cannot supply in the estimated quantities on March 24, 20XX.

All orders will be considered complete upon delivery. In the event shortages do occur, we do request prompt notification from our vendors as to those items which will not be available on your previously scheduled delivery day. These items will then be reevaluated as to their availability from other sources and new orders will be placed in an effort to obtain these products as soon as possible.

Where specific quality specifications have not been listed the specifications will be assumed to be no less than a U.S.D.A. score of 90 points. Where brand names are specified they must be supplied. The specified pack is the required pack. Do not bid on those items which you carry in other than the specified pack.

Please direct any questions to the writer.

Product	Pack	USDA Grade Min. Score	Min. Net Min. Drained Wt.	Size or Count	Remarks
ASPARAGUS Spears	6/#5	Fancy 92/103 oz. net	50½ oz.	54/80	All Green, Mammoth.
BEANS, BAKED Vegetarian	6/#10	Fancy 85			White beans, in tomato sauce.
BEANS, GREEN N. West Blue Lake	6/#10	XStd. 82/101 oz. net	61 oz.	4 or 5 sieve	Pound. Variety.
BEETS Sliced	6/#10	Fancy 90/104 oz. net	68 oz.	2½″ Max. Diam.	
BEETS Whole	6/#10	Fancy 90/104 oz. net	68 oz.	74/124	No softness, peel or black spots.
CARROTS, WHOLE "BELGIAN"	6/#10	Fancy 90/105 oz. net	69 oz.	290/350	Orange-Yellow; no green.
CORN, WHOLE Kernel	6/#10	Fancy 90/106 oz. net	70 oz.		Golden. Brine pack.
MUSHROOMS Whole, Button	24/#8Z	Fancy 85/16 oz. net	8 oz.		Formosa; cream colored.
OKRA, CUT 85/99 oz. net	6/#10	XStd.	60 oz.		½″ to 1″ pods.
PEAS Sweet	6/#10	XStd. 82/105 oz. net	71 oz.	4 or 5 sieve	
PEPPERS Green Diced	6/#10				
PEPPERS Red Diced	6/#10				
POTATOES, DEHY. Pearls	6/#10				Without milk, vit. C added. Packed by Amer. Potato Comp. only. (Whip brand).
POTATOES, DEHY. Slices	6/#4				Packed by the Amer. Potato Comp. only.
POTATOES, INST.	6/#10				
POTATOES Sweet	6/#10	A 85	72 oz.		Golden type Brix 25°.

Figure 1.2 Invitaton to Bid

PAGE 1 **COMPANY:** ____________________

Pack	Item	Estimated Order Quantity	Computer Code Number	Your Bid
6/10	ASPARAGUS SPEARS FCY. 54/80 CT	0	33035	
6/10	BEAN GREEN STD. 4/5 SVE 61/ZD/WT	3	33095	
6/10	BEAN DARK RED KIDNEY/SAUCE FCY.	0	33185	
6/10	BEAN SPROUTS	3	33215	
6/10	BEETS SLD. FCY. 2.5 IN MAX.	7	33245	
6/10	CABBAGE RED	3	33299	
6/10	CORN CREAM STYLE FCY. GLON	0	33335	
6/10	MUSHROOM PIECES	2	33395	
6/10	PEPPERS GREEN DICED	4	33560	
6/10	PEPPERS RED DICED	0	33562	
6/10	PEPPERS GREEN HALVES	3	33565	
24/2$\frac{1}{2}$ lb	PIMENTOS BROKEN FCY.	2	33575	
6/10	SAUERKRAUT FCY. 2.5 PCT SALT MAX.	4	33725	
6/10	TOMATO PUREE FCY. 12 PCT SDS	9	33780	
6/10	TOMATO PASTE FCY. 33 PCT SDS	2	33780	
6/10	TOMATO WHOLE X STD 68/ZD/WT	0	33785	

Figure 1.3 Company Food Quote

Whatever the mode of taking bids, there must be a means of placing the order. Throughout the foodservice industry it is common practice to place orders by telephone. There is no argument that this is efficient, but what is wrong with this practice is that telephone orders give no written record (or confirmation) of orders against which food received can be checked for quantity and quality. When telephone orders are given, the purveyor should write, on the invoice, the specification that was bid. Some salespeople still call on operators and receive an order, although many foodservice operations are now placing their orders through the computer, with a direct link to their supplier. This method, or faxing your order in, will give you a written record of your purchases.

It is also essential that the invoice or a substantial copy of the invoice, called a shipping ticket, accompany the delivery. The invoice must be checked against the buyer's quotation sheet to verify quantity and price. The food received must be checked against the invoice, the quotation sheet, and the specification book to verify that the quantity billed is the quantity received and that the quality received is identical to that specified. Never accept a delivery without verifying weight, quality, and price.

A better method of placing orders is with a purchase order (Figure 1.4). The purchase order names the product, gives the specification or a reference to its number in the buyer's specification manual, and includes the price that has been agreed upon. A good purchase order has at least three parts: a copy for the purveyor, a receiving copy (Figure 1.5) to be used by the buyer in receiving the order, and a copy that is attached to the invoice (Figure 1.6) and used to pay the purveyor.

CHARGE ACCOUNT NO.

DATE OF ORDER	TO BE DELIVERED BY	TERMS

TO BE SHIPPED F.O.B.

IMPORTANT—ADDRESS ALL SHIPMENTS, INVOICES & COMMUNICATIONS TO:

P.D.Q. CAFES
LEWIS REED, BUYER
123 STATE STREET
LOS ANGELES, CA 90667

	QUANTITY	UNIT	ARTICLE		PRICE	PER	AMOUNT
	2	cs.	Pimentos broken fcy.	#33575	32.02	cs.	
	1	cs.	Waterchestnuts sl.	#33830	22.96	cs.	
	1	cs.	Apple rings spiced fcy.	#34045	13.20	cs.	
	4	cs.	Cherries dark sweet pitted	#34245	25.28	cs.	
	3	cs.	Fruit cocktail fcy.	#34455	14.59	cs.	
	12	cs.	Peach cling sliced heavy syrup	#34705	13.04	cs.	
	3	cs.	Pear sliced heavy syrup fcy.	#34786	15.72	cs.	
OUT	3	cs.	Pineapple chunks fcy.	#34815	15.39	cs.	*B/O*
	3	cs.	Pineapple slice mini fcy. hvy syr.	#34885	17.62	cs.	
	12	cs.	Pineapple juice unswt fcy.	#35550	6.48	cs.	
	1	cs.	Chives dehy.	#36065	17.37	cs.	
	6	cs.	Rice wild	#41880	23.39	cs.	
	5	cans	Chili powder	#43140	1.61	can	
	5	qts.	Extract Vanilla pure bourbon	#43400	7.14	qt.	
	2	cs.	Pure honey	#48165	26.65	cs.	
OUT	1	cs.	Strawberry jelly	#48252	46.24	cs.	**B/O**
	1	cs.	Cherry dark water pack	#51010	16.96	cs.	
	3	cs.	Cherry Royal Ann water pack	#51015	16.17	cs.	
	6	cs.	Pear Bartlett half water pack	#51050	17.93	cs.	
	2	cs.	Plums purple water pack	#51060	13.01	cs.	
	10	cs.	Asparagus spears salt free	#51100	21.68	cs.	
	3	cs.	Chick peas, Garbanzo beans		9.60	cs.	
	3	cs.	Kumquats		27.76	cs.	
	5	cs.	Oranges Japanese Mandarin		21.63	cs.	
	3	cs.	Instant chicken bouillon, Individual packets		16.49	cs.	
	3	cs.	Instant beef bouillon, Individual packets		16.49	cs.	
	3	cs.	Soup Minestrone		12.39	cs.	
	1	qt.	Pure lemon flavoring		3.98	qt.	
	2	cs.	Mixed nuts		27.56	cs.	
	1	cs.	Strawberry pie filling		23.83	cs.	
	1	cs.	Apple pie filling		18.28	cs.	
	1	cs.	Cherry pie filling		22.75	cs.	
	1	cs.	Lemon pie filling		17.76	cs.	
	1	cs.	Strawberry glaze		14.86	cs.	

Figure 1.4 Purchase Order

CHARGES COLLECT PREPAID #	CARRIER	CONDITION GOOD	DAMAGED	SHORT	NO. PKGS	ORDER: PARTIAL ❑ COMPLETE ❑	DATE
DATE OF ORDER		TO BE DELIVERED BY					TERMS
RECEIVED FROM:							

REC'D	QUANTITY	UNIT	ARTICLE		PRICE	PER	AMOUNT
2	2	cs.	Pimentos broken fcy.	#33575	32.02	cs.	
1	1	cs.	Waterchestnuts sl.	#33830	22.96	cs.	
1	1	cs.	Apple rings spiced fcy.	#34045	13.20	cs.	
4	4	cs.	Cherries dark sweet pitted	#34245	25.28	cs.	
3	3	cs.	Fruit cocktail fcy.	#34455	14.59	cs.	
12	12	cs.	Peach cling sliced heavy syrup	#34705	13.04	cs.	
3	3	cs.	Pear sliced heavy syrup fcy.	#34786	15.72	cs.	
OUT	3	cs.	Pineapple chunks fcy.	#34815	15.39	cs.	
3	3	cs.	Pineapple slice mini fcy. hvy syr.	#34885	17.62	cs.	
12	12	cs.	Pineapple juice unswt fcy.	#35550	6.48	cs.	
1	1	cs.	Chives dehy.	#36065	17.37	cs.	
6	6	cs.	Rice wild	#41880	23.39	cs.	
5	5	cans	Chili powder	#43140	1.61	can	
5	5	qts.	Extract Vanilla pure bourbon	#43400	7.14	qt.	
2	2	cs.	Pure honey	#48165	26.65	cs.	
OUT	1	cs.	Strawberry jelly	#48252	46.24	cs.	
1	1	cs.	Cherry dark water pack	#51010	16.96	cs.	
3	3	cs.	Cherry Royal Ann water pack	#51015	16.17	cs.	
6	6	cs.	Pear Bartlett half water pack	#51050	17.93	cs.	
2	2	cs.	Plums purple water pack	#51060	13.01	cs.	
7	10	cs.	Asparagus spears salt free	#51100	21.68	cs.	
3	3	cs.	Chick peas, Garbanzo beans		9.60	cs.	
3	3	cs.	Kumquats		27.76	cs.	
5	5	cs.	Oranges Japanese Mandarin		21.63	cs.	
3	3	cs.	Instant chicken bouillon, Individual packets		16.49	cs.	
3	3	cs.	Instant beef bouillon, Individual packets		16.49	cs.	
3	3	cs.	Soup Minestrone		12.39	cs.	
1	1	qt.	Pure lemon flavoring		3.98	qt.	
2	2	cs.	Mixed nuts		27.56	cs.	
1	1	cs.	Strawberry pie filling		23.83	cs .	
1	1	cs.	Apple pie filling		18.28	c s.	
1	1	cs.	Cherry pie filling		22.75	cs.	
1	1	cs.	Lemon pie filling		17.76	cs.	
1	1	cs.	Strawberry glaze		14.86	cs.	

DELIVER TO	NOTIFY	REFERENCE	TOTAL

RECEIVING CLERK	DELIVERED TO	RECEIVED	DATE
Laura	***Dietary***		

Figure 1.5 Receiving Report

INVOICE DATE & NUMBER			PAGE		
03 12		24748	1		SP
1972	11204	40	3		SB
ROUTE	ACCOUNT	LOAD	STOP	. SEG.	

SHIP TO

SOLD TO

SPECIAL INSTRUCTIONS:
PO 58372 WED AM

DUP INV. DOCK OFF HARVEY AVE

CALL ON OUTS

AISLE	SLOT	QUAN	CNTR.	DESCRIPTION	PACK	SIZE	PRICE	EXTENSION	PROD. CODE	
A	06-10	3	CS SB	FRUIT COCKTAIL	6	10	14.59	43.77	02931	141
A	82-10	12	CS SB	PEACH AMB Y C SLICED	6	10	13.04	156.48	04119	564
B	27-10	1	CS SR	APPLE RINGS SPICE RED	6	10	13.20	13.20	02089	50
B	55-10	5	CS SR	ORANGES MANDARIN	6	10	21.63	108.15	36822	240
B	83-10	12	CS SR	PINEAPPLE JUICE	12	46 oz.	6.48	77.76	18614	528
C	64-10	3	CS SB	PINEAPPLE 115/120 MAY	6	10	17.62	52.86	06445	144
C	STOCK OUT		CS SR	PINEAPPLE CHUNKS XHS	6	10	15.39	STOCKOUT	3	
D	08-20	2	CS SG	PIMENTOS BROKEN IMP	24	28 oz.	32.02	64.04	37770	100
D	31-20	1	CS SR	STRAWBERRY PIE FILLING	6	10	23.83	23.83	07641	46
D	34-10	3	CS SB	PEARS BART SLICED	6	10	15.72	47.16	05595	141
D	38-10	1	CS SR	LEMON PIE FILLING RTU	6	10	17.76	17.76	03616	50
D	40-20	1	CS SR	APPLE PIE FILLING RTU	6	10	18.28	18.28	00083	49
E	07-10	4	CS SB	CHERRIES BING PITTED	6	10	25.28	101.12	01545	188
E	07-20	1	CS SR	CHERRY PIE FILLING	6	10	22.75	22.75	01743	56
E	16-10	3	CS SR	KUMQUATS 70/75	12	5	27.76	83.28	03590	138
E	40-10	3	CS SR	BEANS GARBANZO	6	10	9.60	28.80	08896	141
F	01-20	2	CS SR	PLUMS PRUNE DIETETIC	6	10	13.01	26.02	16287	90
F	28-10	3	CS SR	BEEF INST BOUIL PKT	6	100	16.49	49.47	33779	27
F	28-20	3	CS SR	CHICKEN INST BOUL PKT	6	100	16.49	49.47	33797	27
F	40-20	6	CS SR	PEARS BART DIET 30/35	6	10	17.93	107.58	16345	288
F	67-10	7	CS SR	ASPARAGUS DIET M/L GN	24	300	21.68	151.76	17020	217
	PART OUT		CS SR	ASPARAGUS DIET M/L GN	24	300	21.68	PART OUT	3P	
F	71-20	1	CS SR	CHERRIES BING DIETIC	24	303	16.96	16.96	16139	30
F	77-20	3	CS SR	CHERRIES R ANN DIET	24	303	16.17	48.51	16154	90
H	07-20	1	CS SR	WATER CHESTNUTS SLICD	24	30 oz.	22.96	22.96	44917	55
H	17-10	2	CS SR	NUTS SALTED MIXED	12	1 lb.	27.56	55.12	72702	34
H	18-20	6	CS PL	RICE WILD MIX U BEN	6	36 oz.	23.39	140.34	73445	96
J	30-10	1	CS SR	GLAZE STRAWBERRY	6	10	14.86	14.86	87734	45
L	11-10	1	CS SR	CHIVES FREEZE DRIED	12	1 oz.	17.37	17.37	31211	5
L	34-20	3	CS SR	SOUP MINESTRONE	12	5	12.37	37.11	91488	135
	STOCK OUT		CS SR	JELLY STRAWBERRY	6	10	46.24	STOCK OUT	1	
M	08-40X	5	EA SR	CHILI POWDER	1	16 oz.	1.61	8.05	29041	5
M	10-10	5	EA SR	EXTRACT VANILLA PURE	1	QT.	7.14	35.70	82347	15
M	14-10	1	EA SR	EXTRACT LEMON PURE	1	QT.	3.98	3.98	82313	3
			REPACK	CASE						
	INVOICE SUB-TOTALS									
PCS	WGT		AMT.							
107	3814			$697.80 PROCESSED FOODS						

TOTAL PIECES		TOTAL WEIGHT	PLEASE PAY ➡ THIS AMOUNT		TERMS
103		3814		697.80	NET

DUPLICATE INVOICE

NOTE: ALL SHORTAGES AND DAMAGES MUST BE NOTED AT TIME OF RECEIPT

Figure 1.6 Company Invoice

SYSTEMS CONTRACTING

It is the buyer's duty to obtain the specified product at the best possible price from purveyors who can be relied upon. Food passes through many hands before it reaches the food-service operator. The different owners of the product, at one time or another, may include the farmer, the preprocessor (slaughterhouse), the processor (butcher, canner), the regional or national distributor, and the local distributor.

Each time the food changes hands, the new handler (owner) must make a profit. The farther back on this line the food buyer can purchase, the fewer add-on costs will be reflected in the price. Buying "closer to the source" usually means having to buy very large quantities of a single item, either immediately or over a predetermined period of time. In many areas there are individual operators who have formed buying organizations (cooperative buying) to purchase for the group. Frequently, individual buyers can buy in quantities larger than they can stock, at guaranteed prices, by soliciting bids from several local purveyors. They can also make a purchase from a national distributor and then make financial arrangements with a local distributor to warehouse and ship the product on a routine as-needed basis.

Many processing companies maintain warehouses and also sell to local distributors. It often takes no more than a phone call to eliminate one layer of handling costs. The difference can be 10–20 percent in buying price.

Such items as milk, ice cream, and bread are not easily bought on day-to-day or week-to-week bids. In selecting purveyors for these daily-delivery products, the buyer should solicit fixed-price bids for a period of six months or a year and award "sole-purveyor" contracts. Such contracts should be in writing and should specify any procedure for raising or lowering the contracted prices.

Other dairy products, such as cheese, butter, and eggs, can also be contracted for longer periods than week-to-week bids. Most major marketing areas have a publication that gives the wholesale prices being quoted on these items. A contract for cheese, butter, eggs, or even poultry may be bid for and taken on a basis of a percentage above the prices quoted on the sheet. Such a contract may specify that the eggs purchased would be priced $0.06 per dozen higher than the highest price quoted for the particular grade and size on the market sheet. The contract may also allow for cheaper prices for larger quantity purchases, such as $0.06 over market price when ten cases or fewer are purchased, dropping to $0.05 over market price for 11 cases or more, and so forth. Contracts like this simplify an operator's business because they guarantee the source of supply, fix the buy-price to the wholesale market, and limit the number of purveyors. For the purveyor, such a contract guarantees that he or she will receive a price that includes that purveyor's purchase and handling costs and a fair profit. It is also possible to write a systems contract with meat purveyors based on such reports as this.

There are several market reports that wholesale purveyors use as supply and pricing guidelines. Some, such as that of the U.S. Department of Agriculture (USDA), are free; others are available only by subscription. The volume food buyer should learn which publications local purveyors use as market guides and obtain subscriptions to them.

ONE-STOP SHOPPING

One-stop shopping means purchasing all food and supply needs from one purveyor. The advantages of such an arrangement are several:

1. One order to place
2. One delivery to receive
3. One bill to pay

One-stop shopping can reduce the purchasing, receiving, and accounting functions enough to generate substantial savings. However, buying from a one-stop purveyor has some disadvantages:

1. No other sources of supply if the one-stop purveyor has a disaster or labor difficulties
2. No alternative source should goods delivered be other than specified
3. No price competition

Nevetheless, many operators have found that the financial advantages outweigh the disadvantages.

One-stop food shopping may be the accepted and prevailing mode of purchasing. One-stop companies are likely to develop from the merger of several small purveyors. This trend can be seen in the merger of fresh produce with frozen produce houses, which may pick up the distribution of frozen entrées, baked goods, and meat lines and then merge with a general groceries and canned goods purveyor. In almost any city of more than 250,000 people, it is possible to purchase 75 percent or more of your needs from any of several large general variety purveyors. Currently, large companies are realizing the potential of one-stop shopping and developing competitive one-stop shopping services in cities across the country.

COOPERATIVE BUYING

A cooperative has been defined as an association of persons who have voluntarily joined together to achieve a common economic end through the formation of a democratically controlled business organization. The members make equitable contributions to the capital required and accept a fair share of the risks and benefits of the undertaking.

Is buying through a central agency a practical possibility? A central agency is a source through (or from) which many or most of the items you purchase may be obtained, be they perishable or durable.

Research

Setting up a central agency entails a great deal of research and organization to achieve the most suitable type of agency through which member institutions can purchase their requirements. The appointment of a subcommittee to investigate all aspects of such an agency appears to be the logical approach. It must be stressed that the most important ingredient necessary to achieve the most efficient and profitable results is loyalty.

If the agency is to function in the best interests of all, then every constituent member must, at all times, give the highest degree of loyalty to the aims and objects of the agency. Otherwise, its future can be jeopardized.

The agency should be set up as a cooperative. The desire to see a central agency brought into existence is only the first step. Beyond that, there are fundamental requirements to achieve this end.

The first and most important requirement is finance. A cooperative trading agency requires substantial initial capital. Consider, for instance:

Premises have to be acquired, whether purchased, leased, or rented.
Warehousing, showrooms, and office areas have to be created.
Goods have to be purchased for resale, and these goods have to be paid for.
Wages have to be paid.
Vehicles have to be purchased or leased, with subsequent delivery costs to be incurred.
And, of course, there are the ongoing operating expenses to be met.

All these items require initial capital. Therefore, the first consideration is the financial structure. In the cooperative system, all those interested become shareholders and subscribe for an agreed number of shares. If 30 members subscribe for one $5,000 share each, then the actual initial capital will be $150,000 to get the project off the ground.

A preliminary formation meeting would then be held. Here the name of the agency can be decided upon, rules adopted, directors elected, and the various other officers appointed. With the agency subject to trading only with its own members, the net profits of such an agency would be subject to federal income tax only on the amount remaining after deducting the rebates on purchases. The directors decide the basis of markup on the goods purchased for resale. Obviously, the difference between the purchase price and the selling price must be sufficient to pay wages and all operating expenses and still have something left over, if the agency is to function. If the margin between buying and selling is steady, the agency should prosper. As sales rise, the proportion for overhead should grow less, which brings us back to the question of loyalty. For the agency to be effective, its members must ensure that as large a share as possible of their requirements is purchased through their own agency.

TECHNIQUES FOR PURCHASING FOOD

Foodservice purchasing of food, equipment, and supplies is unique. Purchasing policies can affect food and labor costs, the sanitation and safety of the operation, nutritional dependability, and the quality of the program—not to mention production, serving, cleaning costs, and bottom-line profit.

There are many pitfalls to be avoided in purchasing. The following paragraphs discuss some erroneous practices the food buyer must avoid to perform well.

Overbuying

Most often overbuyers are "guessers." They guess instead of plan. To be sure they have enough of everything on hand; they usually have far too much. As a result, overbuying is extremely wasteful and expensive.

The greatest waste occurs with perishable items, such as fruits, vegetables, meats, and all grain products, for which overbuying results in a twofold waste—financial and nutritional. That which spoils is lost and is sheer waste; any that can be utilized, even though it is comparatively old, has lost much of its nutritional value, so it is practically useless. Because fresh produce deteriorates so quickly, it is good practice to pay a little more, if need be, for frequent deliveries than to load up at lower prices.

Underbuying

Like overbuyers, underbuyers do not know the average consumption, so they guess. In addition, they are overcautious. As a result, they frequently run out of items. This causes no end of confusion and brings on justifiable criticism.

It is better to have a little too much than not enough, and in large-quantity foodservice, waste cannot be prevented entirely. A little waste, occasionally, is normal. The food buyer must study and know the operation's needs over a reasonable period and buy accordingly.

Price Buying

A price buyer bases the buying decision solely on price and usually purchases the cheapest products. A buyer can easily make this mistake unless he or she is familiar with correct buying principles. The lowest-priced item is not always the cheapest. In purchasing any food, but particularly perishables, the institution usually gets only what it pays for and no more. Inferior merchandise consumes initial savings, and the institution suffers in the "bargain." Knowledge of merchandise is indispensable in evading this erroneous practice. The buyer must choose the quality best suited for the institution's needs and then shop for the best price.

Quality Buying

The quality buyer buys only the best of everything without considering price. This person is a prime target for salespeople who might otherwise have to shade prices or offer inducements. The quality buyer also leaves him- or herself open to the unscrupulous.

This type of "one track" buying is extravagant and expensive. Moreover, such buyers actually do not always get the high quality they think they are getting. Quality buying is frequently associated with "one house" buying. The buyer is completely sold on a distributor or label that has a reputation for high-quality merchandise. He or she concludes, erroneously, that every item this company carries must be the best when the buyer may, in fact, get a better quality at a lower price from a moderately priced house. The remedy for "quality buying" is to compare brands as well as prices.

Bargain Buying

Bargain buyers insist on a price reduction on every purchase. As long as they get the cut rate, they think they have saved money. Such a buyer is concerned with gross costs and is blind to the net, or overall, costs. This buyer is a victim for inferior merchandise and very often gets trimmed. In contrast, the smart buyer knows that there are few exceptional bargains, especially in food.

The bargain buyer frequently overbuys because he or she is captivated by seemingly getting something for nothing. Nothing is a bargain unless it will be used within a reasonably specified period. Moreover, a price reduction means nothing in itself; there has to be a standard of comparison. The buyer must know the current market price.

Pressure Buying

Pressure buyers cannot say no. They have no sales resistance whatsoever. They overstock, duplicate items, and deal with too many concerns. The savvy food buyer is skeptical of fast-talking pressure specialists with their "great bargains," particularly in perishables.

Cleaning supplies is another area in which the buyer can easily yield to pressure selling. Representatives of reputable houses usually avoid such practices. It is good practice never to make an on-the-spot decision to switch products of this kind; it better to think it over and weigh all the considerations.

Panic Buying

An otherwise capable buyer can be panicked into buying more than he or she needs or paying prices that are too high. Emergencies—real or imagined—are the basis for panic buying. Imaginary emergencies are those that sales representatives and periodicals tend to exaggerate beyond reality, such as droughts, frosts, strikes, shortages, or anything that might affect the flow of merchandise.

Unlike commercial restaurants, institutions need not stock certain items listed as scarce. Usually, it is unwise to rush into buying quantities of an item because it is said to be in danger of becoming scarce or increasing in price. Frequently, the prediction does not materialize. The food buyer must learn to avoid getting overconcerned about being short of certain items.

Extravagant Buying

Extravagant buyers do not appreciate the value of a dollar. They have no norm or guide; moderation is not in their vocabulary, so costs mean little to them. Often they pay high prices for foods that are scarce or luxury items. These buyers want almost everything they see and cannot understand why a superior questions their judgment. They are easily captivated by what is new or a little different, even though what they have is perfectly satisfactory.

To avoid this problem, the food buyer must thoroughly compare prices at all times. He or she must likewise be certain of the actual necessity of the purchase.

Miserly Buying

As stated earlier, a buyer does not always need to have the best. Yet one should not buy inferior products just to save a few dollars or cut expenses.

The food buyer must not descend to purchasing perishables, such as poultry, for example, that cannot be sold over the counter because it has started to deteriorate. It is a waste of money and time to invest in produce that has begun to decay with the intent of sorting out the good stuff. Food buyers must try to avoid being tagged as "cheap." Their responsibility is to provide for the normal needs of the institution, and they do not need to be overconcerned about the cost of ordinary operating expenses.

Personality Buying

Some buyers are influenced too much by the personality of sales representatives. To give most of the orders to one person simply because he or she is more likable than others is contrary to good business practice.

Because this type of buying eliminates competition, it can be extremely costly. Showing favor in this way should prevail only when an individual's product measures up in every respect with that of competitors. In other words, the quality, price, and service must be as good as those of other companies.

Friendship Buying

Friendship is a common sales approach. "I happen to know so-and-so," or "Your friend, so-and-so, sent me to see you," are typical lines. Sometimes the salesperson has almost been assured of an order before he or she calls. A buyer has to be everlastingly on guard lest he or she be unduly influenced in this manner. We usually want to give a friend—or a friend of a friend—a break, but sound business principles must be kept uppermost in mind.

In developing friendships with sales representatives, food buyers must beware of compromising themselves or the institution. Buying can be dangerous for buyers who allow friendship to insinuate itself into their work. Human nature being what it is, this mistake has even jeopardized careers.

Sentiment Buying

It is easy to let the heart rule the head. A buyer must be on guard against any sales approach based on sentiment. The answer to such approaches is simply, "I'd like to buy from you. I'd like to buy from everyone who calls, but, of course, that's impossible. So I must make my choice as I see it at the time. You are welcome to call back, but I cannot assure you of any business unless something unforeseen turns up."

The Satisfied Buyer

Satisfied buyers operate as though no possible improvements can be made under any circumstances. They put unqualified trust in those with whom they deal, without even comparing quality or price. This dangerous procedure can lead to considerable monetary loss.

The term "satisfied" is absent from the vocabulary of smart, progressive buyers who are on the alert to every possible savings or improvement. They see as many sales representatives as possible and are constantly improving their ability. This effort pays big dividends in the long run. Likewise, it alerts sales representatives that they can take nothing for granted.

CONSTRUCTIVE PROCEDURES

In the preceding discussion of buying mistakes, some of the positive measures for correct buying have been pointed out. The following are some more constructive factors that should be part of a food buyer's policies.

Sales Representative

It is usually a sign of a buyer's ignorance if he or she is inconsiderate of, harsh with, or suspicious of sales representatives in general. Most sales representatives are honest and hardworking, know their products, and have many useful tips about purchasing. For the most part, they are educated, trained, and capable.

The food buyer should view the sales rep as being in a position to help in many ways. There is always the possibility that the sales rep may have a money-saving proposition. Selling is highly competitive, and companies constantly generate new inducements to increase sales. Such advantages will never be realized unless the buyer's policy is to try to see every sales representative if at all possible.

Quality

Certainly, the more a food buyer knows, the more he or she will get out of every dollar spent, because price is directly related to quality and quantity.

Price

The food buyer must keep abreast of current prices for specific grades and qualities. It is impossible to study all the market reports and government and consumer bulletins, but at least the buyer should keep some kind of a record so that he or she can readily compare prices when necessary, particularly for items that are purchased frequently. For purchases made at rare intervals, such as equipment, food buyers should shop around until they are satisfied they are getting the best combination of price and service.

Average Consumption

Food buyers should always strive to purchase in accordance with their needs. Before making any purchase, they should compute the average consumption of the item over a definite period. The amount to order then follows logically.

Competitive Buying

If food buyers fail to capitalize on the competition between companies, they lose financially. Instead of being buyers, they are solely order givers. However, the buyer who capitalizes on competition will avoid such pitfalls as satisfied, one-house, sentiment, and friendship buying.

Quantity Buying

In general, buying in quantity means lower costs. Usually a distributor is willing to reduce the cost per unit if it can get volume business. The cost difference may amount to only $0.25 to $0.35 a case, but in lots of 10, 20, or more cases, even small savings add up to a considerable amount. This is why it is so important for the buyer to know the average consumption of each item over a definite period.

Because there is the element of chance in most quantity buying, the food buyer must consider all factors before making such a purchase. What is the predicted supply of the item? If the supply is above normal, the price is likely to drop after a short time. Conversely, if there is a shortage, the price will no doubt rise as the supply dwindles. A normal supply means the price will remain stationary, barring unforeseen circumstances.

In normal times, many food distributors contract for future delivery over a period of time. Suppose a food buyer wants 75 cases of an item. After getting the best price for the quality desired, the buyer arranges to take delivery in 10-, 15-, or 25-case lots and to pay as delivered. This procedure is called "futures"—order the item now and take delivery as needed.

Seasonal Buying

Even with modern transportation and frozen foods, seasonal buying has not been eliminated completely. In fact, a food buyer can make costly mistakes simply because many fresh foods cannot be purchased year-round, in or out of the regular season. Seasons directly affect canned and frozen foods, so the buyer must be alert and familiar with the terms "old pack" and "new pack" and all that they signify.

Supply and Demand

Fundamentally, the price of all merchandise is controlled by the demand for it. Here a smart buyer's knowledge can pay high dividends. Some items are more affected than others, with perishables of all kinds heading the list.

At times, for instance, excellent buys of wholesale cuts of meat are available and the food buyer is far more likely to reap such an advantage from a jobber than directly from a packer. When a packer is overstocked on an item, the packer calls the jobber to take the item off its hands at a greatly reduced price. The jobber can pass these savings along because of the desire for a quick turnover.

Service

A buyer pays not only for merchandise but also for service. To the experienced buyer, service includes such things as timely deliveries, an occasional emergency delivery, exactness in filling orders without padding or substitution, merchandise delivered in perfect condition, willingness to make adjustments or take back items for credit if necessary, procuring articles not in stock, efficient billing, and congenial relations.

In return, the buyer should be reasonable in his or her demands or complaints. A practical way of avoiding unreasonable demands is to obtain all the facts before pushing a complaint. Even though a vendor is obligated, patience pays higher dividends than a harsh or threatening attitude.

Deliveries

The responsibility of the food buyer does not end when the order has been placed. Unless the receiving clerk is vigilant, considerable loss is inevitable. In transporting foodstuffs, problems such as shortages, breakage, and the wrong merchandise being delivered are not rare exceptions. Certain merchandise ought to be weighed, so a large floor scale and a small table scale are indispensable.

To ensure routine adjustments, delivery agents should list mistakes and sign the invoices. The same applies to anything picked up for return and credit. Any faults on the part of the delivery service must be brought to the attention of the main office, and if the condition persists, more drastic action should be considered.

Technological Advances

Although the time-tested method of handwriting an order for the sales representative to pick up is still being used, there have been many advancements in the ordering procedure. Computers and modems can make a sales rep obsolete. The foodservice distributor's order guide can be uploaded to the customer's computer. When customers have finished their inventory and forecasted sales for the upcoming period, they can place their orders when they are ready. The distributor will usually have "cutoff" times for ordering. This is a specific time by which an order or product has to be transmitted. In the case of fresh seafood, it may be 10:00 A.M. for the next day delivery. For fresh cut steaks, it may be 2:00 P.M. for the next day delivery. These cutoff times vary by proximity to the regional fresh markets and the products the distributor carries. Office computers that are tied to full-service food distributors can also assist with inventory control, recipes, and menu costing and can provide prices daily on highly fluctuating commodities, such as lobster tails or filets mignon.

Some private niche industries are adopting Internet technologies for business-to-business commerce between suppliers and restaurants. As of 1997, most of the $150 billion in food ordered annually by restaurants, foodservice organizations, and hotels is handled via phone or fax. Some restaurants that used a version of the Internet service found that they cut in half the time they spent preparing and sending orders. Furthermore, some distributors found that order-filling errors were cut by as much as 60 percent.

Foodservice distributors can save both time and money if the customer places his or her own order. The days of the salesperson traveling from restaurant to restaurant just to take an order may be gone. Many large foodservice companies offer online food purchasing programs. In addition to food, they offer many links to help restaurateurs to become more successful. Among these companies are the following:

- U.S. Foodservice (www.usfoodservice.com/) Large broad-line foodservice distributor. Markets and distributes food and related products nationwide. Corporate offices located in Columbia, Maryland. The company distributes food and related products to more than 250,000 customers, including restaurants, health care facilities, lodging establishments, cafeterias, schools, and colleges. It markets and distributes more than 43,000 national, private label, and signature brand items and employs more than 29,500 foodservice professionals.
- Sysco Corporation (www.sysco.com/) Large foodservice distributor. The site contains information on locations, products, and information of interest. Headquartered in Houston, Texas. The company operates from 152 locations throughout the continental United States and portions of Alaska, Hawaii, and Canada. Beginning in 1970, with sales of $115 million, SYSCO—an acronym for Systems and Services Company—has grown to $26.1 billion in sales for fiscal year 2003.
- Food Services of America (www.fsafood.com/) This is a broad-line food distributor, serving all sectors of the foods industry. Headquartered in Seattle, Washington.

Furthermore, inventory tracking can be done using point of sale (POS) terminals. This eliminates many of the purchasing mistakes that can be made by someone who is basing decisions on human judgment alone.

Food Cost

Foodservice products have two timetables for perishability, indicating perishable in the raw state and perishable again in the processed state. This is part of the complexity of controlling food costs. Here is a list of "40 Thieves,"* or causes of high food costs. There are many more, though some do not apply to all foodservice operations, because of operating differences.

FORTY THIEVES

Purchasing

1. Purchasing too much
2. Purchasing for too high a cost
3. No detailed specifications—quality, weight, type
4. No competitive purchasing policy
5. No cost budget for purchasing
6. No audit of invoices and payments

Receiving

7. Theft by receiving person
8. No system of credits for low-quality or damaged merchandise or goods not received
9. Lack of facilities and/or scales
10. Perishable foods left out of proper storage

Storage

11. Foods improperly placed in storage (e.g., fats, eggs, milk near strong cheese or fish)
12. Storage at wrong temperature and humidity
13. No daily inspection of stored goods
14. Poor sanitation in dry and refrigerated storage areas
15. Prices not marked in storeroom
16. No physical or perpetual inventory policy
17. Lack of single responsibility for food storage and issuing

Issuing

18. No control or record of foods issued from storeroom
19. Sending food or supplies without a requisition or the same amount daily

*Robert C. Petrie, "Food Costs and the 40 Thieves," *Cooking for Profit*, November 1972. Reprinted with permission.

Preparation

20. Excessive trim of vegetables and meats
21. No check on raw yields
22. No use of end products for production of low-cost meals

Production

23. Overproduction, overproduction, overproduction!
24. Wrong methods of cooking
25. Cooking at wrong temperatures
26. Cooking too long
27. No scheduling of foods to be processed (too early, too late)
28. Not using standardized recipes
29. Not cooking in small batches

Service

30. No standard portion sizes
31. No standard-size utensils for serving
32. No care of leftovers
33. No record of food produced and leaving production area
34. Carelessness (spillage, waste, cold food)

Sales

35. Food taken out of building
36. Unrecorded sales and incorrect pricing; "not charging customers" or cash not turned in
37. No food popularity index or comparison of sales and inventory consumption
38. No sales records to detect trends
39. Poor pricing of menu items
40. Employee meal costs—overproduction or unauthorized meals

AN UNCOMPLICATED ACCOUNTING SYSTEM

This section includes a simple-to-keep set of internal records. Though developed for a small operation, this system is equally applicable to large and small hospitals, schools, colleges, in-plant, and industrial foodservices, as well as most commercial operations. The purpose of this section is simply to present a method of tracking and controlling food by accounting methods. A good system makes record keeping simple while providing both control of foods and information for easy future utilization.

Invoices should be recorded for a standard accounting period. The example used here is for a calendar month. The length of the accounting period should be determined by the particular operation. The shorter the period, the more control the operator has. If the labor is available to provide one-week accounting periods, then that period is recommended. The earlier a problem or an out-of-control situation is identified, the earlier it can be resolved.

Invoice Record

The system presented here (Figure 1.7) shows invoices recorded on one continuous record. Each invoice is charged against the specific accounting areas the operator wishes to control. A shorter and perhaps more common grouping may be six to eight categories, as follows:

1. Meat—beef, veal, lamb, pork, sausage, and meat products; chicken, turkey, and other fowl and products made from them. Fresh, frozen, or canned
2. Seafood—all fish, crustaceans, and the like, whether fresh, frozen, or canned
3. Dairy products—butter, cheese, eggs, milk, ice cream, and substitute dairy products, such as margarine
4. Staples—canned, dried, and dehydrated fruits, vegetables; grain products, jams, jellies, and so forth
5. Baked goods—items purchased as finished products, such as bread, cake, and pie
6. Wines and liquors—may be included in food accounting by the institutional operator, but should be kept as a wholly separate accounting category by restaurants and hotels
7. Produce—vegetables and fruits, fresh and frozen

The purpose of these breakdowns is to establish percentage consumption guidelines that then can be used to determine overexpenditure or theft. These breakdowns also aid in predicting future expenses or the effect of menu changes on expenditures. Keeping these records forces close examination of invoices and a working knowledge of what goods are in the house at any given time.

Order/Receiving/Inventory Record

Often, the only record of receipt of food is a daily invoice record (Figure 1.8). It is also desirable to have a central record from which one can determine shortages and amount of product in the house and from which one can refer to past consumption.

Inventories should be taken at least monthly. There are two kinds of inventories—perpetual and physical.

A perpetual inventory can tell an operator how much of any product (or all products) should be on hand at any time. It is done by adding the amount of product received and sub-

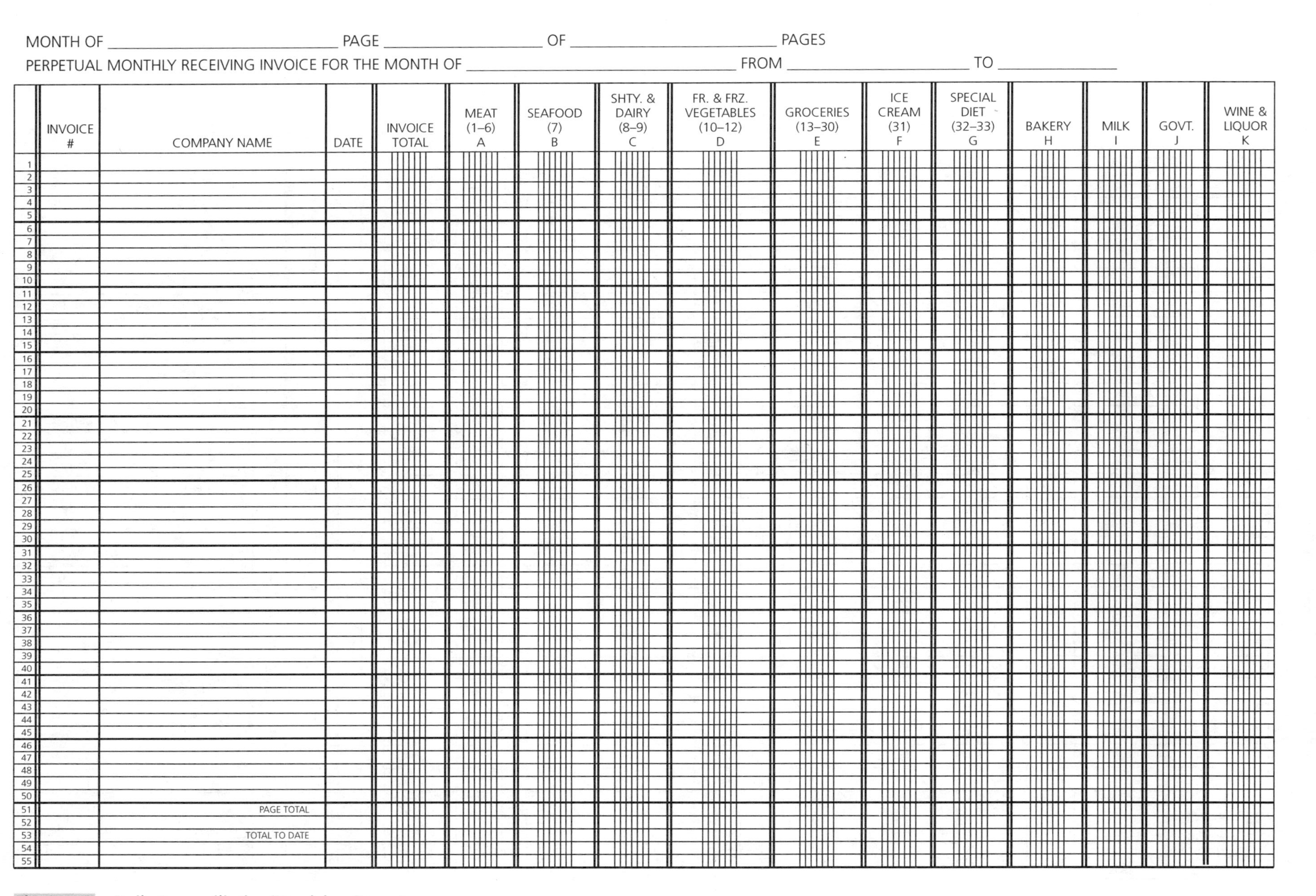

MONTH OF ______________ PAGE ______________ OF ______________ PAGES

PERPETUAL MONTHLY RECEIVING INVOICE FOR THE MONTH OF ______________ FROM ______________ TO ______________

	INVOICE #	COMPANY NAME	DATE	INVOICE TOTAL	MEAT (1–6) A	SEAFOOD (7) B	SHTY. & DAIRY (8–9) C	FR. & FRZ. VEGETABLES (10–12) D	GROCERIES (13–30) E	ICE CREAM (31) F	SPECIAL DIET (32–33) G	BAKERY H	MILK I	GOVT. J	WINE & LIQUOR K
1															
2															
3															
4															
5															
6															
7															
8															
9															
10															
11															
12															
13															
14															
15															
16															
17															
18															
19															
20															
21															
22															
23															
24															
25															
26															
27															
28															
29															
30															
31															
32															
33															
34															
35															
36															
37															
38															
39															
40															
41															
42															
43															
44															
45															
46															
47															
48															
49															
50															
51		PAGE TOTAL													
52															
53		TOTAL TO DATE													
54															
55															

Figure 1.7 Daily Reconciliation/Receiving Report

ITEM	UNIT	BEG. INV.	M	T	W	Th	F	END	S/S	M	T	W	Th	F	END	S/S	M	T	W	Th	F	END	S/S	M	T	W	Th	F	END INV.	@	–	CON-SUMED	
BEEF																																	
1 Liver, Steer	lb.																																1
2 Round, MGB No. 158	lb.																																2
3 Round, MGB No. 158 corned	lb.																																3
4 Round, MGB No. 1168r	lb.																																4
5 Round, diced 1 in. MBG No. 135	lb.																																5
6 Round, diced 1 in. mBG No. 135	lb.																																6
7 Round, ground 80/20	lb.																																7
8 Flk Stk MBG 193	lb.																																8
9 Skt Stk 1-1/2 in	lb.																																9
10 Strip Loin MBG 178	lb.																																10
11 Strip Stks MBG 1178	lb.																																11
12 Bologna 100% Beef	lb.																																12
13 Salami 100% Beef	lb.																																13
14 Franks 8 -100% Beef	lb.																																14
15 Pastrami, Pickles Romanian	lb.																																15
16 Patties 4 oz 80/20	lb.																																16
17 Salibury Steak, 5 oz 80/20	lb.																																17
18 Stuffed Cabbage	lb.																																18
19 Stuffed Peppers	lb.																																19
20 Knockwurst, 4 to Skinless	lb.																																20
21 Beef Burgandy	lb.																																21
22 Beef Stew	lb.																																22
23 Top Round — Cooked	lb.																																23
24 Bttn Round — Cooked	lb.																																24
25 Corned Beef — Cooked	lb.																																25
26																																	26
27																																	27

PAGE 1 TOTAL 868.01

Figure 1.8 Foodservice Monthly Inventory/Order/Receiving Record

tracting the amount of product used since the last physical inventory. Thus, if there was one box of lobster tails on hand on February 28, six boxes received during March, and five boxes issued before March 31, the perpetual inventory would be 1 + 6 = 7 − 5 = 2 boxes. A perpetual inventory is important because it tells what should be on hand at any given time.

A physical inventory must be taken in order to know what is actually on hand. To take a physical inventory, one must count every item in the storage areas. If, using the earlier example, the physical inventory reveals only one box of lobster tails on hand, then the operator must decide whether:

(a) 1 box of lobster tails was stolen, or
(b) 1 box of lobster tails was never received, or
(c) last month's physical inventory was wrong, or
(d) an error has been made in bookkeeping.

Having been alerted, the operator can track down the missing box and take corrective action.

Performing an inventory also tells the operator the dollar value of the food on hand. Because there is commonly more than one price on food items in a weekly or monthly accounting period, the operator has a choice of dollar values to put on each product. He or she may choose to use the average price, the first price (FIFO [first in, first out] accounting), or the last price paid (LIFO [last in, first out] accounting). Because food prices fluctuate frequently, the most common and the easiest method is to use the last price paid.

A good inventory record categorizes all food items into the same groupings as used in the invoice record. Figure 1.8 is a page from an inventory/order/ receiving book that allows an operator to keep a record of items received and disbursed, the perpetual inventory, the physical inventory, and consumption. As orders are written, they are entered above a split line under the day of the month on which delivery is anticipated. On the day the goods are delivered, the receiving clerk enters the actual quantity received under the split line. This makes it easy to identify shortages immediately. The latest price paid is entered in the price column at the right side of the book.

At the end of the month, a physical inventory is taken and recorded directly into the "end inventory" column. This figure is then multiplied by the latest price and becomes the dollar inventory. The quantity consumed is determined by adding the beginning inventory (last month's ending inventory) to the quantities received and subtracting the end inventory. There is a column for this figure at the far right of the page.

The consumption figure is transferred to a master consumption book that holds the consumption figures for a three-year period. This way, the operator can trace the long-term consumption of each item. This record is a tremendous aid in making future contract purchase commitments. The Order/Receiving/Inventory Record has sections for food supplies, which are the same categories included in the Invoice Record. A second part of the Order/ Receiving/Inventory Record is for supplies. On the back of the Order/Receiving/ Inventory Record is a page that recaps the total dollar value of all goods by their respective categories.

Budget Reports

Every foodservice operator has an obligation to supply to management all relevant statistics that can be derived from his or her operation. Figure 1.9 is typical of a form that might be used in an institutional operation. A commercial operation would expand the form to allow for such additional costs as rent, utilities, payroll taxes, and insurance.

Cost Analysis

The cost analysis section of the form (Figure 1.10a) allows an analysis of where the food dollar was spent during the past month. An explanation of the categories and the reason for the seven-section breakdown is discussed in the preceding section on the Monthly Invoice Record. Note that the column on the far left details the performance for the same month last year. This figure is used rather than a 12-month average because the numbers of meals served in the same month of different years are comparable, and therefore a better comparison than a moving average of unlike months. The current month's consumption in each category, as well as an average for the year-to-date (per month), is calculated to give a broad general measure of dollars currently being spent. The amount of money spent in each category should remain about the same. If there is a large difference, there should be a good explanation.

For example, in 2004 the operation was spending 3 to 5 percent more of its food dollar on meat than in 2000. This was because of its increased use of convenience meats and a decrease in the amount of seafood used. Kosher had dropped because of a decrease in kosher customers. Bakery increased because the operation had reduced its on-premise baking. Milk decreased both because of the availability of other beverages and because the operation converted from packaged half pints to bulk milk dispensers.

Frequently, these percentage changes represent a trade-off. Bakery costs increase with more outside expenditures, but the costs of raw products used in baking, such as flour, eggs, and shortening, decrease. Only with these percentage calculations can one accurately measure how much new programs are costing.

	This Month		This Year to Date			
	Actual Expenditures	Budgeted Expenditures	Actual Expenditures	Budgeted Expenditures	%	12 Month Moving Average
Foodservice Personnel	$14,211	$16,274	$85,719	$95,992	89.5	
Clerical	716	713	4,401	4,214	104.4	
620.9 Total Salaries	$14,927	$17,057	$90,120	$100,206	89.9	$15,444
620.30 Equip. Repairs	124	150	933	1,000	93.3	190
620.51 Personnel Expenses		25	48	75	64.0	
620.59 Food Supplies	14,892	14,250	86,485	86,750	99.7	13,953
620.60 Kitchen Supplies	2,195	1,500	14,021	10,640	131.8	2,035
620.75 Training	250	400	799	850	94.0	76
Total Supplies & Expenses	$17,461	$16,325	$102,286	$99,315	103.0	$16,254
Subtotal Expenditures	$32,388	$33,382	$192,406	$199,521	96.4	$31,698
Cutback		(1,242)		(7,459)	(3.73)	
Total Expenditures	$32,388	$32,140	$192,406	$192,062	100.2	$31,698

	Customers	Employees	Total	12 Month Moving Average
This Month:				
Meals Served	13,142	104	13,246	XXXXX
Salary Expense	$12,098	$2,829	$14,927	XXXXX
Salaries per Meal Served	$.92	$.40	$.71	$.76
Food Expense	$9,183	$4,966	$14,149	XXXXX
Food Cost per Meal Served	$.70	$.70	$.70	$.71
Supplies Expense	$1,821	$427	$2,248	XXXXX
All Other Expenses	$861	$203	$1,064	XXXXX
Food & Supplies Cost	$11,004	$5,393	$16,397	XXXXX
Food & Supplies Cost per Meal Served	$.84	$.76	$.81	$.80
Total Direct Costs	$23,102	$8,222	$31,324	XXXXX
Direct Costs per Meal Served	$1.76	$1.17	$1.55	$1.57
This Year to Date:				
Meals Served	77,353	43,036	120,389	XXXXX
Salary Expense	$72,851	$17,269	$90,120	XXXXX
Salaries per Meal Served	$.94	$.40	$.75	$.76
Food Expense	$55,061	$30,623	$85,684	XXXXX
Food Cost per Meal Served	$.71	$.71	$.71	$.71
Supplies Expense	$11,007	$2,581	$13,588	XXXXX
All Other Expenses	$2,000	$470	$2,470	XXXXX
Food & Supplies Cost	$66,068	$33,204	$99,272	XXXXX
Food & Supplies Cost per Meal Served	$.86	$.77	$.83	$.80
Total Direct Costs	$38,649	$50,473	$189,392	XXXXX
Direct Costs per Meal Served	$1.79	$1.17	$1.57	$1.57

Figure 1.9 Foodservice Budget Report

June 20XX

	%	FOOD	totals	meat	seafood	dairy	vegetables	groceries	ice cream	kosher	bakery	milk	govern-ment	liquor
12,866		Purchases	13,216	4,518	420	1,128	2,246	1,720	178	190	1,743	1,047	43	(17)
11,291		+ Beg. Inven.	11,182	1,682	83	435	352	8,314	53	65	33	12	—	153
24,157		= Subtotal	24,398	6,200	503	1,563	2,598	1,034	231	255	1,776	1,059	43	136
10,736		– End. Inven.	10,249	1,536	207	574	547	6,950	50	82	129	26	—	148
13,421		= Total Cons.	14,149	4,664	296	989	2,051	3,084	181	173	1,647	1,033	43	(12)

This Month Last Year	%	FOOD EXPENDITURE	$ this month	$ this %	year	avg.	%
4,691	35.0	Meat	4,664	33.0	28,120	4,687	32.9
497	3.7	Seafood	296	2.1	2,371	395	2.8
837	6.2	Shrtg & Dairy	989	7.0	6,360	1,060	7.4
1,535	11.4	Fr. & Froz. Veg.	2,051	14.5	10,519	1,753	12.3
3,065	22.8	Groceries	3,084	21.8	19,263	3,210	22.5
—	—	Ice cream	181	1.2	793	132	0.9
341	2.5	Kosher	173	1.2	1,473	246	1.7
1,147	8.6	Bakery	1,647	11.6	8,423	1,404	9.8
1,171	8.9	Milk	1,033	7.3	7,470	1,245	8.7
77	0.6	Government	43	0.3	479	80	0.5
60	0.5	Wines & Liquor	(12)	(0.1)	413	69	0.5
13,421	100.0	Total	14,149	100.0	85,684	14,281	100.0

This Month Last Year	%	SUPPLIES	Totals	paper	clng.	service
1,221		Purchases	1,907	580	848	479
4,184		+ Beg. Inventory	4,157	1,866	1,442	849
6,105		= Subtotal	6,064	2,446	2,298	1,328
4,005		– End. Inventory	3,816	1,367	1,562	887
2,100		= Total Cons.	2,248	1,079	728	441

	%	SUPPLIES EXPENDITURE	$ this month	%	$ this year	avg.	%
980	46.7	Paper Goods	1,079	48.0	5,781	964	42.5
258	12.3	Cleaning Goods	728	32.4	3,363	561	24.8
862	41.0	Serviceware	441	19.6	4,444	741	32.7
2,100	100.0	Total	2,248	100.0	13,588	2,265	100.0

Figure 1.10a Food and Supplies Expenditures

Statistical Analysis

This section of the form (Figure 1.10b) contains the statistical analysis of the major phases of the operation. The record is meaningful because it gives a thumbnail sketch of everything going on during the particular month and year, as well as the long-term trend. To understand this section and its figures and indications, a knowledge of industry-wide standards of performance is necessary.

"Number of patient days" is the total number of occupied beds for the month. In the hospital used as an example, there is a large group (44) of daycare patients who eat only one meal per day at the hospital. To calculate the patient days for foodservice purposes, the number of daycare patient days is divided by 3. This figure is then added to the inpatient census to produce the total number of patient days.

As explained later, this figure is essential in forecasting budget needs for the coming year. "Cost per patient day" is achieved by dividing the total direct costs by the total number of patient days. Note from the example that the cost per patient day is less than three times the direct cost per patient meal. This is because not every patient eats every meal.

Here is an example of how these figures aided this operation: A closed ward was opened, the daycare program was expanded, and the policies governing weekend passes were changed. It was essential to know what the effect on the foodservice budget would be. With statistics on past cost per patient day, the operator could calculate cost per patient day for a full three-meal per day census. By interpolating these figures and using a percentage factor for the gain in census, he was able to calculate the increased costs in minutes. Without the ongoing figures on record, the increased costs might have taken days to forecast.

	This Month	This Year	12 Month Average
No. of Patient Days	6,084	36,879	XXX
Cost per Patient Day	$3.80	$3.76	XXX
Patient Meals per Patient Day	2.16	2.10	XXX
Total Meals per Patient Day	3.33	3.26	3.27
No. of Man-Hours Worked	3,421	20,797	3,538
No. of Meals Served	20,246	120,389	XXX
Meals Served per Man-Hour Worked	5.9	5.8	5.6
Salary Expenses	$4,297	$90,120	XXX
Cost per Man-Hour Worked	$4.18	$4.33	$4.28
Cafeteria Income	$3,886	$23,218	$3,717
Check Average	$.70	$.68	n.a.
Income per Staff Meal	$.48	$.54	n.a.

Figure 1.10b Cost Analysis: Statistical

"Patient meals per patient day" is derived by dividing the total number of patient meals served by the total number of patient days. The "total meals per patient day" figure is arrived at by dividing the total meals served by the total number of patient days. The difference between patient meals and total meals is the number of staff meals per patient day.

Knowing these three figures enables the operator to see the relationship between them, as well as the trend of patient and staff eating habits month by month, season by season, and year by year. Fluctuations in these figures signal such things as a changing program or a changing attitude toward menus.

"Number of employee hours worked" is the actual number from payroll records. Knowing this figure is important in controlling labor costs. Once the operator has done a labor analysis, it becomes possible to carefully regulate vacation schedules according to workload needs and to keep the man-hours worked close to peak productivity.

"Meals served per man-hour worked" is calculated by dividing the number of meals by the number of man-hours worked. This figure or its reciprocal, man-minutes worked per meal served, is the primary index of productivity. Knowledge of the average meals served per man-hour worked for similar operations would be useful in measuring cost-effectiveness. "Cost per man-hour worked" is calculated by dividing total payroll expense by the number of man-hours worked. This figure does not indicate the average wage per hour paid to employees; it is relatively meaningless, and it is hard to understand why government and industry use it in graphing labor indexes.

The relevant figure is cost per man-hour worked. This figure takes into account the cost of such "free" fringe benefits as vacations and holidays. Seeing this figure every month emphasizes the importance of finding new and better labor-saving work methods and equipment.

"Cafeteria income" is the amount of cash revenue derived from cafeteria sales to staff. The "check average" is the average amount paid for a meal if the staff pays for meals. "Income per staff meal" is the average amount of cash revenue for all staff meals. This includes groups that eat free, such as resident physicians, foodservice staff, and official visitors. The income per staff meal figure deducted from the cost per staff meal (total cost divided by number of meals served) indicates the amount of subsidy for staff food.

Budget or Pro-Forma Development

It is usually the responsibility of each department head to submit a proposed annual budget to management for approval. Justifications for present staff expenses, and any increases, must also be presented. The intent is to provide an honest estimate of needs.

The management should also be knowledgeable of departmental needs. Any padded requests or improper justifications to make the coming year's budget easier or to look good by having money left at the end can be shaken out quickly.

In developing estimates on the funds to be needed, the operator must call upon many resources. General economic conditions reveal many trends. Knowledge of such things as customer trends, health trends, weather trends, large union wage settlements, and more rigid federal controls on various foods all aid in making estimates. Similarly, reading the *Wall Street Journal* reports on distant commodity futures, talking to middle managers of food contracting companies and to buyers who work for purveyors, and reading trade magazines, the FDA papers, and the daily USDA Market Reports will help in estimating.

If they have done these things all year, operators will have better knowledge than a guess about price trends for the coming year. In addition, they should keep records on the actual price trends for important items, as well as for the 11 broad categories on the daily reconciliation/receiving report for reviewing these records, they can see the average price rises of previous years. They know that menu changes are planned for the next year, and now they can estimate the effect these will have on expenses in each category.

With all the aforementioned factors at hand, an operator can make a final estimate of the anticipated price rises for the coming year. When these figures are multiplied by the per-meal expenditures of the current year, the operator arrives at a per-meal cost for the next year.

With knowledge of the customer counts for the past year, and using the "food costs" and "moving averages" calculations mentioned earlier, the operator can now simply multiply to get a food expenditure estimate. This figure, as indicated by past experience, will usually end up within 2 percentage points of the next year's actual expenses.

In "salaries and wages," estimates can be nearly exact because of the knowledge of the next year's scheduled wage gains, vacation and holiday time due, and past records of sick time taken and overtime needs.

This system of budget development works well, because the department head who must live with the budget plays a large part in its development and is the person who has the expertise in that area. A cost accounting line budget is recommended because that format makes it easy to locate specific areas of excess expenditures.

INTERNET RESOURCES

www.ism.ws

Institute for Supply Management (ISM), founded in 1915, is the largest supply management association in the world. ISM's related areas include standards of excellence, research, promotional activities, and education. ISM provides opportunities for the promotion of the profession and the expansion of professional skills and knowledge.

stats.bls.gov/oco/ocos023.htm

The *U.S. Department of Labor Bureau of Labor Statistics Occupational Outlook Handbook* for purchasing managers, buyers, and purchasing agents, includes the nature of the work, working conditions, employment, training, qualifications, and advancement. There are also sections on earnings, job outlook, and related occupations.

CHAPTER

2

Fruits, Fresh and Processed

In 2003, foodservice operations accounted for approximately 45 percent of the total crop of fresh produce. Americans consumed 195 lb of fresh vegetables and 128 lb of fresh fruit per person. The purchasing of fresh produce requires constant attention, owing to several factors: high perishability, different market practices, and the daily changing of supply, demand, and quality. The handling of produce has improved, and fruits and vegetables may be purchased in many stages of processing. Methods of packing include the following:

Fresh: Fruits and vegetables that have been picked and packed for delivery in the fresh state. Generally, these have been freed of soil and insects and cooled before packing.

Canning: The process of putting food into a container to which enough heat is applied to kill all bacteria. In addition to using high heat to kill bacteria, there are processes in which a combination of pressure and lower heat can be applied to reduce the boiling point needed to sterilize the product.

Freezing: A process whereby fruits and vegetables are kept at a low enough temperature to prevent spoilage.

Dehydrating: A process by which fruits and vegetables have most of their water extracted. Drying: A process by which fruits and vegetables, mostly peas, beans, and cereal, are allowed to dry naturally or are machine-dried of most of their water content.

Freeze-drying: A process in which fruits and vegetables are frozen and then dried in a vacuum so that the water content changes from ice to vapor, bypassing the liquid state. There is less damage to products that are freeze-dried than to those that are dried or dehydrated.

Dehydro-freezing: A process by which foods are dehydrated, then frozen.

Irradiation: A process by which the bacteria in foods are killed through exposure to radiation.

All of these processes compromise the quality of fresh fruits and vegetables to some extent. However, it is usually not economically possible to use only fresh produce, and the various types of processing make it possible to serve almost any kind of fruit or vegetable out of its growing season. There are USDA standards for fresh, canned, frozen, dehydrated, and dried fruits and vegetables.

WRITING THE SPECIFICATIONS

An attempt has been made in this book to present pertinent information on states of processing for fruits and vegetables. The essential grade information on products in all its processing states has been included after the general information on each item. The grade information of product examples, and the details of the specification are covered in the following section.

FRESH PRODUCE

A large variety of fresh fruits and vegetables are available every season. It is not unusual for 100 or more distinctly different kinds of fresh produce (not counting the several varieties of each) to be on the market at the same time. Even in midwinter, a shopper in one of the large supermarkets can probably buy 47 different fresh fruits and vegetables, plus a number of varieties. The variety of fresh produce available in some markets that have extensive produce departments will be even greater. The extension of the period of availability is due to such factors as increased production of truck crops in southern areas during the winter and early spring, planting of varieties that have been developed to produce earlier or later than previously, increased imports, improvement in storage methods, and improvements in transportation and terminal distribution, which permit the transfer of tender produce for long distances, during which they are maintained in good condition.

SPECIALIZED PRODUCTION

The growing complexity of producing and marketing fresh fruits and vegetables has caused a trend away from the very small market garden or small home orchard and toward large commercial operations where soil, climate, and water are most favorable. Large specialized farms and orchards, run by well-trained operators, are able to take advantage of the most modern equipment and methods. Because the costs of labor, material, and capital are rising, increased efficiency in production, packing, and marketing is essential.

Although farms of very small acreage far from market are no longer economical, there are still many truck farms of 40 to 60 acres, more or less, that are operated with much success. Orchards somewhat smaller than this are also successful. Many are located close to city markets. The trend in fruit and vegetable growing is in line with the general trend in agriculture—that is, toward fewer and larger farms in selected areas. Organic produce is also more prevalent, and smaller farms are changing to supply the need for such items.

PACKING

There has been a revolution in packing, and it is continuing. One trend was the elimination of waste parts, such as carrot tops, radish tops, cauliflower jacket leaves, celery tops, and inedible parts of greens. Transportation, material, and labor are so costly that it has become essential not to ship inedible parts when they can be removed at the shipping point. Another trend is toward automatic assembling, filling, and closing of shipping containers. This is coupled with mechanized washing, sorting, and conveying. Automatic checking of color by electronic means is used in some cases. Almost all bagging of fruits and vegetables at the shipping point and at repackaging houses at terminals is done more or less automatically. Commodities are weighed by various mechanical means and are poured into the bag, which is then closed mechanically.

There is also a trend toward packing fruits and vegetables in consumer units at the shipping point, then placing the consumer units in a master container for shipment. Another trend apparent in the last few years is a shift from large, heavy shipping packages to smaller ones that are more easily handled and can be recycled. The size and weight of a package are important considerations in restaurants and other institutions, as well as in retail stores where packages may be lifted by hand. A great deal of research and experimentation with various kinds of containers, liners, and trays is continuing. The type of package used is important. It is intended to protect the packed commodity from crushing or bruising, permit easy stacking and securing in a rail car or truck, permit refrigeration as required, and perform other functions, such as, in some cases, providing an atmosphere low in oxygen or providing chemical inhibition of the development of molds. The container needs to be designed for easy assembling, filling, and closing; its weight should be as low as possible; and costs need to be as low as attainable, while still permitting requirements to be met.

TRANSPORTATION

Rapid transportation is of major importance in the fresh fruit and vegetable industry. Fresh produce is a living organism and will loose quality rapidly if not handled properly. It cannot stand long delays, wide variations in temperature, or rough handling. Smooth, rapid movement at proper, well-controlled temperatures is necessary. Transportation has improved considerably since the turn of the century, both as to speed and protection of products from damage. On average, it takes about eight days to haul a carload of fresh produce by rail freight from the West Coast to the East Coast. Express transport is faster, but it costs more. Trucks are used widely for hauls up to 1,500 miles or even farther. Transportation is one of the major costs of marketing fresh produce and may range from a few percent of the consumer's dollar spent for these products to 20 percent or more. Transportation rates have gone up in line with prices of other services. One of the newer developments is "piggybacking." A truck is loaded on a special train flat car and unloaded at its destination ready to roll to its delivery point. Or a truck body or other container, minus wheels, may be loaded in the same way, then placed on a wheeled chassis at the receiving terminal. The Interstate Commerce Commission (ICC) says trailer-on-flat-car operations are continuing to grow. Progress is also being made with air transport of fresh fruits and vegetables. Because air transport, in most cases, is more expensive than other types of carriage, it is applicable especially to commodities of relatively high value for the bulk and weight. Development of new jet transports designed for cargoes has stimulated air shipment of highly perishable commodities.

TEMPERATURE CONTROL

The use of refrigeration or other methods of temperature control is on the increase at all stages of fresh fruit and vegetable marketing, from shortly after harvest right through to the consumer. Although most produce is shipped to a central "produce market" and sold from there, some quantities of produce are marketed near where they are grown and are not refrigerated en route. Yet vast amounts are moved long distances and must be maintained at a suitable temperature to keep them in good condition. This involves cooling or warming of the load, depending on outside temperatures. At the same time, high relative humidity, about 90 percent, is desirable for most commodities to prevent wilting or moisture loss. Humidity in both iced and mechanically refrigerated cars is usually high. Ice, on top of the load or in the package, is used for some commodities, especially the leafy vegetables, to provide added moisture and refrigeration. Evaporation also is reduced by the use of film box liners and film bags for both fruits and vegetables. Important applications of refrigeration to fresh fruits and vegetables include precooling before shipment; cooling in short or long storage; air-conditioning of wholesale produce warehouses and retail produce departments; refrigeration of retail displays; walk-in refrigeration for storing produce in retail stores; and refrigeration in the home.

Precooling

As the term implies, precooling is the rapid initial removal of field heat prior to shipment or storage. This is done as soon after harvest as possible. Tests show that precooling slows the ripening process in fruit and the breakdown process in vegetables, checks the development of molds and bacteria, and prolongs the market life of the produce. Various studies have shown that a reduction of 15°–18° F in the temperature of deciduous fruits slows the ripening and the respiration rate by approximately one-half. Although the ratio of reduction of temperature to slowing of respiration is not the same for all commodities, it is sufficient to justify precooling of perishables before shipment. Methods of precooling include:

1. Placing packages in an ordinary refrigerated storage room (a slow process, and therefore not as effective in increasing market life as some other methods)
2. Hydro-cooling by means of an ice-cold water spray or water bath (a rapid process)
3. Forced-air cooling in a tunnel or specially built room (a fairly rapid means)
4. Cooling in refrigerated rail cars or trucks (fast or slow, depending on the process used)
5. Vacuum cooling (a rapid process)

Hydro-Cooling

The most common form of hydro-cooler is a tunnel constructed of metal sheets, through which the produce to be cooled is passed on a conveyor. Icy water is showered over the unlidded containers of produce. Often, a chemical is put into the water as a sterilizing agent. A typical hydro-cooler 31 ft long by 7 ft wide, operating efficiently, can reduce the pulp temperature of peaches from 85° to 50° F in 12–14 minutes. It can cool 350 to 450 bushels an hour. A well-refrigerated storage room would take about 48 hours to reduce the temperature to the same extent.

Vacuum Cooling

Vacuum cooling can reduce the temperature of lettuce from 73° to 35° F in 30 minutes. This period of time will cover the entire cycle of operations, from placing the lettuce cartons in the vacuum tank to removing them. One plant at Watsonville, California, has a cooling capacity of 75 carloads per day. In some plants, the tank will admit a railroad car. The lettuce is loaded ready to ship, and the entire load is cooled in one operation. The vacuum cooling system is based on two facts: first, that water evaporating from a surface cools the surface; second, that as pressure is decreased, the boiling point of water is reduced. At the boiling point, evaporation is rapid. A vacuum cooler consists of a gastight chamber into which the produce is loaded. Pressure in the chamber is then reduced by exhausting air, resulting in rapid evaporation and cooling. A carton of lettuce (40–45 lb net) loses about 2 to 2.5 percent of its weight. Vacuum cooling can be applied to any of the leafy vegetables as well as to some other commodities to which a little water can be applied for evaporation.

Icing

One of the common methods of refrigeration of fresh produce is the use of ice, such as for broccoli, which provides both moisture and heat absorption. The ice may be placed in bunkers and air circulated through the load; or ice may be placed in the produce container or blown over the top of the load; or a combination of these methods may be used.

Continuous Refrigeration

Fruits and vegetables that have been refrigerated at the shipping point and en route should, for best results, continue under refrigeration at the receiving point. Wholesalers generally have cold rooms that are insulated and mechanically refrigerated. These are known as "holding rooms." They also have special ripening rooms for such commodities as bananas and tomatoes that need to be kept relatively warm and humid until the correct stage of ripeness is reached.

Long Storage

Items such as apples, pears, and potatoes may be held for several months in specially built storage cellars, where temperature and other conditions are controlled. The operators have the special knowledge and equipment needed to hold the product and still maintain quality. This type of storage is usually located at the shipping point. Apples may be held for as long as a year before being shipped out.

DISTRIBUTION

Distributors, the so-called middlemen, carry out the function of placing foods where people who want them can buy them. The functions of the distributor are not as well understood as the functions of the producer. This leads to a misunderstanding as to the value that marketing and distribution add to a product. The distributive chain in a simple form is grower-to-shipper-to-transportation company-to-wholesaler-to-retailer-to-consumer. The are many important variations:

- The produce may be packed, shipped, and sold by a cooperative agency owned by the growers.
- A broker may buy at the shipping point for a distant receiver; the commodities may be serviced by a packing and cooling agency; the produce may be stored for later sale either in a general storage warehouse or in a specialized house; it may be packaged in consumer units at the shipping point or shipped for repacking at the terminal.
- The produce may be shipped directly to a buyer in a terminal market, or it may occasionally be shipped as a "roller," to be sold while en route and diverted to the buyer.
- The produce may be shipped on consignment to a receiver who sells it and then deducts a service charge or commission.
- The produce may be shipped to an auction house that sells it to the highest bidder.
- The produce may be handled through a broker at its destination. Produce may also be shipped to a service wholesaler that buys for its own account and then warehouses, stores, ripens, repacks as necessary, and delivers to retail stores.
- The ultimate distributor, for example, a retail chain, may buy commodities in the field or orchard, or at the packing house, and handle them all the way through the stores, even transporting them long distances in its own trucks.
- Repackers may buy commodities by the carload for delivery to their warehouses, or buy at auction or at the terminal market, and then repack items in consumer-size packages for retailers.

These marketing channels have undergone considerable changes since the late 1980s. Prior to 1987, fresh fruit and vegetable markets were more fragmented; most transactions took place between produce grower-shippers and wholesalers on a day-to-day basis, based on fluctuating market prices and quality levels. Today, a typical produce sale may take place between a multiproduct grower-shipper and a large supermarket retailer under a standing agreement or contract specifying various conditions and terms, including marketing services provided by the grower-shipper, volume discounts, and other price adjustments and quality specifications. Changes in these marketing services coincided with the growth of value-added and consumer-branded products, increasing variety, consolidation of food wholesalers and retailers, the expansion of the foodservice sector, and the greater role of produce imports and year-round supply.

In 1997, $1.1 billion worth of produce was sold directly to the consumer, $34.3 billion in retail stores, and $35.4 billion through foodservice establishments. As consumption has increased, so has the demand for variety, convenience, and quality, as evidenced by the explosion in produce department offerings. Many products (for example, lettuce and tomatoes) are available year-round, produce is precut, and more packaged and branded products are available. The share of branded produce increased from 7 percent in 1987 to 19 percent in 1997, and fresh-cut produce and packaged salads rose from 1 percent to 15 percent of total sales.

Three percent of the farms in the United States supply 75 percent of the nation's food. Ninety percent of all fresh vegetables consumed in the United States are now grown in the San Joaquin valley in California.

Licensed Dealers

About 25,000 licensed dealers largely carry on the wholesale distribution of fresh produce. Under the Perishable Agriculture Commodities (PAC) Act, all commission merchants, brokers, and dealers buying or selling fresh fruits and vegetables in wholesale or jobbing quantities in interstate commerce must have a license from the U.S. Department of Agriculture. All licensed dealers must carry on their business under the requirements of the Act. Those who violate this law can be called to account under the PAC Act (PACA). If necessary, the license of a dealer can be suspended or revoked by the secretary of agriculture. In the absence of a license, the dealer cannot continue in business.

Trading Rules

Under the PACA, trading rules are laid down. Because dealers are operating under known rules enforceable by law, they can deal quickly at a distance. Marketing terms are defined under the Act so that a specific and binding agreement for the sale of large quantities of fresh produce can be made with a few words. The rules and the enforcement machinery have helped to speed up the buying and selling of fresh produce. This is important because delays can cause great loss of perishable products. Reliable firms in the trade have expressed their approval of the PACA and the way it is administered by the USDA. The Act has helped raise business standards in the fresh fruit and vegetable industry to a relatively high level.

Brokers

The broker is a middleman who does not take title to merchandise. Rather, his or her principal function is to bring buyer and seller together. Brokers transact business in the name of this principal function. They deal in large units, which in the case of fruits and vegetables is the carload or truckload. Brokers are generally classified as buying brokers and selling brokers. The broker takes a commission, for instance $100 or $200 per car, or so much per package, depending on value.

Auctions

Fresh fruits and vegetables are sold at auction at some shipping points and at some terminal markets. The auction operates at high speed and in a dramatic way. The auctioneers call for bids, and interested buyers respond. They know each buyer in the room and instantly recognize his or her bid signal, which may be a word or a gesture. The hammer bangs down to sell the lot to the highest bidder. If bids are slow in coming on any lot, the auctioneer may pass it and go on to another, returning later to the slow lot. Everything possible is done to sell fast. Onlookers unfamiliar with auctions usually cannot make heads or tails of the proceedings because no merchandise is in sight.

The buyers work from lists of the merchandise that they have previously inspected. The seasoned buyers have no difficulty in following the rapid-fire offers. New York City, Chicago, Philadelphia, Boston, Cleveland, St. Louis, and Detroit have produce auctions that operate all year.

Wholesaling

Food wholesaling in the United States is a $589 billion business. It consists of that part of food marketing in which goods are assembled, stored, and transported to customers, including retailers, foodservice operators, other wholesalers, government agencies, and other types of businesses. Leading the way with sales of $422 billion is the retail food store sector, followed by the $358 billion foodservice sector. Sales to other wholesalers represent a significant portion (more than 25 percent) of total wholesale sales. These include small specialty wholesalers that purchase goods from larger wholesalers rather than manufacturers.

The Census of Wholesale Trade broadly classifies three basic types of wholesalers: merchant wholesalers, manufacturers' sales branches and offices, and agents and brokers. Merchant wholesalers' sales account for the largest percentage of food wholesale sales, at 56 percent. These wholesalers (also referred to as third-party wholesalers) are firms primarily engaged in buying groceries and grocery products from processors or manufacturers and reselling to retailers, institutions, and other businesses. Manufacturers' sales branches and offices are wholesale operations maintained by grocery manufacturers or processors to market their own products. Brokers and agents are wholesale operators that buy or sell as representatives of others for a commission and typically do not physically handle the products. They may serve as representatives of manufacturers or processors, but do not take title to the goods.

The Census of Wholesale Trade also broadly classifies grocery wholesalers as general line, specialty, or miscellaneous. General line distributors (also referred to as broadline or full-line distributors) handle a broad line of groceries, health and beauty aids, and household products. Specialty distributors are primarily engaged in the wholesale distribution of items such as frozen foods, dairy products, poultry products, fish, meat and meat products, or fresh fruits and vegetables. Miscellaneous distributors are primarily engaged in the wholesale distribution of a narrow range of dry groceries such as canned foods, coffee, bread, or soft drinks.

Among the three groups, specialty grocery wholesalers account for the highest percentage of sales (43 percent).

Service Wholesaler

A merchant wholesaler whose functions are not well understood is the service wholesaler. The word *service* indicates that this wholesaler does more than just buy and sell. It also warehouses, delivers, extends credit, provides market information, may send out salespersons, and may, when necessary, aid retail customers with their merchandising. The customers of service wholesalers are varied. They include independent retailers, voluntary and cooperative associations of retailers, chain stores, and institutions such as restaurants and hospitals. Ordinarily, a service wholesaler has both rail and truck docks; receives merchandise in carload or truckload lots; has common and refrigerated storage areas, banana ripening rooms, tomato ripening rooms; and often has facilities for packaging produce in consumer units. The service wholesaler maintains large and small trucks and may serve an area within a radius of 100 or more miles. A firm may serve several hundred stores and institutions.

RESTAURANTS AND MASS-FEEDING INSTITUTIONS

According to the Bureau of Labor Statistics, in 1998 about 38 percent of all food was consumed away from home, and that figure has continued to grow. Most of this consumption is done in restaurants or other eating places and large-scale feeding establishments such as in-plant cafeterias, and in public and private institutions such as hospitals and schools. Large restaurants and institutional eating places buy their fresh fruits and vegetables from service wholesalers specializing in these commodities or from restaurant and hotel supply houses that handle an extensive line of both fresh and processed commodities of all kinds. Smaller restaurants often buy from these retailers.

WASTE AND LOSS

Despite all the advances made in growing, harvesting, storing, and distributing fresh fruits and vegetables, a large amount of waste still occurs. Some waste is inevitable, as in eliminating culls and in trimming vegetables for marketing, but much waste can be prevented or reduced. Great strides have been made in combating losses in growing due to pests, such as insects, weeds, rodents, nematodes, fungi, and bacterial and viral organisms. Losses in marketing have also been greatly reduced by improved temperature control, better packages, faster transportation, and the use of chemicals to inhibit the development of molds and bacteria. Waxing some items has cut down the loss of moisture and preserved them longer. Loss in the marketing process, aside from the normal trimming of vegetables for display, is still heavy. The USDA has estimated that of the produce sent to market, 11 percent of all fruit and 8 percent of all vegetables is lost during this process. Although all general loss figures are necessarily estimates, various detailed studies of losses of particular items indicate that these estimates are conservative.

STANDARD GRADES AND INSPECTION

Buyers can purchase a carload, or many carloads, of fresh fruits and vegetables in a few minutes from someone thousands of miles away and know what they are getting after a phone call or a letter on a fax machine. This can be done because there are federal standard grades and packs, as well as state standards, and federally approved trade terms and definitions that apply to these perishable products. In addition, a system of federal and state inspection is provided both to protect the public and to avoid the innumerable disputes that might otherwise arise between buyer and seller as to the grade, quality, and condition of fresh produce.

CONTRACTS

Each time a buyer and a seller of fresh produce close a deal, they have made a binding and enforceable contract, even though no detailed contract has been written and signed. An exchange of short confirmatory messages by telegram or fax accomplishes the same purpose as a lengthy document. This is possible because a brief statement such as, for example, "U.S. Extra Number 1 topped Carrots" has a definite and detailed meaning known to buyer and seller and stated in the standards published by the U.S. Department of Agriculture (USDA). A federal or state inspector, examining a shipment of carrots, can determine whether they actually are of the stated grade. About 40 percent of all shipments of fresh fruits and vegetables are officially inspected, mostly at shipping point. In many cases, inspection is a routine procedure. The party ordering the inspection pays the fee for inspection. The fee is based on the quantity and number of kinds of commodities to be inspected. Arrangements are also made, if desired, for continuous inspection of fresh fruits and vegetables at the packing plant. Such packages may then be labeled with the U.S. shield with the wording "Packed under continuous inspection of the U.S. Department of Agriculture." The inspection fees cover the government's costs for providing this service.

VOLUNTARY GRADING

Use of the U.S. standards and use of the inspection service are generally voluntary, but may be compulsory under certain circumstances. Under provisions of the Marketing Agreement Act, authority is granted for restriction of shipments of produce by grade, size, or maturity where marketing agreements are in effect. These agreements are adopted by vote of those concerned. Grading and inspection of certain products has been made compulsory in areas that have adopted marketing agreements and orders.

COMPULSORY GRADING

The laws of some states require grading of various fresh fruits and vegetables in accordance with official U.S. standards or state grades. A few states also require inspection of some products. The Export Apple and Pear Act makes it unlawful to ship apples or pears in the fresh state to foreign countries in car lot quantities, unless they meet certain minimum grades prescribed by the secretary of agriculture.

The grades prescribe and define quality terms such as "U.S. Fancy," "U.S. Extra No. 1," "U.S. No. 1," "U.S. Combination," and "U.S. No. 2." Grades may specify size, maturity, color, cleanliness, shape, freedom from specified injuries or damage, interior structure of the fruit or vegetable, absence of seed stems, or any other factor of quality. The grades also set up tolerances, which are percentages of permissible variations from any specification. Tolerances are necessary because in commercial practice it is impossible to pack fruits or vegetables to meet specifications 100 percent. As a rule, "U.S. No. 1" grade represents good, average quality that is practicable to pack under commercial conditions. Usually, under normal growing conditions, more than half of a crop will be of U.S. No. 1 grade. The designation "U.S. No. 2" ordinarily represents the quality of the lowest grade that is deemed practicable to pack under normal conditions. Superior products are packed "U.S. Fancy" or "U.S. Extra No. 1." The USDA in response to requests of those interested works out the grades. The policy of the department is not to issue standards for official use until they are considered practicable and workable. Congressional action in 1913 paved the way for the declaration of the first standards for fruits and vegetables in 1917. Potatoes were the first products for which standards were adopted. The inspection service started in 1918 when Congress provided for inspection at receiving markets. In 1922, Congress extended the service to shipping points. Inspection offices are now maintained in 78 of the larger cities throughout the continental United States, one in Puerto Rico, and one in Hawaii.

INSPECTION PROCEDURE

Trained inspectors are thoroughly conversant with grades and with all fruits and vegetables. Some inspectors specialize in certain products, but all are licensed to inspect any fresh fruit or vegetable. The inspector uses a random sampling procedure. He or she selects a number of boxes at random from the sorting line, or the car, or lot, and from each box examines a representative number of items, also selected at random. The sampling is large enough to make it reasonably certain that it reflects the quality of the entire lot. The inspector notes findings as to each item inspected. If the samples are found to lack uniformity, the inspector will inspect more of the merchandise than he or she would otherwise. Upon completion of the examination, the inspector totals the scores of various defects and calculates percentages of defects. The inspector makes up a worksheet and report, and from the report makes up an inspection certificate. He or she notes the condition of the car, how it is loaded, how it is iced, what the pack is, and the quality and condition of the merchandise. Federal inspection certificates are prima facie evidence in any court of law.

GRADE, CONDITION, AND QUALITY

A distinction should be noted in regard to the terms *grade*, *condition*, and *quality*. *Grade*, as used by the USDA, is the sum of the characteristics of a commodity at the time it is graded, and includes both quality and condition.

Quality denotes those characteristics that are relatively permanent, such as shape, solidity, color, maturity, and freedom from insect damage.

Condition relates to factors that may change, such as decay and firmness. If a commodity is graded, for example, "U.S. Extra Fancy" when packed, it means that it not only has the quality characteristics of the grade, but is also in good condition within the meaning of the grade. That is, it has no factor of poor condition outside the tolerances permitted. A product that is graded "U.S. Extra Fancy" when packed may, after a long journey or careless handling, be found to be out of grade owing to its condition. The following are important factors to specify for fresh fruits.

Grade

When you specify a U.S. grade, such as U.S. No. 1, it has a definite, detailed meaning that makes it unnecessary to write out the details. A full set of U.S. grade standards for fresh fruits and vegetables can be obtained free by writing the Fresh Products Standardization Section, Agricultural Marketing Service, Washington, DC 20250. In using grades, however, it should be noted that there can be quite a range of quality within a grade, owing to tolerances provided in the grade. Note also that U.S. No. 1, though it is the highest grade for some commodities, is a lower grade for others. Higher grades include U.S. Extra Fancy, U.S. Fancy, and U.S. Extra No. 1. The buyer should at all times specify that fruit must be up to the desirable grade at delivery, and not merely at the time it was shipped.

The U.S. and state standard grades are handy tools for the buyer. Instead of detailing lengthy specifications, he or she can ask for U.S. No. 1 celery or U.S. 1 cauliflower. The buyer should specify that the item must meet the grade at time of delivery to his or her receiving room. A vegetable that met the grade at the shipping point may be below grade later. The purchaser can add any special requirements; for example, he or she may want Pascal celery of a particular size from a preferred area, or may want to specify film-wrapped cauliflower minus the jacket leaves and ribs.

Quality

Grade alone does not necessarily define all the quality factors the restaurateur is interested in when buying a fruit for a certain purpose. The buyer should specify any additional factors as necessary.

Variety

In almost every instance it is necessary to specify variety, because there are considerable differences that make one variety suitable for one purpose and another for a different purpose. Thus, Delicious apples might be suitable for table service or salads but would not be suitable for baking.

Quantity

Quantity should be stated in precise terms appropriate to the commodity. In the case of strawberries, it may be thirty 12-pt trays. In the case of watermelons, it may be 12 melons, average 30 lb each. In the case of bananas, it may be ten 40-lb cartons, also specifying the size of the bananas as small, medium, or large.

Brand

It is advisable to specify brands when possible, because fruits packed under some labels are consistently good. Judging labels requires experience, but the suppliers can help with this.

Growing Area

Fruit from one area may be much different from fruit from another area, so it is often practical to specify the source desired. For example, because of climatic conditions, pears of the far western United States are superior to those of the eastern regions. Florida produces mainly juice oranges, while California produces eating oranges.

Other Factors

Some other factors may also be of importance in ordering specific items. Case weight or item weight can vary greatly for some products (e.g., cases of lettuce). Degree of ripeness will affect when the produce can be used, (e.g., avocados or bananas). Precut produce is gaining popularity in the food-service industry and may be a consideration. Packaging may vary between suppliers; a case may be two 5-lb bags or four 5-lb bags of bell peppers.

PRICE

Price is important, but the buyer shopping on the basis of price only is likely to be a poor buyer. If the buyer gets consistently low prices, he or she is also going to get consistently low-quality product and service. Fruit and vegetable dealers are as smart as the buyers are, and they are not going to give something for nothing. Moreover, the price-only buyer tends to buy a little from one dealer and a little from another, splitting up the purchases so much that he or she may not be a profitable customer for anyone. Thus, this buyer does not merit the attention and cooperation that others may deserve and get. It is well known in the trade that produce offered as a bargain is likely to be anything but a bargain.

The best buyer is not necessarily the one who gets everything for the least money. Instead, he or she is the one who buys vegetables best suited to the particular use to which they are to be put and which please customers. The proper quality at the right price is the value the buyer is looking for. Experience shows that the cheapest vegetables often may not be the best value. Price needs to be balanced against such factors as freshness, tenderness, shape and appearance, size, trim loss, and total waste.

INSPECTION

When fruit is delivered, it should be inspected at once to ensure that it is of the kind, quantity, variety, grade, size, pack, and of any special quality that has been ordered. It is not necessary to remove and check every piece to make an adequate inspection. Professional inspectors of the USDA do not do that. A random spot check of a reasonable number of samples from a reasonable number of packages is all that is necessary. Overinspection means overhandling, which increases damage, and is time-consuming. If fruit has been ordered by brand, check the trademarks to see if they conform. Check for proper ripeness. It may be necessary to change plans as to when certain items will be served. Some fruits, such as pears, bananas, cantaloupes, and avocados, if too firm, can be ripened by holding them at room temperature until they are at the proper stage for eating.

ADEQUATE CARE

When good money has been spent to buy high-quality fruits to serve discriminating customers, it is only good sense to spend some time and effort to give the fruits the right care. Here are some general rules:

Insist on gentle handling: It is not unusual for those who deliver fruits, as well as those who handle the packages in the foodservice operation, to be rough. It takes supervision, explanation, and a rather firm policy and follow-up to induce all concerned to place, rather than drop, fruit packages. They should never be thrown or pushed roughly across a floor. They should be handled the same as a crate of eggs. It will be easier to persuade workers to handle the fruit carefully if backbreaking labor is eliminated by the provision of suitable materials-handling equipment, such as two-wheel hand trucks, a four-wheel, bar-handle truck, semilive skids and jack, a conveyor, or whatever the particular situation and volume handled call for.

Handle as little as possible: Attention should be given to receiving and routing fruit efficiently so that it need not be moved repeatedly. This takes planning. No rule can be given, but certainly new fruit should not be stacked on top of older fruit or in such a way as to block efforts to reach and use the older fruit first. The rule should be first in, first out. Dating each package on receipt is a help.

Avoid letting fruit stand in danger areas: When fruit is received, it is sometimes allowed to stand almost any place before it is put in storage, and damage can result. A bad place, for example, is near a hot radiator, on a wet floor, or on a receiving platform in extreme cold, heat, or wind. If fruit is to be refrigerated, it should be moved into the cold room without delay.

Stack packages according to their shape and weight: Stack so that the pressure is on the structure of the package, not on the fruit. Some packages are properly stacked on their bottoms, some on their sides, and some on end. Bulge packs should not be stacked on the bulge, because that results in crushing pressure on the fruit. In general, keep stacks low for ease of handling and to avoid excessive pressure on the lower layers.

Maintain proper temperature, humidity, and ventilation: Fresh fruits are alive. In a sense, they "breathe," taking up oxygen and giving off carbon dioxide and other products such as ethylene. They generate heat, which must be removed to maintain low temperatures. Fruits should be stacked so that air can circulate around the packages. No one rule as to temperature can be given. It is recognized that most foodservice operations do not have facilities for keeping different kinds of produce at different temperatures and humidities. In most cases, the choice is simply between keeping a commodity refrigerated or at room temperature around 45° F. For brief periods, such as one or two days, neither exact temperature nor exact humidity is important, but for long periods, proper control of both is essential. However, it should be noted that it is unwise to expose some fruits, such as bananas or avocados, to such low temperatures as 32°–40° F for even a short time. Quality will be adversely affected.

FRESHNESS

Specify and insist on freshness. If vegetables are wilted or stale, they should be rejected regardless of "bargain" prices. Check how well the original characteristics of the vegetable have been preserved, such as bright, deep color; crispness; good weight for size; good shape; lack of mechanical damage; and absence of decay. They gradually will lose water. Retention of freshness requires lowering their temperature to retard the life processes and, in most cases but not all, keeping humidity high to conserve moisture. Vegetables that are warm on arrival should be considered suspect even if visible wilting has not yet occurred. The useful life of most vegetables is greatly shortened by allowing them to stay warm even for a few hours. (Exceptions include sweet potatoes, white potatoes, tomatoes, and some others.)

SEASONALITY

Vegetables bought very early or very late in their season need to be bought with extra care, with the exception of sweet corn, which is often at its best in early and late periods. These products will most certainly be high in price but will not necessarily be of high quality. Generally, vegetables are lower in price and are likely to be of better quality and flavor when they are in season. Check the availability chart. (A precise, detailed guide to the availability and sources of each commodity by months is the annual *USDA Report on Fresh Fruit and Vegetable Unload Totals for 41 Cities*. Any large-scale buyer should have this guide at hand. It is free from the U.S. Department of Agriculture, Washington, DC 20250.)

VARIETAL DISTINCTIONS

For most vegetables, varietal distinctions are unimportant and are little used in institutional buying. Type, but not variety, is of considerable importance. Thus, "Pascal" is not a varietal name for celery but a type designation. "Danish" cabbage designates a type, and so does "domestic round." Variety is more important for potatoes, but even so, type and origin are of more consequence. Even experts have trouble picking varieties out of a jumbled pile of many varieties of potatoes. In instances where variety is important, the purchaser can find a discussion of varieties in Chapter 3.

SIZE

If size is important for a particular use, as it often is, it should be definitely specified. Such terms as *small* or *large* should be used only if these are defined in a standard grade specified in your order. Otherwise, size should be stated in terms such as *length* or *diameter*, or in terms of the number of units in a standard pack, such as 24s for lettuce, or weight, such as 20–25-lb watermelons.

The end use will determine what size is needed. The medium sizes generally cost more per package than either the very small or very large, because mediums are more in demand. However, if a whole fruit is to be served, obviously a large fruit will cost more than a much smaller one. Size is usually indicated by the count in a standard container, the lowest counts meaning the largest sizes.

PACKAGING

Vegetables are available in many kinds of containers and packs, with different net weights, as well as different degrees of protection. The buyer should specify a container suitable for his or her purpose. For example, an institution will probably want Brussels sprouts in a 25-lb drum or carton rather than in a crate or tray containing pint cups. Information on containers and net weights and other data on weights and measures for many products is available in Statistical Bulletin 362, *Conversion Factors and Weights and Measures*, available from the USDA.

PARTIALLY PREPARED VEGETABLES

Local produce suppliers, in many cases, offer foodservice buyers a number of fresh vegetables in partially prepared form. Most universally available are prepeeled potatoes, often marketed as peeled, whole; peeled and cut to various French fry sizes and styles; peeled and sliced for cottage fries; peeled, cut, and blanched for French fries. Other items offered by some suppliers include washed, cut, and mixed salad greens; coleslaw mixes; peeled, diced, sliced, or shredded carrots; washed and cut spinach; peeled, sliced, and chopped onions; peeled and sliced or diced turnips and parsnips. These partially prepared vegetables are currently available everywhere at most produce dealers.

HANDLING CARE

Although most vegetables need low temperature and high humidity during any holding period, there are exceptions, so each vegetable needs to be considered carefully and separately. All have in common the need for careful and knowledgeable handling. None should be banged around, thrown, or dropped. The term *hardware* is sometimes applied to such vegetables as potatoes and cabbage, but it is a false term. Damage and loss result from the mistaken idea that some items can be handled roughly without harm. Handlers should be instructed about the need for gentleness, and this requirement should be enforced. Vegetables should not be allowed to stand in frigid winter or high summer temperatures or in high wind on a loading dock, or to be placed temporarily next to a hot radiator or allowed to stand in a puddle of water. Upon receipt and inspection, all should be placed promptly in storage with suitable temperature and humidity. All should have some ventilation space and should not be tightly stacked, because fresh vegetables generate heat and need air movement to carry it away. Green leafy vegetables, in particular, generate a lot of heat. The following guidelines are important for handling fresh vegetables:

Stacking: Crates, cartons, and other vegetable containers should be stacked properly to avoid pressure on the produce itself and to prevent toppling. Packages that have a bulge should be stacked to keep weight off the bulge. The height of stacks should be low enough to prevent crushing weight on the bottom packages. If vegetables are to be stacked high, then palletizing with suitable support for upper stacks is required.

Odors: Some vegetables, such as onions, give off odors that can penetrate and incorporate with other products such as butter, eggs, and cheese. Products that pick up odors readily should not be stored with vegetables even if their temperature requirements happen to be compatible.

Life and Storage: A booklet on storage temperatures (such as *USDA Agriculture Handbook 66*) gives data on the length of storage under certain conditions of temperature and humidity. However, these figures cannot be applied to vegetables as received at a foodservice operation. Their life has been shortened by their trip through the marketing process. Unless it is known that vegetables have been rapidly cooled immediately after harvest and have been kept at satisfactory temperatures, long storage is not desirable—except in the case of vegetables that do not have critical temperature requirements, such as potatoes, or in the case of vegetables that naturally have a long keeping period, such as topped carrots. It is recognized that foodservice operations, in most cases, cannot provide finely adjusted temperatures or humidity ranges required for different vegetables, and this is not necessary for brief periods. However, optimum temperature and humidity for commercial storage are quoted in the USDA handbook to give the reader a yardstick for measuring the storage conditions he or she can provide.

SAMPLE SPECIFICATIONS FOR FRUITS—FRESH AND PROCESSED

Apples

Packaging: 3 and 5 lb poly bags, bushel baskets, 38 to 42 lb boxes
Grades: U.S. Extra Fancy, U.S. Fancy, U.S. No. 1, U.S. Utility, Combination
Sizes (count): 64, 72, 80, 88, 100, 113, 125, 138, 150, 163

Varieties

There are innumerable varieties of apples, and more than 7,000 have been named. However, there are 300 varieties harvested, and actually only 17 clonal varieties make up 91 percent of total production in the United States. (A clonal variety is one that is produced directly from a portion of the plant, such as a bud or shoot, and not from seed. All standard Delicious trees, for example, have been propagated from a single tree that showed a deviation that was desirable.) More Delicious apples (Red and Golden) are produced (55 percent) than any other variety. McIntosh is second at 7 percent. These two varieties are almost two-thirds of the total. There are a great number of "sports." (A sport is a sudden spontaneous deviation or variation of an organism from type, beyond the usual limits of individual variation.)

Apple growers are on the lookout for tree limbs that show fruit with favorable variations, such as more red color or color that comes out earlier. Many new sports are announced every year; now sports of sports (sometimes called supersports) are coming out, and there are already sports of supersports. It would take far more space than is available here to list and describe the sports. For example there is Starkrimson, which is a supersport of Red Delicious, and there are already many super-supersports of Starkrimson. Some of the sports are spur-type trees on which fruit spurs are formed along scaffold limbs, and fruit is produced on practically all parts of the tree. The rank of the varieties and each variety's percentage of total production are as follows:

1.	Delicious	38 percent
2.	McIntosh	7 percent
3.	Golden Delicious	17 percent
4.	Rome Beauty	7 percent
5.	Jonathan	4 percent
6.	Winesap	2 percent
7.	York	5 percent
8.	Stayman	3 percent
9.	Yellow Newtown	2 percent
10.	Cortland	2 percent
11.	Northern Spy	2 percent
12.	ll others	1 percent

Percentages are rounded to nearest whole unit. Following are data by variety, including all four major varieties, the harvesting and marketing seasons, the average crop, main growing areas, and a description of the fruit.

Summer

Gravenstein. Harvest July into September, with main marketing season July 15 to September 15; grown mainly in California but also in New England and Appalachia; fruit above medium to large; skin thin, tender, slightly rough, greenish yellow to orange yellow overlaid with broken stripes of light and dark red; flesh whitish yellow, firm, moderately fine, crisp, moderately tender, juicy, sprightly, medium acid, aromatic; good for table use, cooking, and salads; not a storage apple and must be sold quickly.

The Gravenstein is said to have been found originally in the Duke of Augustinberg's garden at Gravenstein in Holstein, Germany. Another theory is that it derived its name from being found in the garden of the castle of Grafenstein in Schleswig. In any event, it is a common apple throughout Germany and Sweden and was received by English growers from those countries. The Gravenstein may have been received in the United States in the vicinity of Albany, New York, prior to 1826.

Others There are a great many summer apple varieties available from late June well into September. Among the more popular varieties are:

Lodi. From June on into early August; size medium to large; skin greenish yellow, similar to the Yellow Transparent of which it is a relative and which it has largely replaced; flesh medium firm; flavor acid; good for sauce and pies.

Astrachan (Red Astrachan). Size medium to sometimes large; very perishable; skin rather thin, moderately tender, smooth, pale yellow or greenish, often nearly or quite overspread with light and dark red, splashed and irregularly striped with deep crimson or carmine, and covered with rather heavy bluish bloom, numerous small whitish dots; flesh white, often strongly tinged with red, rather fine, crisp, tender, juicy, brisk medium acid, aromatic, sometimes slightly astringent; good cooking apple before it becomes fully ripe; desirable for dessert when fully ripe and mellow.

Starr. Large to very large; skin rather thick, tough, nearly smooth, green becoming yellowish green, sometimes with an indication of a faint blush, numerous small and large pale or russet dots; flesh tinged with yellow, moderately fine, very tender, crisp, very juicy, sprightly medium acid, aromatic; good for sauce and pies.

Summer Rambo. Large to very large; skin thick, smooth, attractive, clear bright yellow or greenish tinged with carmine striping; flesh yellowish green, firm, tender, very juicy, medium acid, aromatic; a distinctive "sauce" apple.

N.W. Greening. Large to very large; skin smooth, entirely green, turning to yellow with advanced ripeness; flesh greenish white, very firm, juicy, acid; generally used in the premature stage for "green apple" pies and other dishes calling for tartness and shape-retaining firmness.

Williams Red. Medium to large; skin moderately thick, rather tender, nearly smooth, pale yellow overlaid with bright deep red, indistinctly striped with dark red or crimson, numerous small grayish or russet inconspicuous dots; flesh sometimes tinged with red, firm, a little coarse, moderately crisp, tender, rather juicy, becoming dry when overripe, mildly acid, aromatic; good for dessert.

Yellow Transparent. Size medium to above medium, sometimes large; skin thin, tender, smooth, waxy, pale greenish yellow changing to attractive yellowish white, numerous greenish and light-colored dots, often submerged; flesh white, moderately firm, fine grained, crisp, tender, juicy, sprightly medium acid with pleasant but not high flavor; good cooker.

Fenton. This is a red sport of Beacon; ripens in August; size medium, shape oblate; skin fairly thick and tough and handles well; flesh crisp, juicy, and sprightly to sub-acid; Miller Red is similar.

Fall

Grimes Golden. Harvest September 1 to early October, but mainly in September; main marketing season September 10 to December 1; grown mainly in Virginia, West Virginia, Pennsylvania, and Ohio; fruit medium to large; skin tough, somewhat rough, clear deep yellow with scattering of pale yellow or russet dots; flesh yellow, very firm, tender, crisp, moderately coarse, moderately juicy, medium acid, rich, aromatic, sprightly; food for cooking or dessert.

Jonathan. Harvest generally September 1 to mid-October, but mainly September 15 to October 1; may be marketed until May; average crop 8.3 million bushels; grown mainly in Michigan, Washington, Illinois, Appalachian states, Ohio; fruit medium to small, rarely large; skin thin, tough, smooth, pale bright yellow overlaid with lively red faintly striped with carmine; well-colored fruit almost completely covered with red, which deepens to purplish on sunny side; less-colored fruit has green to red-toned appearance; prevailing effect attractive, lively, deep red; flesh whitish or somewhat yellow, sometimes tinged with red, firm, moderately fine, crisp, tender, juicy, very aromatic, sprightly medium acid; excellent for dessert or cooking; fine for baking.

Wealthy. Harvest generally mid-August to early October, mainly late August into mid-September; main marketing season September to October; average crop 1.5 million bushels; grown mainly in New York, Michigan, Wisconsin; size above medium to large when well grown but often small on old trees; skin thin, tough, pale yellow or greenish, blushed and marked with narrow stripes and splashes of red, deepening in highly colored specimens to brilliant red, numerous small inconspicuous pale or russet dots; prevailing effect bright red; flesh whitish, sometimes stained with red, moderately fine, crisp, tender, very juicy, agreeable medium acid, sprightly, somewhat aromatic; excellent for fresh eating; extra good for pies, sauce, and baking.

Winter

Baldwin. Harvest mid-September through October, mainly September 25 to October 25; marketing season mainly October through December, but some fruit marketed to June 1; grown mainly in New York, New England, Michigan, and Ohio; size above medium and sometimes large to very large; skin tough, smooth, light yellow or greenish blushed and mottled with bright red, indistinctly striped with deep carmine, prevailing effect bright red; flesh yellowish, firm, moderately coarse, crisp, rather tender, juicy to very juicy, agreeably medium acid, sprightly, somewhat aromatic; good for dessert and salad; extra good for pie and sauce; about 75 percent are sold to processors.

Ben Davis. Harvest generally September 25 through October, mainly October 10 to 25; marketing season mainly October to mid-November with some supplies to April 1 and later; grown mostly in New York and Virginia; production declining steadily; size usually above medium to large; skin tough, waxy, bright, smooth, usually glossy, clear yellow or greenish, mottled and washed with bright red, striped and splashed with bright dark crimson, inconspicuous light, whitish, or brown dots, prevailing effect deep, bright red or red striped; flesh whitish, slightly tinged with yellow, firm, moderately coarse, not very crisp, somewhat aromatic, juicy, mildly medium acid; usually used for cooking; virtually all go to processors.

Cortland. Harvest September 10 to October 20, mainly September 25 to October 10; marketing season mainly October to mid-January; grown mostly in New York, New England, Ohio, Michigan, Pennsylvania, and Wisconsin; size medium to large; skin thin, moderately tender, smooth, shiny, red and striped indistinctly with deep dark carmine, prevailing effect red to deep red; size medium to large; flesh snow white, fairly firm, tender, delicate in texture, mildly medium acid, mildly aromatic, does not turn brown on exposure to air; excellent for eating out of hand, in salads, and in fruit cocktails where flesh holds snowy white color; fine for all cooking purposes.

Delicious. Harvest late August into November, depending on area, but mostly September 1 through October 25; marketing season mainly September 20 to May 20 with some available into June; grown in substantial quantities in 19 states, mainly Washington, Virginia, California, Michigan, New York, Pennsylvania, Oregon; size medium to large; skin thin, tough, smooth, brilliant red over yellow with areas of lighter and darker color, striped red to solid red, prevailing effect deep red with lighter red areas, sometimes with small areas of yellowish green; calyx end has five characteristic low ridges or knobs; flesh white, fine grained, tender, crisp, juicy, moderately low acid, mildly aromatic; excellent for eating raw in salads; not generally used for cooking; named accidentally by Lloyd C. Stark, governor of Missouri, who bit into one at a fair in 1894 and exclaimed, "That's delicious!"

Golden Delicious. Harvest late August to November 1, depending on area, but mostly September 10 to October 20; marketing season mostly September 25 to April and some into June; grown mostly in Washington, Illinois, Virginia, Pennsylvania, West Virginia, Michigan, as well as in 12 other states; size medium to large; skin bright yellow or golden, rather tough, and may be shiny, velvety, or russeted; flesh white, often with yellowish tinge, crisp, fine grained, juicy, moderately low acid, mildly aromatic; fine for eating out of hand; especially good for salads because it does not turn brown; excellent for cooking.

McIntosh. Harvest August 20 to October 30, but mostly September 10 to October 10; marketing season mostly September 25 to May 1, but some may be available to June 30; grown mostly in New York, New England, Michigan, and in seven other states; size above medium, sometimes large; skin thin, moderately tender, smooth, readily separating from the flesh, clear whitish yellow or greenish washed and deeply blushed with bright red and faintly striped with carmine; highly colored specimens become dark, almost purplish red, with carmine stripes obscured or obliterated; prevailing effect bright red; flesh white or slightly tinged with yellow, sometimes veined with red, firm, fine, crisp, tender, very juicy, characteristically and agreeably aromatic; perfumed, sprightly, medium acid becoming mild and nearly sweet when ripe; excellent for dessert and salad; good for cooking and baking but less cooking time is needed than for most other varieties.

Newton (Yellow Newtown, Albermarle Pippin). Harvest September 10 to November 25, but mostly September 20 to October 20; marketing season mostly September 20 to January 30, but may extend to June 1; grown mostly in California and Oregon, with some in Washington and Virginia; size medium to very large; skin rather tough, smooth, or slightly roughened with brownish russet dots; when fully mature the more highly colored apples are bright yellow, sometimes with distinct pinkish blush, especially about the base; less highly colored fruit is greenish yellow shaded more or less with duller brownish pink; flesh slightly tinged with yellow, firm, crisp, moderately fine grained, juicy, sprightly, medium acid, highly aromatic; excellent for dessert; good cooker; one of the best keepers.

Northern Spy. Harvest October 1 to November 5, mostly October 15 to 31; marketing season mostly October 10 to May 1 with some available to June 30; grown mostly in Michigan, New York, and New England; size usually large to very large; skin thin, tender, smooth, glossy; clear pale ground color is nearly concealed with bright pinkish red, mottled and splashed with carmine, and overspread with thin, delicate bloom, prevailing effect bright red or striped red; flesh yellowish, rather firm, moderately fine grained, very tender, crisp, very juicy, sprightly, aromatic, medium acid; excellent for dessert or cooking.

Rhode Island Greening. Harvest May 12 to October 25, mostly September 10 to October 10; marketing season mostly October 1 to February 1, extending sometimes to May 1; grown mostly in New York, with some in Michigan and New England, few in Ohio; size above medium to large or very large; skin moderately thick, tough, smooth, waxy, grass green varying to rather yellow, sometimes blushed with brownish red, which rarely deepens to a distinct bright red, greenish white or russet dots often submerged, prevailing effect green or yellowish; flesh yellowish, firm, moderately fine grained, crisp, tender, juicy, rich, sprightly, medium acid; not generally used for dessert or salad; extra good for pie, sauce, and good for baking.

Rome Beauty. Harvest September 1 to November 10, mostly October 1 to October 30; marketing season mostly October 1 to April 1 with some supplies available to July 1; grown mostly in New York, Pennsylvania, New Jersey, West Virginia, Washington, California, Virginia, Ohio, North Carolina, and in eight other states; size medium to very large; skin thick, tough, smooth, yellow or greenish more or less mottled with bright red, which in highly colored specimens deepens to almost carmine, numerous small whitish or brown dots, prevailing effect red or red mingled with yellow; flesh nearly white with slight tinge of yellow or green, firm, moderately fine grained to a little coarse, rather crisp, juicy, slightly aromatic, agreeably mild, medium acid; excellent for baking because it holds its shape; good for pie, sauce, and all cooking purposes; not generally used for eating raw but is rather good raw when fully ripe.

Stayman. Harvest September 1 to November 1, mostly October 1 to November 1; marketing season mostly October 1 to April 30 and some supplies may be available later; grown mostly in Virginia, Pennsylvania, West Virginia, New Jersey, Ohio, North Carolina, and in seven other states; size medium to large; skin smooth, rather thick, often nearly completely covered with rather dull mixed red or rather indistinctly striped with dull carmine, light russet dots often rather large and conspicuous, striped effect more noticeable in less highly colored specimens; flesh tinged with yellow or slightly greenish, firm, moderately fine grained, tender, crisp, juicy to very juicy, aromatic, sprightly, pleasantly medium acid; fine all-purpose apple.

Winesap. Harvest September 15 to November 15, mostly October 5 to November 1; marketing season mostly November 15 to June 30 with some supplies available as late as August; grown mostly in Washington and Virginia, also in 12 other states; size medium to small; skin medium thick, tough, smooth, glossy, bright red indistinctly striped and blotched with very dark purplish red over a distinctly yellow ground color, or green if not fully mature, overspread with faint bloom; dots rather small, scattering, whitish; prevailing effect bright deep red; flesh tinged with yellow, veins sometimes red, very firm, rather coarse, moderately crisp, very juicy, sprightly, medium acid; fine all-purpose apple, good for eating fresh; one of the best keepers.

York Imperial. Harvest September 20 to October 30, mostly October 1 to October 25; marketing season mostly October 1 to March 30 but some supplies sometimes available to June 1; grown mostly in Virginia, Pennsylvania, West Virginia; size medium to large; skin tough, bright, smooth, blushed to solid light red or pinkish red, with dots pale or russet, often conspicuous; flesh yellowish, firm, crisp, a little coarse, moderately juicy; at first sprightly medium acid but becomes mild medium acid or nearly sweet, somewhat aromatic; good for eating fresh and a fine cooking apple, holding its shape and flavor under heat. This fruit has peculiar oblique lines, as though each apple was leaning a little to one side.

DWARF ROOTSTOCKS

Thousands of trees of many varieties are now being planted on rootstocks that control the ultimate tree size. These are known as dwarfing stocks. The use of such stocks is not new and, in fact, was familiar to Roman agriculturists. Recently, however, after 100 years of study and experiment in the United States, and for centuries before that in Europe, the interest of U.S. growers in these stocks has revived. There are many kinds of dwarf rootstocks, and they produce different degrees of dwarfing. In addition, the effect depends to some extent on the variety that is grafted or budded onto the rootstock. The main object of dwarfing is to improve growing efficiency, inasmuch as smaller apple trees cost less to prune, thin, spray, and harvest. More trees can be planted per acre; they often come into production earlier than regular trees, and it is said that better-quality fruit is possible owing to ease of spraying and harvesting and because of greater exposure to light, which produces better color. Such trees are generally on Malling or Malling Merton rootstocks. There are many kinds, identified by number, and each has its good and bad characteristics and needs to be selected to fit individual needs. On the negative side is the inability of the present standard dwarf stocks to stand extreme cold, their susceptibility to being blown over by a high wind, the requirement for strong bracing of many trees, and the requirement for more intensive horticultural practices than needed for regular trees.

INTERNET RESOURCES

www.pma.com

The Produce Marketing Association, founded in 1949, is a trade association that serves members who market fresh fruits, vegetables, and related products worldwide. Members are involved in the production, distribution, retail, and foodservice sectors of the industry. Their core purpose is to sustain and enhance an environment that advances the marketing of produce and related products and services.

www.ams.usda.gov/howtobuy/canned.htm

How to Buy Canned and Frozen Fruit, by the U.S. Department of Agriculture Agricultural Marketing Service (USDA AMS).

CHAPTER

3

Vegetables

Fruits and vegetables continue to be an important part of the American diet. The new federal guidelines suggest that consumers eat at five to nine servings of fruits and vegetables each day. Providing foodservice operations with safe, wholesome fruits and vegetables should be a main priority of purveyors. Fruits and vegetables can pick up dirt and soil as they are being harvested, handled, packed, and shipped. They may also have trace amounts of chemicals and bacteria on the outer areas that can be removed by washing. Germs can adhere to the surface of produce and can be passed to the flesh when cut (cross-contamination). Wash the produce just before you plan to use it, not when you put it away. Washing in slightly warm water brings out the flavor and aroma of the fruit or vegetable you are preparing. When receiving produce, look for fresh-looking fruits and vegetables that are not bruised, shriveled, discolored, or slimy. Do not purchase anything with an off odor or off color.

BECAUSE OF THE NATURE OF FRESH PRODUCE, THERE ARE SEVERAL ITEMS TO BE CONSIDERED

Fresh vegetables are highly perishable; buy only what the operation will use within a few days.

Put produce away promptly. The most important thing foodservice operations can do is wash all vegetables in clean water before preparing.

Dry with a paper towel. Greens, such as spinach, chard, kale, and collards, should be cooked while wet, immediately after washing.

Produce used in salads, such as lettuce, radishes, carrots, and the like, should be washed in the coldest water available to maintain crispness.

Peel and discard outer leaves or rinds, such as the rinds of cucumbers, which may have a protective waxy

spray on them. Scrub vegetables such as potatoes and carrots if you intend to prepare and serve them with the skin on.
Cover and refrigerate produce you have cut.
Read and follow label instructions such as "Keep Refrigerated" or "Use by" (a certain date).
Keep prepared fruit salads or other cut vegetable items in the refrigerator until just before serving.
Discard all cut produce items if they have been out of the refrigerator for more than four hours.

SAMPLE SPECIFICATIONS FOR VEGETABLES

Anise
Packaging: 45 lb crate
Grade: U.S. No. 1

Marketing Season

Anise is scarce at all times, but is more likely to be available October through April, with the peak period in November and December.

Containers

In California, where most anise is grown, the crop continues to be harvested by hand, hauled to a packing shed, cleaned, trimmed, cut to length, and packed. The container most used is the Sturdee nailed 15½ in. crate or a wirebound crate holding 40–50 lb, averaging 45 lb, according to the California Department of Agriculture. The Western Growers Association (WGA) crate, holding 80–85 lb, is still used to a small extent. Other crates are seen on the market from other areas, such as a ⅘ bushel crate from Virginia and a ⅖ bushel crate from Arizona.

Grades

The U.S. standards for "sweet anise" (finocchio), adopted in 1930, provide one grade, U.S. 1. This grade consists of stalks of sweet anise that are firm, tender, well trimmed, and fairly well blanched; that are free from decay and from damage caused by growth cracks, pithy branches, wilting, freezing, seed stems, dirt, discoloration, disease, insects, or by mechanical or other means.

Unless otherwise specified, the minimum diameter of each bulb should be not less than 2 in. In order to allow for variations other than size not related to proper grading and handling, not more than 10 percent by count of any lot may be below the requirements of this grade, but not to exceed ⅒ of this amount, should be allowed for decay.

Uses

Anise bulbs may be quartered and eaten raw with salt or dressing. The inner stalks can be served raw with salt or dressing. Or you can cut the bulb and stalks in strips, crisp them in ice water, and serve as you would celery. Served hot, anise makes an enticing vegetable dish. Try it sautéed lightly in butter, then baked in a casserole with a rich cheese sauce. Or if you want to braise it, cut two stalks in quarters lengthwise, then sauté in butter or olive oil until lightly browned, add salt and pepper, cover the pan, and simmer until tender. Serve with the juice. Another idea is to chop the stalks and add them to tomato sauce or other sauces for fish or meat to give a delightfully different fragrance. Anise can also be added to soup stock or to tossed green salads. Cut the anise into slices, drop the slices into boiling salted water and simmer until tender, then serve with salt, paprika, or melted butter.

Artichokes
Packaging: 20 and 25 lb crates
Grades: U.S. No. 1, U.S. No. 2
Size: (each) Small (less than 8 oz), Medium (8 oz), Large (10 oz), Extra Large (14 oz), Jumbo (20 oz)

Marketing Season

Artichokes are available throughout the year, with the peak of supply occurring from March to May and the fewest available in July and August.

Varieties

The only variety of artichoke produced commercially in important quantities is the Green Globe, grown in California. This variety has deep green heads, 3 or 4 in. diameter, round but slightly elongated. Other globular varieties, none of which are in commercial production, include the White Globe, Red Dutch, and Giant Bud.

Containers

The main container is the wooden box or fiberboard carton, 7 in. deep, 11 in. wide, and 20⅝ in. long. The net weight of the contents is 20–25 lb. A few shippers use a box 9¾ in. deep and of the same length and width as the more standard containers.

Quality

The most desirable artichokes are compact, plump, heavy in relation to size, somewhat globular, and with large, fresh, fleshy, tightly clinging, green leaf scales. Freshness is indicated by the green color. Overmature artichokes have hard-tipped leaf scales that are opening or spreading; the center formation may be fuzzy and dark pink or purple in color. Leaf scales on such overmature specimens are tough and woody when cooked and may be undesirably strong in flavor. Seriously discolored artichokes are usually bruised or lacking in freshness. Bruises may appear as dark off-colored areas at the points of injury. They may also show mold growth. Bruised or seriously discolored artichokes usually turn grayish black or black when cooked. The size of the artichoke is not important in relation to quality or flavor.

Grades

U.S. grade standards for artichokes provide two grades, U.S. No. 1 and U.S. No. 2. U.S. No. 1 consists of artichokes that are properly trimmed, fairly well formed, fairly compact, not overdeveloped, and free from damage caused by worms, snails, bruising, freezing, disease, insects, or other means. To allow for variations incident to proper grading and handling, not more than 10 percent, by count, of any lot may be below the requirements of this grade, but no part of this tolerance is allowed for decay.

U.S. No. 2 consists of artichokes that are not badly spread or overdeveloped, and that are free from serious damage caused by worms, bruising, freezing, disease, insects, or other means. To allow for variations incident to proper grading and handling, not more than 10 percent, by count, of any lot may be below the requirements of this grade, but no part of this tolerance is allowed for decay.

Uses and Preparation

The artichoke has a delicate, nutty flavor, causing it to be prized in salads or hors d'oeuvre. It can be eaten in its entirety, or each leaf can be pulled off and dipped into a sauce. It can be served as a hot vegetable, with butter or special sauce, or served cold.

How to cook: Wash artichokes, trim stems, and pull off and discard tough outside leaves at the base. Cut off the top and spread the leaves open. Remove the fuzzy thistle and tiny inner leaves with the tip of a teaspoon. This can be done before or after cooking. Stand artichokes upright in a deep saucepan just big enough to hold them snugly, or tie them with a string so they will retain their shape. Add 1 teaspoon salt, 1 tablespoon lemon juice, and boiling water to cover. Cook, covered, 45 to 60 minutes, or until base is soft. Lift out with two spoons and let drain.

How to serve: If a whole artichoke is served to each person, place each on an individual salad plate. A half artichoke may be served on the dinner plate. Serve sauce, if thin, in small paper cups or in tiny bowls placed on the salad plates beside the artichokes. If sauce is thick, it may be served in lettuce cups.

How to eat: This is one vegetable that can be eaten properly with the fingers. Just pluck off each petal and dip it in a savory sauce—a smooth hollandaise, spicy vinaigrette, béchamel, or melted butter with a dash of fresh lemon juice. When the prized heart is reached, after all the leaves have been removed, eat it with a fork.

Asparagus
Packaging: 15, 16½, and 30 lb pyramid crates
Grades: U.S. No. 1, U.S. No. 2
Size: (varies from state to state)

Marketing Season

Ninety-two percent of fresh asparagus is available from February through June. Peak months are April and May. California produces the most asparagus (55 percent), followed by Washington state (12 percent), and imported asparagus is also used in this country (16 percent).

Varieties

Asparagus varieties are of two general types, based on the color of the spears. The most important group includes the varieties Mary Washington, Martha Washington, Reading Giant, Palmetto, and Argenteuil. Spears of this group become dark green in sunlight. The less important group includes such varieties as Conover's Colossal and Mammoth White. These produce light green or whitish spears. These light-colored varieties should not be confused with white (blanched) asparagus as grown for canning. The older, light green varieties—Conover's Colossal and Mammoth White—and the dark green variety—Palmetto—have been largely replaced by the more rust-resistant Mary Washington and Martha Washington varieties. Reading Giant and Argenteuil are also largely replaced by the Washington varieties. In the early 1900s, asparagus rust threatened the U.S. asparagus industry. In response to this threat, J. B. Norton developed the Washington varieties. He crossed an unknown American male plant with a Reading Giant female from England. The male plant lent remarkable vigor and rust resistance to the resulting fruit, the plant of which was called Martha Washington. Since Norton's time a number of new selections from Washington have been released. The newer improved selections replace the older strains. None of these are immune to rust, but some are more resistant than others. Recently, Waltham and Roberts have been planted widely in the east, and 711 and 500 W in the west. Currently, two new strains, 66 and 72—resistant to Fusarium wilt—are being planted in the west.

Besides having rust resistance, the Washington varieties are of high commercial quality, fully equal or superior to the best other varieties on the market in earliness, vigor of growth, and size and quality of shoot. They are also more uniform in size, shape, and color than the old standard varieties and are very productive of large spears. The Martha

Washington variety is the more rust resistant, but the Mary Washington variety is slightly earlier, more vigorous, and resistant enough to make it more popular for general planting. Three new selections from the Washington varieties have been introduced. These are Paradise, Mary Washington 500, and Mary Washington 499. These are all reported to be rust resistant.

Containers

Market News Service of USDA lists the following containers as those used in the more important producing areas: the pyramid crate holding 26–32 lb net; fiberboard box or carton holding twenty-four 1½ lb consumer packages, total of 36 lb; pony crate, 12 lb; and 8 qt basket, loose, 10 lb. Other packages listed by USDA used in some areas are the bushel basket, 30–35 lb net; squares, 12 lb; ½ crate, 12 lb; used citrus crates or boxes, 48–50 lb; crate holding 7 in. cut spears, 23–25 lb; and Los Angeles lug for tips, 17–18 lb.

Asparagus is shipped in different forms, that is, bunched, loose, or overwrapped. USDA *Market News* reports show that California growers generally use pyramid crates containing sixteen 1½ lb bunches or twelve 2 lb bunches. New Jersey growers use pyramid crates containing twelve 2¼ lb bunches. The pyramid crate has two compartments. Its tapered shape and solid ends prevent shifting during handling and shipping. There are various forms and sizes of pyramid crates. The bottoms are usually lined with paper that is covered with a layer of damp moss or other material. The butts of the bunches are placed on the moss. When packed in this way and kept at a temperature just above freezing, asparagus can be shipped a long distance and arrive in good condition. A package of special interest is a master container of wood and fiberboard that holds 16 consumer units, open-faced cartons each holding 1½ lb of asparagus. Each unit carton contains a water-absorbent pad in the bottom. The package is suitable for going through a hydro-cooler. Increased airfreight shipment of asparagus has brought the development of special containers for such shipment. These containers are composed of 45 percent wax and 55 percent cardboard, weigh 17 oz, and hold 15 lb of asparagus.

Servings and Weights

A pound of asparagus as purchased provides 3.38 portions of four medium cooked spears. For such portions for 25 persons, 7.5 lb are required as purchased, and for 100 persons, 29.75 lb. A pound as purchased provides 2.61 portions of 3 oz of cut, cooked spears. For such portions for 25 persons, 9.75 lb are required, and for 100 persons, 38.5 lb. The yield in cooked, cut spears is 49 percent of the purchased asparagus. A crate of 28 lb net at 49 percent yield gives 73.17 portions of 3 oz of cooked cut spears, and 1.5 crates are required for 100 portions.

Quality

Asparagus should be fresh and firm with closed, compact tips and the entire green portion tender. Asparagus ages rapidly after cutting; the tips become partially open, spread, or wilted, and the stalks become tough and fibrous. Tender asparagus is brittle and easily punctured. Slightly wilted stalks may sometimes freshen in cold water, but are usually undesirable. Angular or flat stalks are apt to be tough and woody.

Asparagus is cut a few inches below the surface of the ground when the spears have developed the desired length aboveground. If growth is rapid, a green shoot 6–10 in. long may be obtained before any part of it has become tough. After a few inches of the tip are green, the white portion below the ground begins to toughen. Thus, the white portion of asparagus as commonly displayed in markets is generally tough, but can be used for flavoring or for soup. At the time of harvest, asparagus has some tough fibers toward the butt end, the extent of the fibrous portion depending on the length and maturity of the spears.

Grades

The U.S. standards for asparagus provide two grades, U.S. No. 1 and U.S. No. 2. U.S. No. 1 consists of stalks of asparagus that are fresh, well-trimmed, and fairly straight; free from decay; and free from damage caused by spreading or broken tips, dirt, disease, insects, or other means. (For definitions of such words as *fresh* and *damage*, see the following section.)

Size: Unless otherwise specified, the diameter of each stalk is not less than ½ in.
Color: Unless otherwise specified, not less than ⅔ of the stalk length is of a green color.

Definitions

Fresh means that the stalk is not limp or flabby. *Well trimmed* means that at least two-thirds of the butt of the stalk is smoothly trimmed in a plane approximately parallel to the bottom of the container and that the butt is not stringy or frayed. *Damage* means any defect, or any combination of defects, that materially detracts from the appearance, or the edible or marketing quality, of the stalk. *Diameter* means the greatest thickness of the stalk measured at a point approximately 1 in. from the butt. *Fairly well trimmed* means that at least one-third of the butt of the stalk is smoothly trimmed in a plane approximately parallel to the bottom of the container, and that the butt is not badly stringy or frayed. *Badly misshapen* means that the stalk is so badly flattened, crooked, or otherwise deformed that its appearance is seriously affected.

Serious damage means any defect, or any combination of defects, that seriously detracts from the appearance, or the edible or marketing quality, of the stalk.

Asparagus—Canned

Grade A

Color: Spears, tips, and points possess good characteristic green, light green, or yellowish green color typical of well-developed asparagus, and the bottom portion of not more than 10 percent or one unit, whichever is larger, may possess typical white or yellowish white color not to exceed 1/8 of the length of the unit.

GREEN TIPPED—Unit possesses a good characteristic green, light green, or yellowish green color with typical white or yellowish white color at the base ends typical of well-developed asparagus; and not more than 20 percent by count may possess typical white or yellowish white color in excess of 1/2 of the length of the unit, or may be all green.

GREEN TIPPED AND WHITE—Possess good characteristic white or yellowish white color, and may possess green, light green, or yellowish green heads and adjacent areas, not to exceed 1/2 of the length.

WHITE—May possess good characteristic white or yellowish white color typical of well-developed asparagus; and not more than 10 percent or one unit, whichever is larger, may possess green, light green, or yellowish green heads and adjacent areas, not to exceed 1/2 of the length.

Defects: SPEARS, TIPS, AND POINTS: GREEN AND GREEN TIPPED—Not more than 10 percent, and GREEN TIPPED, GREEN TIPPED AND WHITE, AND WHITE—Not more than 15 percent should have shattered heads, misshapen, poorly cut, damaged, and seriously damaged units, provided that not more than 3 percent may be seriously damaged; or one unit in a single container may be seriously damaged if such unit exceeds the allowance of 3 percent, provided that in all of the containers comprising the sample such damaged units do not exceed an average of 3 percent of the total number of units.

Character: Not less than 85 percent are well developed and the remainder at least fairly well developed. Tough Units—Maximum tolerance SPEARS, TIPS, AND POINTS: GREEN and GREEN TIPPED 10 percent and WHITE 20 percent.

Asparagus—Frozen

Grade A

Is of good similar varietal characteristics; has a good flavor and odor; has no grit or silt present that affects the appearance or edibility, in which no more than 5 percent, by weight, of loose material may be present; and has an attractive appearance and eating quality within the limits specified for the various quality factors.

Grade B

Is of similar varietal characteristics; has a good flavor and odor; has no more than a trace of grit or silt present that slightly affects the appearance or edibility.

INTERNET RESOURCES

www.pma.com

The Produce Marketing Association, founded in 1949, is a trade association that serves members who market fresh fruits, vegetables, and related products worldwide. Members are involved in the production, distribution, retail, and foodservice sectors of the industry. Their core purpose is to sustain and enhance an environment that advances the marketing of produce and related products and services.

www.ams.usda.gov/howtobuy/fveg.htm

How to Buy Fresh Vegetables, by the U.S. Department of Agriculture Agricultural Marketing Service (USDA AMS).

CHAPTER

4

Dairy Products

MILK AND MILK PRODUCTS

Every state, and most municipalities, operate a rigid milk regulation program. The Grade A Pasteurized Milk Ordinance recommended by the U.S. Public Health Service, first published in 1924 and revised periodically, sets forth in detail the type of regulation that is desirable. This ordinance is formulated as a guide for state and local communities and represents recommendations of federal and state Departments of Agriculture; national, state, and local health authorities; and officials of the dairy industry. The milk supplies of much of the United States are covered by the ordinance.

State and local governments exercise legal responsibility for health protection of milk supplies for the market, for the most part. The dairy industry works cooperatively with both states and municipalities to maintain the standards for the milk supply. The industry is constantly interested in technological advances, in procedures and equipment, and in research to make sure any technological change produces quality milk.

Supervision of Milk Supply

The guidelines that govern a quality milk supply can be summarized as follows:

1. Inspection and sanitary control of farms and milk plants
2. Examination and testing of herds for the elimination of bovine diseases related to public health
3. Regular instruction on desirable sanitary practices for persons engaged in production, processing, and distribution of milk
4. Proper pasteurization of milk
5. Laboratory examination of milk
6. Monitoring of milk supplies by federal, state, and local health officials to protect against unintentional additives

Grades of Milk

Grade A designates quality fluid milk and is the grade purchased in retail stores and delivered to consumers. Milk used for manufacturing milk products—butter, cheese, and ice cream—is designated "Manufacturing Grade." These grades are based on the conditions under which the milk is produced and handled, and on the bacterial count of the final product. The grades and their meanings vary only slightly, according to local regulations. Where the U.S. Public Health Service Grade A Pasteurized Milk Ordinance has been adopted, uniform standards are prescribed.

In different parts of the world, the milk of various species of animals is used for food. In the United States, however, cows furnish virtually all the available market milk. Therefore, unless otherwise specified, the term *milk*, as used in this book, refers to cows' milk.

This food is among the most perishable of all foods. Milk, as it comes from the cow, provides an excellent medium for bacterial growth. It is subject to many possible flavor changes unless it is constantly protected from contamination each step of the way from the cow to the consumer. Milk is relied on as an important source of many of the nutrients known to be necessary for proper development and maintenance of the human body. Maximum retention of these nutrients must be ensured as milk is stored, processed, transported, and distributed in its many forms.

The rules for the care of milk are these: Keep it clean, covered, and cold, and consume it within a reasonable period of time. Milk should be protected from sunlight to preserve its riboflavin and its good flavor. Off-flavor in milk does not make it unsafe, merely unpleasant to the taste. Milk should be stored promptly in the refrigerator or consumed quickly. After a milk container is opened, it should be covered before storage. Milk removed from the original container should not be returned to it.

The Attributes of Quality Milk

Quality milk has been described as milk that has a low bacterial count, good flavor and appearance, satisfactory keeping quality, and high nutritive value. It is free of disease-producing bacteria, toxic substances, and foreign material.

Progress in dairy technology and public health measures results in milk that can be depended on as a safe, nutritious, pleasing food, even though it may be produced hundreds or thousands of miles away. Vigilance is continuously exercised in maintaining this quality as new challenges arise in the environment. Pasteurization is a basic safeguard in the processing of all milk.

Pasteurized Milk

Proper pasteurization is the only practical commercial measure today that destroys all disease organisms in fluid milk. Examination of cows and milk handlers can be done only at intervals. Disease organisms may enter milk accidentally from sources such as flies, contaminated water, and utensils. Pasteurization is by no means a substitute for sanitary dairy practices, but is rather an additional safeguard for the consumer's health.

How Milk is Pasteurized

Pasteurization is defined as the process of heating milk to at least 145° F and holding it at, or above, this temperature continuously for at least 30 minutes in approved and properly operated equipment; or heating it to at least 161° F and holding it at, or above, this temperature continuously for at least 15 seconds in approved and properly operated equipment. The latter method is predominant today. Following this treatment, the milk is cooled promptly to 45° F or lower.

Effects of Pasteurization

Pasteurization destroys all pathogenic organisms and most nonpathogenic bacteria, so that milk may be safely consumed. This process also improves the keeping quality of milk. The food value is not changed significantly. There is no apparent undesirable effect on the protein, fat, carbohydrate, or mineral content of milk, nor on vitamins A, D, E, riboflavin, niacin, pyridoxine, or biotin. Both vitamins A and E are subject to oxidative deterioration, but they appear to be quite stable in fluid dairy products. Only slight losses, if any, occur in vitamins B_{12} and K and pantothenic acid. Greater losses of ascorbic acid and thiamine have been reported, but the level of ascorbic acid in milk is not of special importance. A varied American diet includes other sources rich in ascorbic acid. Although pasteurization reduces thiamine somewhat, milk still supplies a significant amount of it in the daily diet. Carefully controlled, high-temperature, short-time pasteurization permits maximum retention of ascorbic acid and thiamine.

Homogenized Milk

Pasteurized milk is also homogenized. Today most whole milk is homogenized immediately after it is pasteurized. The milk is treated mechanically to break up the fat into smaller globules and then to disperse these globules permanently in a fine emulsion throughout the milk. The heated milk is channeled to the homogenizer, where the milk, under high pressure, is forced through very tiny openings. Nothing is added or removed.

Effects of Homogenization

Physically, homogenized milk differs from ordinary whole milk in that there is no separation of cream, and the product remains uniform throughout. Small differences in color and viscosity may be noted in the homogenized milk, and there is a slightly richer taste. Homogenization lowers the curd tension, which results in the formation of a softer curd during human digestion.

Kinds of Milk

Today milk appears on the market in many forms to appeal to the varied tastes of consumers and to satisfy their demands. A wide range of products has been developed to improve keeping quality, facilitate distribution and storage, make maximum use of by-products, and preserve surplus. The processing involved in producing each form of milk is designed and controlled to protect the health of the consumer. Some forms of milk are available in all communities; others may be found in only a few communities. State and local governments generally establish standards of composition for all fluid milk products. Federal standards of identity have been established only for evaporated, condensed, and nonfat dry milk.

Whole Fresh Fluid Milk

Composition Although the composition of milk is variable between cows and seasons of the year, minimum standards for the composition of whole milk have been set by individual states. Many states define whole milk as milk that contains not less than 3.25 percent milk fat and not less than 8.25 percent milk solids-not-fat. At the milk plant, the milk from different farms is pooled and "standardized" to meet or exceed the minimum legal requirements.

The standards of composition, however, vary from state to state. Even within a state, the milk composition may be well above the minimum standard. Milk of higher milk solids-not-fat and milk fat is available in most markets. This may be a premium product or milk from certain breeds of cows that is sold by breed name in many communities.

Basic Processing Most of the whole fluid milk marketed in this country is pasteurized and homogenized. Milk that receives no heat treatment is called raw milk. Unless it is certified milk, raw milk should be pasteurized at home, or boiled, before it is consumed.

Certified Milk

Certified milk originated in 1893 in response to a need for safe milk. Its certification on the container means that the conditions under which it was produced and distributed conform to the high standards for cleanliness set forth by the American Association of Medical Milk Commissions. These standards have been recognized in many state and local laws. The Grade A Pasteurized Milk Ordinance provides for the sale of certified pasteurized milk derived from certified raw milk.

Certified milk continues to be available in only a few communities. Where it is available, it may be either raw or pasteurized, though most certified milk is now being pasteurized. It may also be homogenized and may be fortified with vitamin D.

Soft Curd Milk

For certain uses, it is considered desirable to modify cows' milk so that the curd tension is considerably less, and the curd formed in digestion is softer. In digestion, the curd formed from soft curd milk tends to leave the stomach more quickly than the curd of ordinary milk.

Commercial Preparation Soft curd milk may be produced by homogenization, enzymatic treatment, sonic vibration, ion exchange, or the addition of various salts. Now that almost all fresh fluid milk sold is homogenized, and thus has a soft curd, the product labeled "soft curd milk," prepared by the other processes, is seldom seen.

Low-Sodium Milk

A process of ion exchange can remove 90 percent or more of the sodium that occurs naturally in milk. Fresh whole milk is passed through an ion-exchange resin to replace the sodium it contains with potassium. The milk is pasteurized and homogenized.

During the ion-exchange process, some B vitamins and calcium are lost. Despite this loss, low-sodium milk has special use in certain sodium-restricted diets. It permits the inclusion of milk and other protein foods that may otherwise have to be severely limited because of their high sodium content. Low-sodium milk is available in various parts of the country as a dry, canned, or fresh product.

Fortified Milks

Fortified milks are those containing added amounts of one or more of the essential nutrients normally present in milk.

Vitamin D Milk The Council on Foods and Nutrition of the American Medical Association recognized the fortification of milk with vitamin D as being of significance to public health. Food in general does not contain appreciable quantities of vitamin D. The primary source for this vitamin is the action of sunlight on the skin. A small but not physiologically significant amount occurs normally in milk. However, milk is the only food the Council on Foods and Nutrition of the American Medical Association has approved for fortification with vitamin D. Milk provides the proportion of calcium and phosphorus that must be present with vitamin D for normal calcification of bones and teeth.

To meet the requirements for acceptance by the Council, vitamin D milk must contain 400 International Units (IU) of vitamin D per quart, usually added in the form of a concentrate.

Multiple Fortified Milk Milk can be fortified with substances such as vitamins A and D, multivitamin preparations, minerals, lactose, and nonfat dry milk. The substance added, and the amount of fortification, will vary, depending on the dairy company. The dairy company, in turn, conforms to the state standards for multiple fortified milk, where these standards exist. Usually the products and amounts added must be declared on the label. Fortified milks also vary in fat content. They can be made with whole, partially skimmed, or skim milk.

Concentrated Milks

Concentrated milks may be fresh, frozen, evaporated, condensed, or dried. Milks are concentrated by the removal of varying amounts of water, under carefully controlled heat and vacuum conditions. All may be reconstituted by the addition of appropriate quantities of water.

Concentrated Fresh Milk Fresh whole milk is concentrated by first pasteurizing and homogenizing, and then removing two-thirds of the water under vacuum. This 3:1 concentrate, standardized to about 10.5 percent milk fat, is rehomogenized, repasteurized, and packaged. Although perishable, concentrated fresh milk may retain its flavor and sweetness for as long as six weeks when stored at near-freezing temperature. This milk is available in only a few communities.

Concentrated Frozen Milk To increase the keeping quality of fresh concentrated milk, it may be quickly frozen and held at –10° to 20° F for several months. Like other frozen foods, it must be used soon after defrosting. This milk, too, is available in only a few communities.

Concentrated Canned Milk Considerable research is under way on the production of concentrated, sterilized milk, aseptically packaged in cans. The product will keep three months on the shelf or six months if refrigerated.

Evaporated Milk

In the manufacture of evaporated milk, slightly more than half the water is removed by heating pasteurized whole milk at 122°–131° F in vacuum pans. After evaporation, the milk is homogenized, and usually vitamin D is added to provide 400 IU per reconstituted quart. The evaporated milk is sealed in cans and sterilized at about 239° F for 15 minutes, thus preventing bacterial spoilage. A can of evaporated milk requires no refrigeration until opened.

Condensed Milk

In the preparation of condensed milk, sugar is added to the milk before the evaporation process is initiated. This milk contains not less than 28.0 percent milk solids and not less than 8.3 percent milk fat. The sugar, which represents about 40–45 percent of the condensed milk, acts as a preservative. The milk, sealed in cans, can be stored without further heat treatment and without refrigeration.

Nonfat Dry Milk

Nonfat dry milk is made of fresh whole milk from which both water and fat have been removed. After the fat has been removed, the skim milk is pasteurized and about two-thirds of the water removed under vacuum. Spraying it into a chamber of hot, filtered air dries this concentrated skim milk. The resulting product is a fine-textured powder of very low moisture content. A further step, the instantizing process, produces dry milk that dissolves in water instantly. In one method, the dry milk is moistened with steam, then redried. Except for small losses in ascorbic acid, vitamin B_{12}, and biotin, the processing has no appreciable effect on the nutritive value. The presence of the milk fat does require special packaging to prevent oxidation during storage.

Skim Milk

Skim milk is milk from which fat has been removed by centrifugation to reduce its milk fat content to less than that of whole milk. In the skim milk ordinarily available, the fat content is 0.1 percent, although it may be lower or higher. Various states have established standards ranging from 8.0 to 9.25 percent as the minimum for the total solids in skim milk. The product is pasteurized.

With the exception of milk fat and the vitamin A contained in the milk fat, the usual nutrients of milk—the protein, lactose, minerals, and water-soluble vitamins—remain, for the most part, in the skim milk. Because the vitamin A in whole milk is removed with the fat, a water-soluble vitamin A and D concentrate is frequently added to skim milk. Such fortified skim milk often contains 2,000 IU of vitamin A and 400 IU of vitamin D per quart. However, the amounts may vary, as discussed in the earlier section "Fortified Milks." Fortified skim milk may also contain additional milk solids-not-fat.

1 Percent Milk

As its name implies, 1 percent milk contains 1 percent milk fat. Made from fresh whole and skim milks, 1 percent milk is pasteurized and homogenized. Adding milk solids-not-fat and various vitamin and mineral preparations may enrich it.

2 Percent Milk

As its name implies, 2 percent milk contains 2 percent milk fat. Made from fresh whole and skim milks, 2 percent milk is pasteurized and homogenized. Adding milk solids-not-fat and various vitamin and mineral preparations may enrich it.

Cultured Milks

Cultured milks are prepared from pasteurized (or sterilized) milk. Certain desirable bacterial cultures, whose growth under controlled conditions of sanitation, inoculation, and temperature yield a variety of milks, have been added. These fermented milks may exert a favorable influence on the flora of the intestinal tract.

Buttermilk

Commercially produced buttermilk is a cultured product. Today it is not the by-product derived from churning cream into butter. Most of the cultured buttermilk marketed in the United States is made of fresh skim milk. However, cultured buttermilk may be made from fresh fluid whole milk, concentrated fluid milk (whole or skim), or reconstituted nonfat dry milk.

Pasteurized skim milk is cultured chiefly with *Streptococcus lactis* and incubated at 68°–72° F until the acidity is 0.8–0.9 percent, expressed as lactic acid. The result is milk with a characteristic tangy flavor and smooth, rich body. Butter granules are sometimes added in an amount to produce buttermilk testing 1 or less than 1 percent milk fat. The concentration of milk solids-not-fat is similar to that of whole milk.

Acidophilus Milk

Pasteurized skim milk, cultured with *Lactobacillus acidophilus* and incubated at 100° F, is called acidophilus milk. This tart milk, available in only a few communities, is sometimes used to combat excessive intestinal putrefaction by changing the bacterial flora of the intestine. As a therapeutic product, its use may be prescribed after antibiotic treatment to help reestablish a normal balance of bacterial flora in the intestine.

Yogurt

Possessing a consistency resembling that of custard, yogurt is usually manufactured from fresh, partially skimmed milk, enriched with added milk solids-not-fat. Fermentation is accomplished by a mixed culture of one or more strains of organisms, such as *Streptococcus thermophilous*, *Bacterium bulgaricum*, and *Plocamobacterium yoghourtii*. The milk is pasteurized and homogenized, inoculated, and incubated at 112°–115° F. The final product has a tangy flavor and contains between 11 and 12 percent milk solids. It is available in varied flavors.

Effects of Fermentation In fermented milks, changes due to bacterial growth include formation of lactic acid from lactose and coagulation of the milk protein casein. Bacterial enzymatic action on protein and fat constituents, plus the effect of the increased concentration of acid, changes the physical properties and chemical structure of the milk.

A thicker body and a pleasing flavor and aroma are developed in the finished product. These characteristics vary with the type of culture and kind of milk used, the concentration of milk fat and milk solids-not-fat, the fermentation process, and the temperature at which it is carried out. Some alteration in the vitamin concentration may occur, but there is no evidence of major changes in these nutrients. These products are said to promote biological synthesis of vitamins within the small intestine.

Flavored Milks and Milk Drinks

Flavored milk is whole milk with syrup or powder, containing a wholesome flavoring agent, and sugar added. A flavored milk drink, or dairy drink, is skim or partially skimmed milk similarly flavored and sweetened. These milks are pasteurized and usually homogenized.

Chocolate Milk Whole milk flavored with a chocolate syrup or powder is called chocolate milk. Usually, its milk fat content is the same as that of whole milk, and it contains 1 percent cocoa or 1½ percent liquid chocolate, plus 5 percent sugar, and less than 1 percent stabilizers.

Chocolate Dairy Drink Skim or partially skimmed milk flavored with chocolate syrup or powder is called chocolate dairy drink. Frequently, its milk fat content is about 2.3 percent and its milk solids-not-fat about 90 percent of the amount in skim milk. Otherwise, it contains the same ingredients as chocolate milk and is processed in the same manner.

Food Value of Chocolate in Milk Research indicates that the addition of normal quantities of good-grade chocolate has no appreciable effect on the availability of either the calcium or the protein of milk to human beings. Therefore, the nutritive value of milk is not significantly altered by the addition of this flavoring, except in regard to the increased caloric value, chiefly from the added sugar. The sugar and chocolate content brings the caloric value of chocolate dairy drinks made of skim milk to a slightly higher level than that of plain whole milk, but to a lower level than that of chocolate milk.

Canned Whole Milk

Whole milk that is homogenized, sterilized at 270°–280° F for eight to ten seconds, and canned aseptically is available chiefly for use on ships or for export. It can be stored at room temperature until opened, after which it requires refrigeration.

Frozen Whole Milk

Homogenized, pasteurized whole milk can be quickly frozen and kept below –10° F for six weeks to three months. Like concentrated frozen milk, it must be used soon after defrosting. This milk, used on ships and at overseas military installations, is not ordinarily available in retail markets.

Effect of Freezing Freezing does not measurably change the nutritive value of milk. Freezing causes a destabilization of the protein, however, and particles of precipitated protein may be visible on a glass container when the milk thaws. On thawing, there is also a tendency for the fat to separate. Milk that has been accidentally frozen (on the doorstep in winter, for example) is quite safe to use. If, during freezing, the cap has been pushed out of the bottle, it is a wise precaution to boil the milk for all uses.

Federal Regulations for Milk and Cream

The following paragraphs outline the federal regulations on milk and cream. These regulations should constitute the buyer's specifications for milk purchases. The regulations pertain except where state standards exceed the federal standards. There are some definitions of the dairy descriptions to be noted. "Cream" means the liquid milk product, high in fat and separated from milk, which may have been adjusted by adding milk, concentrated milk, dry whole milk, skim milk, concentrated skim milk, or nonfat dry milk. Cream contains not less than 18 percent milk fat. The name "whipped cream" should not be applied to any product other than one made by whipping the cream that complies with the standards of identity for whipping cream. If flavoring or sweetening is added, the resulting product should be so identified.

"Pasteurized," when used to describe a dairy product, means that every particle of such product shall have been heated in properly operated equipment to one of the temperatures specified in the table at the end of this paragraph, and held continuously at or above that temperature for the specified time (or other time/temperature relationship that has been demonstrated to be equivalent thereto in microbial destruction).

TEMPERATURE*	TIME
145° F	30 minutes
161° F	15 seconds
191° F	1 second
204° F	0.05 second
221° F	0.01 second

*If the dairy ingredient has a fat content of 10 percent or more, or if it contains added sweeteners, the specified temperature shall be increased by 5 degrees.

"Ultra-pasteurized," when used to describe a dairy product, means that the product shall have been thermally processed at or above 280° F for at least 2 seconds, either before or after packaging, so as to produce a product that has an extended shelf life under refrigerated conditions.

Milk

Milk is the lacteal secretion, practically free from colostrum, obtained by the complete milking of one or more healthy cows. Milk that is in final package form for beverage use shall have been pasteurized or ultra-pasteurized, and shall contain not less than 8¼ percent milk solids-not-fat and not less than 3¼ percent milk fat. Milk may have been adjusted by separating part of the milk fat therefrom, or by adding thereto cream, concentrated milk, dry whole milk, skim milk, concentrated skim milk, or nonfat dry milk. Milk may be homogenized.

Vitamin Addition (Optional) If added, vitamin A shall be present in such quantity that each quart of the food contains not less than 2,000 International Units thereof, within limits of good manufacturing practice. If added, vitamin D shall be present in such quantity that each quart of the food contains 400 International Units thereof, within limits of good manufacturing practice.

The following safe and suitable ingredients may be used:

1. Carriers for vitamins A and D
2. Characterizing flavoring ingredients (with or without coloring, nutritive sweetener, emulsifiers, and stabilizers, as follows:
   (a) Fruit and fruit juice (including concentrated fruit and fruit juice)
   (b) Natural and artificial food flavorings

The following terms shall accompany the name of the food wherever it appears on the principal display panel or panels of the label, in letters not less than one-half the height of the letters used in such name:

If vitamins are added, the phrase "vitamin A" or "vitamin A added" or "vitamin D" or "vitamin D added," or "vitamins A and D" or "vitamins A and D added," as is appropriate.
The word "vitamin" may be abbreviated "vit."
The word "ultra-pasteurized," if the food has been ultra-pasteurized.
The word "pasteurized," if the food has been pasteurized.
The word "homogenized," if the food has been homogenized.

Lowfat Milk

Lowfat milk is milk from which sufficient milk fat has been removed to produce a food having, within limits of good manufacturing practice, one of the following milk fat contents: ½, 1, 1½, or 2 percent. Lowfat milk is pasteurized or ultra-pasteurized, contains added vitamin A, and contains not less than 8¼ percent milk solids-not-fat. Lowfat milk may be homogenized. Vitamin A shall be present in such quantity that each quart of the food contains not less than 2,000 International Units thereof, within limits of good manufacturing practice. Addition of vitamin D is optional. If added, vitamin D shall be present in such quantity that each quart of the food contains 400 International Units thereof, within limits of good manufacturing process.

The following safe and suitable ingredients may be used:

1. Carriers for vitamins A and D
2. Concentrated skim milk, nonfat dry milk, or other milk-derived ingredients to increase the nonfat solids content of the food, provided that the ratio of protein to total nonfat solids of the food and the protein efficiency ratio of all protein present shall not be decreased as a result of adding such ingredients

3. Characterizing flavoring ingredients (with or without coloring, nutritive sweetener, emulsifiers, and stabilizers) as follows:
   (a) Fruit and fruit juice (including concentrated fruit and fruit juice)
   (b) Natural and artificial food flavorings

The name of the food is "lowfat milk." The name of the food shall appear on the label in type of uniform size, style, and color. The name of the food shall be accompanied on the label by a declaration indicating the presence of any characterizing flavoring.

The following terms shall accompany the name of the food wherever it appears on the principal display panel or panels of the label, in letters not less than one-half the height of the letters used in such name:

1. The phrase "____% milk fat," the blank to be filled in with the fraction ½, or multiple thereof, to indicate the actual fat content of the food.
2. The phrase "vitamin A" or "vitamin A added," or if vitamin D is added, the phrase "vitamins A and D added." The word "vitamin" may be abbreviated "vit."
3. The word "ultra-pasteurized," if the food has been ultra-pasteurized.
4. The phrase "protein fortified" or "fortified with protein," if the food contains not less than 10 percent milk-derived nonfat solids.

The following terms may appear on the label:

The word "pasteurized," if the food has been pasteurized.
The word "homogenized," if the food has been homogenized.

Skim Milk

Skim milk is milk from which sufficient milk fat has been removed to reduce its milk fat content to less than 0.5 percent. Skim milk that is in final package form for beverage use shall have been pasteurized or ultra-pasteurized, shall contain added vitamin A, and shall contain not less than 8¼ percent milk solids-not-fat. Skim milk may be homogenized.

Vitamin A shall be present in such quantity that each quart of the food contains not less than 2,000 International Units thereof, within limits of good manufacturing practice. Addition of vitamin D is optional. If added, vitamin D shall be present in such quantity that each quart of the food contains 400 International Units thereof, within limits of good manufacturing practice.

The following safe and suitable ingredients may be used:

1. Carriers for vitamins A and D.
2. Concentrated skim milk, nonfat dry milk, or other milk-derived ingredients to increase the nonfat solids content of the food, provided that the ratio of protein to total nonfat solids of the food and the protein efficiency ratio of all protein present shall not be decreased as a result of adding such ingredients.
3. When one or more of the optional, milk-derived ingredients, emulsifiers, stabilizers, or a combination of both, is used, they must be used in an amount not more than 2 percent by weight of the solids in such ingredients.
4. Characterizing flavoring ingredients (with or without coloring, nutritive sweetener, emulsifiers, and stabilizers) as follows:
   (a) Fruit and fruit juice (including concentrated fruit and fruit juice)
   (b) Natural and artificial food flavorings

The name of the food is "skim milk" or, alternatively, "nonfat milk." The name of the food shall appear on the label in type of uniform size, style, and color. The name of the food shall be accompanied on the label by a declaration indicating the presence of any characterizing flavoring.

The following terms shall accompany the name of the food wherever it appears on the principal display panel or panels of the label, in letters not less than one-half the height of the letters used in such name:

The phrase "vitamin A" or "vitamin A added," or if vitamin D is added, the phrase "vitamins A and D" or "vitamins A and D added." The word "vitamin" may be abbreviated "vit."
The word "ultra-pasteurized," if the food has been ultra-pasteurized
The phrase "protein fortified" or "fortified with protein," if the food contains not less than 10 percent milk-derived nonfat solids
The word "pasteurized," if the food has been pasteurized
The word "homogenized," if the food has been homogenized

Half-and-Half

Half-and-half is the food consisting of a mixture of milk and cream that contains more than 10.5 percent but less than 18 percent milk fat. It is pasteurized or ultra-pasteurized and may be homogenized.

The following safe and suitable optional ingredients may be used:

1. Emulsifiers
2. Stabilizers
3. Nutritive sweeteners
4. Characterizing flavoring ingredients (with or without coloring) as follows:
   (a) Fruit and fruit juice (including concentrated fruit and fruit juice)
   (b) Natural and artificial food flavoring

The name of the food is "half-and-half" and shall be accompanied on the label by a declaration indicating the presence of any characterizing flavoring.

The following terms shall accompany the name of the food wherever it appears on the principal display panel or panels of the label, in letters not less than one-half the height of the letters used in such name:

The word "ultra-pasteurized," if the food has been ultra-pasteurized
The word "sweetened," if no characterizing flavor ingredients are used, but nutritive sweetener is added
The word "pasteurized," if the food has been pasteurized
The word "homogenized," if the food has been homogenized

Light Cream

Light cream is cream that contains more than 18 percent but less than 30 percent milk fat. It is pasteurized or ultra-pasteurized and may be homogenized.

The following safe and suitable ingredients may be used:

1. Emulsifiers
2. Stabilizers
3. Nutritive sweeteners
4. Characterizing flavoring ingredients (with or without coloring) as follows:
   (a) Fruit and fruit juice (including concentrated fruit and fruit juice)
   (b) Natural and artificial food flavoring

The name of the food is "light cream" or, alternatively, "coffee cream" or "table cream." The following terms shall accompany the name of the food wherever it appears on the principal display panel or panels of the label, in letters shall be not less than one-half the height of the letters used in such name:

The word "ultra-pasteurized," if the food has been ultra-pasteurized
The word "sweetened," if no characterizing flavoring ingredients are used, but nutritive sweetener is added
The word "pasteurized," if the food has been pasteurized
The word "homogenized," if the food has been homogenized

Light Whipping Cream

Light whipping cream is cream that contains more than 30 percent, but less than 36 percent, milk fat. It is pasteurized or ultra-pasteurized, and may be homogenized.

The following safe and suitable optional ingredients may be used:

1. Emulsifiers
2. Stabilizers
3. Nutritive sweeteners
4. Characterizing flavoring ingredients (with or without coloring) as follows:
   (a) Fruit and fruit juice (including concentrated fruit and fruit juice)
   (b) Natural and artificial food flavoring

The name of the food is "light whipping cream" or, alternatively, "whipping cream." The name of the food shall be accompanied on the label by a declaration indicating the presence of any characterizing flavoring.

The following terms shall accompany the name of the food wherever it appears on the principal display panel or panels of the label, in letters not less than one-half the height of the letters used in such name:

The word "ultra-pasteurized," if the food has been ultra-pasteurized
The word "sweetened," if no characterizing flavoring ingredients are used, but nutritive sweetener is added

The following terms may appear on the label:

The word "pasteurized," if the food has been pasteurized
The word "homogenized," if the food has been homogenized

Heavy Cream

Heavy cream is cream that contains more than 36 percent milk fat. It is pasteurized or ultra-pasteurized and may be homogenized.

The following safe and suitable optional ingredients may be used:

1. Emulsifiers
2. Stabilizers
3. Nutritive sweeteners
4. Characterizing flavoring ingredients (with or without coloring) as follows:
   (a) Fruit and fruit juice (including concentrated fruit and fruit juice)
   (b) Natural and artificial food flavoring

The name of the food is "heavy cream" or, alternatively, "heavy whipping cream." The name of the food shall be accompanied on the label by a declaration indicating the presence of any characterizing flavoring.

The following terms shall accompany the name of the food wherever it appears on the principal display panel or panels of the label, in letters not less than one-half the height of the letters used in such name:

The word "ultra-pasteurized," if the food has been ultra-pasteurized
The word "sweetened," if no characterizing flavoring ingredients are used, but nutritive sweetener is added

The following terms may appear on the label:

The word "pasteurized," if the food has been pasteurized
The word "homogenized," if the food has been homogenized

When used in food, each of the ingredients previously specified, in this section shall be declared on the label.

Evaporated Milk

Evaporated milk is the liquid food obtained by the partial removal of water from milk. The milk fat and total milk solids contents of the food are not less than 7.5 and 25.5 percent, respectively. It is homogenized and sealed in a container and so processed by heat, either before or after sealing, as to prevent spoilage.

Evaporated milk contains added vitamin D, which shall be present in such quantity that each fluid ounce of the food contains 25 International Units thereof, within limits of good manufacturing practice. Addition of vitamin A is optional. If added, vitamin A shall be present in such quantity that each fluid ounce of the food contains not less than 125 International Units thereof, within limits of good manufacturing practice.

The following safe and suitable ingredients may be used:

1. Carriers for vitamins A and D
2. Emulsifiers
3. Stabilizers

The name of the food is "evaporated milk." The phrase "vitamin D" or "vitamin D added," or "vitamins A and D" or "vitamins A and D added," as is appropriate, shall immediately precede or follow the name of the food wherever it appears on the principal display panel or panels of the label, in letters not less than one-half the height of the letters used in such name.

Concentrated Milk

Concentrated milk is the liquid food obtained by partial removal of water from milk. The milk fat and total milk solids contents of the food are not less than 7.5 and 25.5 percent, respectively. It is pasteurized, but is not processed by heat so as to prevent spoilage. It may be homogenized. If added, vitamin D shall be present in such quantity that each fluid ounce of the food contains 25 International Units thereof, within limits of good manufacturing practice. Safe and suitable carriers may be used for vitamin D.

The name of the food is "concentrated milk" or, alternatively, "condensed milk." If the food contains added vitamin D, the phrase "vitamin D" or "vitamin D added" shall accompany the name of the food wherever it appears on the principal display panel or panels of the label, in letters not less than one-half the height of the letters used in such name. The word "homogenized" may appear on the label if the food has been homogenized.

Sweetened Condensed Milk

Sweetened condensed milk is the food obtained by the partial removal of water only from a mixture of milk and safe and suitable nutritive sweetener. The finished food contains not less than 8.5 percent by weight of milk fat, and not less than 28 percent by weight of total milk solids. The quantity of nutritive sweetener used is sufficient to prevent spoilage. The food is pasteurized and may be homogenized.

The name of the food is "sweetened condensed milk." The word "homogenized" may appear on the label if the food has been homogenized.

The optional sweetener used shall be declared on the label.

Nonfat Dry Milk

Nonfat dry milk is the product obtained by removal of water only from pasteurized skim milk. It contains not more than 5 percent by weight of moisture, and not more than 1½ percent by weight of milk fat, unless otherwise indicated.

The name of the food is "nonfat dry milk." If the fat content is over 1½ percent by weight, the name of the food on the principal display panel or panels shall be accompanied by the statement "contains ____% milk fat," the blank to be filled in with the percentage to the nearest one-tenth of 1 percent of fat contained, within limits of good manufacturing practice.

Nonfat Dry Milk Fortified with Vitamins A and D

Nonfat dry milk fortified with vitamins A and D conforms to the standard of identity for nonfat dry milk, except that vitamins A and D are added as prescribed in the following paragraph.

Vitamin Addition Vitamin A is added in such quantity that, when prepared according to the label directions, each quart of the reconstituted product contains 2,000 International Units thereof.

Vitamin D is added in such quantity that, when prepared according to label directions, each quart of the reconstituted product contains 400 International Units thereof.

These requirements will be deemed to have been met if reasonable overages, within limits of good manufacturing practice, are present to ensure that the required levels of vitamins are maintained throughout the shelf life of the food expected under customary conditions of distribution.

Dry Whole Milk

Dry whole milk (made by the spray process or the atmospheric roller process) is the product resulting from the removal of water from milk and contains the lactose, milk proteins, milk fat, and milk minerals in the same relative proportions as in the fresh milk from which it is made. The term "milk" means milk produced by healthy cows and pasteurized at a temperature of 161° F for 15 seconds, or its equivalent in bacterial destruction, before or during the manufacture of the dry whole milk.

U.S. Grade The U.S. grades of dry whole milk are determined on the basis of flavor and odor, physical appearance, bacterial estimate, butterfat content, coliform estimate, copper content, iron content, moisture content, oxygen content, scorched particle content, solubility index, and titratable acidity.

Nonfat Dry Milk (Spray Process)

Nonfat dry milk is the product resulting from the removal of fat and water from milk and contains the lactose, milk proteins, and milk minerals in the same relative proportions as in the fresh milk from which it is made. It contains not over 5 percent by weight of moisture. The fat content shall not exceed 1½ percent by weight. The term "milk," when used in this section, means fresh, sweet milk, produced by healthy cows, that has been pasteurized before or during the manufacture of the nonfat dry milk.

U.S. Grade The U.S. grades of nonfat dry milk (spray process) are determined on the basis of flavor and odor, physical appearance, bacterial estimate on the basis of standard plate count, butterfat content, moisture content, scorched particle content, solubility index, and titratable acidity. The final U.S. grade is established on the basis of the lowest rating of any one of the quality characteristics.

Instant Nonfat Dry Milk

Instant nonfat dry milk is nonfat dry milk that has been produced in such a manner as to improve substantially its dispersing and reliquefication characteristics over those produced by the conventional processes.

Nonfat dry milk is the product resulting from the removal of fat and water from milk, and contains the lactose, milk proteins, and milk minerals in the same relative proportion as in the fresh milk from which it is made. It contains not over 5 percent by weight of moisture. The fat content is not over 1½ percent by weight, unless otherwise indicated. The term "milk," when used in this section, means fresh, sweet milk, produced by healthy cows, that has been pasteurized before or during the manufacture of the instant nonfat dry milk.

U.S. Grade The only U.S. grade is U.S. Extra. The U.S. grade of instant nonfat dry milk is determined on the basis of flavor and odor, physical appearance, bacterial estimate on the basis of standard plate count, coliform count, milk fat content, moisture content, scorched particle content, solubility index, titratable acidity, and dispersibility. The final U.S. grade is established on the basis of the lowest rating of any one of the quality characteristics.

Dry Buttermilk

Dry buttermilk (made by the spray process or the atmospheric roller process) is the product resulting from drying liquid buttermilk, derived from the manufacture of sweet cream butter to which no alkali or other chemical has been added. It has been pasteurized either before or during the process of manufacture, at a temperature of 161° F for 15 seconds, or its equivalent in bacterial destruction.

CHEESE

Cheese is a highly nutritious food. It is of value in diet because it contains almost all the protein, essential minerals, vitamins, and other nutrients of milk in concentrated form.

It is difficult to classify the different cheeses satisfactorily into groups. There are two main ways to classify the different cheeses. One method is to group them into the 18 distinct types or families of natural cheese. The following cheeses are typical of the 18 kinds or families:

Brick	Gouda	Romano
Camembert	Hand	Roquefort
Cheddar	Limburger	Sapsago
Cottage	Neufchatel	Swiss
Cream	Parmesan	Trappist
Edam	Provolone	Whey cheeses

No two of these are made by the same method, and each varied procedure produces characteristics and qualities peculiar to the particular kind of cheese.

Cheeses can also be classified by hardness or ripening process, such as:

1. Very hard, hard, semisoft, or soft
2. Ripened by bacteria, mold, surface organisms, or a combination of these, or unripened

Following are examples of cheeses in these classifications:

1. Very hard (grating):
   (a) Ripened by bacteria—Asiago old, Parmesan, Romano, Sapsago, Spalen
2. Hard:
   (a) Ripened by bacteria, without eyes—Cheddar, Granular, Caciocavallo
   (b) Ripened by bacteria, with eyes—Swiss, Emmentaler, Gruyère
3. Semisoft:
   (a) Ripened principally by bacteria—Brick, Munster
   (b) Ripened by bacteria and surface microorganisms—Limburger, Port du Salut, Trappist
   (c) Ripened by mold in the interior—Roquefort, Gorgonzola, Blue, Stilton, Wensleydale
4. Soft:
   (a) Ripened—Brie, Butter, Camembert, Cooked, Hand, Neufchâtel
   (b) Unripened—Cottage, Pot, Baker's, Cream, Ricotta

"Natural" cheese is made directly from milk. Most cheese is made from cow's milk, although cheese can also be made from the milk of camels, mares, buffaloes, and reindeer. Another kind of cheese is "processed cheese"; it is made by combining or blending two or more natural cheeses. Cheese is made by the modification of milk in the following steps:

Curdling: This is usually done by heating the milk to anywhere from 80° F to 135° F (depending on the type of cheese to be made), adding rennet, and waiting for the cheese to curdle into curds and whey. Farm cheese is sometimes made by waiting for the lactic acid in the milk to curdle the milk, instead of adding rennet.

Removal of whey: The curdled mass is broken into small pieces and drained off.

Curing: Hard cheeses are kept in a cool environment for as long as three years. Salt, certain bacteria, and molds may be added to the cheese to help it develop its characteristic flavor.

There are federal standards for many cheeses produced in America. Where these standards exist, the specifications are given in the following section. The cheeses of other countries differ according to the producer, as many different types of milks and curing processes are used. The descriptions of cheeses presented here attempt to cover only the most common varieties. There are at least 600 different varieties on the market in the United States; obviously, space does not allow the classification of all of them. Terms used to describe cheese composition include the following:

Milk. Means the lacteal secretion, practically free from colostrum, obtained by the complete milking of one or more healthy cows. It may be clarified and may be adjusted by separating part of the fat to create concentrated milk, reconstituted milk, and dry whole milk. Water, in a sufficient quantity to reconstitute concentrated and dry forms, may be added.

Nonfat milk. Means skim milk, concentrated skim milk, reconstituted skim milk, and nonfat dry milk. Water, in a sufficient quantity to reconstitute concentrated and dry forms, may be added.

Cream. Means cream, reconstituted cream, and dry cream. Water, in a sufficient quantity to reconstitute concentrated and dry forms, may be added.

Pasteurized. When used to describe a dairy ingredient, means that every particle of such ingredient shall have been heated in properly operated equipment to one of the temperatures specified.

Ultra-pasteurized. When used to describe a dairy ingredient, means that such ingredient shall have been thermally processed at or above 280° F for at least two seconds.

Procedure set forth. The phrase used to describe the process that each individual cheese must go through during manufacture to produce the desired qualities.

Methods of Analysis

Moisture, milkfat, and phosphatase levels in cheeses will be determined by the following methods of analysis from *Official Methods of Analysis of the Association of Official Analytical Chemists*, 13th edition, 1980 (copies are available from the Association of Official Analytical Chemists, 481 N Frederick Ave., Suite 500, Gaithersburg, MD 20877, or available for inspection at the Office of the Federal Register, 1100 L Street, NW, Washington, DC 20408).

Cheese Varieties and Descriptions

Asiago Fresh and Asiago Soft Cheese

Asiago fresh cheese and Asiago soft cheese are the foods prepared from milk and other ingredients specified by the procedure set forth or by any other procedure that produces a finished cheese having the same physical and chemical properties. They contain not more than 45 percent of moisture, and their solids contain not less than 50 percent of milk fat. They are cured for not less than 60 days.

For the purposes of this section, the word *milk* means cow's milk, which may be adjusted by separating part of the fat therefrom or by adding thereto one or more of the following: cream, skim milk, concentrated skim milk, nonfat dry milk, or water in a quantity sufficient to reconstitute any concentrated skim milk or nonfat dry milk used. Asiago fresh and Asiago soft cheese are round and flat and usually weigh between 16 and 22 lb.

Asiago Medium Cheese

Asiago medium cheese conforms to the definition and standard of identity and is subject to the requirements for label statement of optional ingredients for Asiago fresh cheese, except that it contains not more than 35 percent moisture, its solids contain not less than 45 percent of milk fat, and it is cured for not less than six months.

Asiago Old Cheese

Asiago old cheese conforms to the definition and standard of identity and is subject to the requirements for label statement of optional ingredients for Asiago fresh cheese, except that it contains not more than 32 percent moisture, its solids contain not less than 42 percent of milk fat, and it is cured for not less than one year.

Blue Cheese

Blue cheese is the food prepared by the procedure set forth or by any other procedure that produces a finished cheese having the same physical and chemical properties. It is characterized by the presence of bluish green mold, *Penicillium roquefortii*, throughout the cheese. The minimum milk fat content is 50 percent by weight of the solids, and the maximum moisture content is 46 percent by weight. The dairy ingredients used may be pasteurized. Blue cheese is at least 60 days old.

Blue or green color in an amount to neutralize the natural yellow color of the curd may be added. Vegetable fats or oils, which may be hydrogenated, can be used as a coating for the rind. Blue cheese is about 7½ inches in diameter and weighs from 4½ to 5 pounds. It is round and flat like Gorgonzola, but smaller.

Brick Cheese

Brick cheese is the food prepared from dairy ingredients and other ingredients specified in this section by the procedure set forth or by any other procedure that produces a finished cheese having the same physical and chemical properties. The minimum milk fat content is 50 percent by weight of the solids, and the maximum moisture content is 44 percent by weight. If the dairy ingredients used are not pasteurized, the cheese is cured at a temperature of not less than 35° F for at least 60 days. Coloring is optional and may be added. Brick cheese measures about 10 in. long, 5 in. wide, and 3 in. thick and weighs approximately 5 lb.

Brick Cheese for Manufacturing Brick cheese for manufacturing conforms to the definition and standard of identity for brick cheese, except that the dairy ingredients are not pasteurized and curing is not required.

Caciocavallo Siciliano Cheese

Caciocavallo Siciliano cheese is the food prepared from cow's milk or sheep's milk or goat's milk or mixtures of all of these and other ingredients specified in this section. It has a stringy texture and is made in oblong shapes. It contains not more than 40 percent of moisture, and its solids contain not less than 42 percent milk fat as determined by the methods prescribed. It is cured for not less than 90 days at a temperature of not less than 35° F.

For the purposes of this section, the word *milk* means cow's milk or goat's milk or sheep's milk or mixtures of two or all of these. Such milk may be adjusted by separating part of the fat therefrom or (in the case of cow's milk) by adding one or more of the following: cream, skim milk, concentrated skim milk, nonfat dry milk; (in the case of goat's milk) the corresponding products from goat's milk; (in the case of sheep's milk) the corresponding products from sheep's milk; and water in a quantity sufficient to reconstitute any such concentrated or dried products used.

When Caciocavallo Siciliano cheese is made solely from cow's milk, the name of such cheese is "Caciocavallo Siciliano cheese." When made from sheep's milk or goat's milk or mixtures of these, or one or both of these with cow's milk, the name is followed by the words "made from ____," the blank being filled in with the name or names of the milks used, in order of predominance by weight. Each cheese weighs between 17½ and 22 lb and is in the shape of an oblong block.

Cheddar Cheese

Cheddar cheese is the food prepared by the procedure set forth or by any other procedure that produces a finished cheese having the same physical and chemical properties. The minimum milk fat content is 50 percent by weight of the solids, and the maximum moisture content is 39 percent by weight. If the dairy ingredients used are not pasteurized, the cheese is cured at a temperature of not less than 35° F for at least 60 days.

Milk, nonfat milk, or cream, as defined, may be used alone or in combination.

Cheddar cheese accounts for 67 percent of all cheese produced in the United States. In addition to being the name of the cheese, it is also the step in the manufacturing process and the name of the most common style. Each cheese is about $14\frac{1}{2}$ in. in diameter, 12 in. thick, and weighs between 70 and 78 lb.

BUTTER

Grading

The U.S. grade of butter is determined on the basis of classifying first the flavor characteristics and then the characteristics in body, color, and salt. See Figure 4.1. Flavor is the basic quality factor in grading butter and is determined organoleptically by taste and smell. The flavor characteristic is identified and, together with its relative intensity, is rated according to the applicable classification. When more than one flavor characteristic is discernible in a sample of butter, the flavor classification of the sample shall be established on the basis of the flavor that carries the lowest rating. Body, color, and salt characteristics are then noted, and any defects are disrated in accordance with the established classification, subject to disratings for body, color, and salt. When the disratings for body, color, and salt exceed the permitted amount for any flavor classification, the final U.S. grade shall be lowered accordingly.

Figure 4.1 Grade and Quality Shields for consumer packages of dairy products manufactured under the grading and quality control service. AA shield on top-level quality for butter. (Courtesy of the U.S. Department of Agriculture.)

Flavor

The flavor of butter is determined primarily by the senses of taste and smell. To register its full taste sensation, a substance must be soluble so that it can be carried quickly to the taste buds. The sense of smell supplements taste in determining flavor in butter. The warmth of the mouth melts the butter and frees its volatile aromas, which then enter the olfactory chambers, coming forward into the nose. Moisture in the mouth and nasal passages enhances the transmission of flavor sensations.

There are only four primary taste sensations: sweet, sour, salt, and bitter. Sugar produces the sensation of sweetness; lactic acid or a tart apple produces a sour taste; common table salt produces a sensation of saltiness; and quinine produces a bitter sensation. When melted butter comes in direct contact with the taste buds, its sweet and salty characteristics are detected by the taste buds located at the tip of the tongue, its sour characteristics by those on the sides of the tongue, and its bitter characteristics by those at the back of the tongue. The proper procedure in grading butter is first to use the sense of smell and then the sense of taste to confirm and establish the character, probable origin, and degree of development of each flavor present. By discerning carefully the odor or aroma characteristics of the sample, the character, and the degree of the flavor present, the grader is able to identify and classify the flavor properly.

Aroma in butter may be present to a greater or lesser degree. In the higher grades of butter, a pleasing aroma accentuates certain pleasing or desirable flavors. An objectionable aroma or odor is generally associated with flavors present in the lower grades of butter and serves to accentuate the undesirable flavor characteristics. The aroma noted in butter before it is tasted reflects a general indication of its quality.

The temperature of the butter at the time of grading is important in determining the true characteristics of flavor and aroma. The temperature of the butter should preferably range from 40° F to 50° F. A temperature of about 70° F is preferable in the grading room; it should not be below 60° F.

The following is a listing of certain flavor characteristics and their probable causes:

Aged. Associated with short or extended holding periods of the butter above freezer temperature, depending on the actual temperature level; may also occur if high-quality raw materials are not handled and processed promptly.

Barny. Attributable to absorption of barn odors or contamination during milking; may also result from an abnormal condition of cows.

Bitter. Attributable to the action of microorganisms in the cream before churning, certain types of feeds, and late lactation.

Coarse-acid. Associated with moderate acid development in the milk or cream or excessive ripening of the cream.

Cooked (fine). Associated with high-temperature pasteurization of sweet cream.

Cooked (coarse). Associated with high-temperature pasteurization of cream with slight acid development.

Feed. Attributable to feed flavors in milk carried through into the butter. Most dry feeds, such as hay and many of the concentrates, silage, green alfalfa, and various grasses, produce feed flavor in butter. Silage flavors may vary in degree and character, depending on the time of feeding, extent of fermentation, and the kind of silage.

Flat. Attributable to excessive washing of the butter or to a low percentage of fats of volatile acids and other volatile products that help to produce a pleasing butter flavor.

Lipase. Associated with the action of the enzyme lipase; may be accentuated by late lactation, excessive agitation of raw milk and cream at critical temperature of lipase activation (warming cold milk to 80°–90° F and then cooling again to 40°–50° F), or prolonged storage of raw cream at low temperatures that favor the growth of lipolytic organisms.

Malty. Attributable to the growth of the organism *Streptococcus lactis* var. *Maltigenses* in milk or cream. It is often traced to improperly washed and sanitized cream cans or other utensils in which this organism has developed.

Metallic. Attributed to keeping milk or cream on the farm in poorly tinned containers. Contact of cream with iron or copper equipment in the plant may also cause the cream to develop a metallic flavor; the use of can steamings is also a contributing factor.

Musty. Attributable to cream held in a damp cellar; may also result from cows grazing on slough grass, eating musty or moldy feed (hay and silage), or drinking stagnant water.

Neutralizer. Attributed to the excessive or improper use of alkaline products to reduce the acidity of the cream before pasteurization.

Old cream. Attributable to aged cream or inadequate cooling of the cream on the farm. This flavor may be accentuated by unclean cans, utensils, and processing equipment.

Scorched. Associated with the pasteurization, at a relatively high temperature, of cream with developed acidity; pasteurization after holding cream in forewarming vats for long periods; or faulty pasteurization equipment.

Smothered. Attributable generally to improper handling and delayed cooling of the cream on the farm.

Sour. Attributable to the churning of high acid cream, cream not properly neutralized, or to acid development after pasteurization.

Stale. Attributable to holding cream for extended periods, usually without adequate cooling, and may be associated with partial oxidation of the fat. This flavor may be accentuated by faulty sanitation.

Storage. Associated with holding butter in storage at freezer temperatures for several months or longer.

Utensil. Attributable to a faulty physical condition of utensils and equipment, improper sanitary care of such items as milking machines, pails, farm bulk tanks, producer cans, or a possible lack of proper washing and bactericidal treatment of processing equipment.

Weed. Attributable to weed-infested pastures and weedy hay eaten by cows.

Whey. Attributable to use of whey cream, or a blend of whole milk cream and whey cream.

Wild onion or garlic. Attributable to cows grazing on wild onion- or garlic-infested pastures.

Woody. Attributable to new wooden churns or other wooden equipment that has not been properly treated before being used; may also be caused by action of microorganisms.

Yeasty. Usually attributable to cream being held at high temperatures, which causes yeast fermentation.

Body

The factor of body in butter is considered from the standpoint of its characteristics or defects. Defects in body are disrated according to degree or intensity. Milk fat in butter is a mixture of various triglycerides of different melting points and appears in the form of fat globules and as free fat. In both these forms, a part of the fat is crystalline and another part is liquid. Some fats are solid at temperatures up to 100° F or even higher; others are still liquid at temperatures far below the freezing point. Because of this, butter at the temperatures at which it is usually handled is always a mixture of solid (crystallized) and liquid fat. The variations in the composition of milk fat thus have a great influence on the texture and spreadability of butter.

In summer, when milk fat contains more liquid fat, butter tends to be weak and leaky. In winter, when its solid fat content is high, butter tends to be hard and brittle, resulting in unsatisfactory spreadability. The ratio between the crystalline and the liquid particles depends on the composition of the butterfat (varying with the season of the year), manufacturing methods, and the temperature of the butter.

There may be a wide range of variation in the percentage of globular fat, also due presumably to differences in the composition of fat and manufacturing practices. Close attention should be given to the proper relationship between the temperatures of heating and cooling, and rate of cooling, of the cream, as well as the temperature of churning, washing, and working of the butter, at different seasons of the year when there are different fat conditions. This is important in maintaining a uniformly firm, waxy body possessing good spreadability.

The temperature of the cream after holding is a very important factor that determines the hardness of the butter. Butter with either a poor or excellent body may be made from cream containing either a hard or soft fat, depending on the methods of processing. The state of the water droplets and air in butter also plays a vital role in the body of the butter.

Butter with a firm, waxy body has an attractive appearance, has granules that are close knit, and cuts clean when sliced, with good spreadability. The trier sample from such butter will show this clean-cut, smooth, waxy appearance.

The temperature of the butter at the time of grading is important in determining the true characteristics of body, and should be between 40° and 50° F.

The following is a listing of certain body defects and their probable causes:

Crumbly. Attributable to a high proportion of fat crystals in the free fat. Such a condition is associated with higher-melting-point fats that result from feeding certain dry feeds like cottonseed meal, and is also associated with cows in late lactation. Cooling cream rapidly helps to form small globules of particles. If enough liquid fat is available, the butter will not crumble. It will crumble if crystals are large and there is no liquid fat. Cooling cream too rapidly may also cause crumbliness. Lower wash water temperature (10°–20° F below the temperature of the buttermilk) will help to correct crumbliness.

Gummy. Attributable to the presence of a high percentage of high-melting-point fats. Feeding cottonseed meal or whole cottonseed in quantities large enough to supply the bulk of the protein in a ration will result in a high proportion of high-melting-point fats and a hard-bodied butter. Such cream requires slower cooling, higher churning temperature, higher temperature wash water, and longer working time.

Leaky. Attributable generally to insufficient working, resulting in incomplete incorporation of the water. The water droplets are not sufficiently reduced in size to be well distributed throughout the mass of the butter. When the fat is soft and the granules are not sufficiently firm at the start of the working process, they mass together too quickly and do not offer enough resistance to break up the water droplets and obtain a minute and uniform distribution of the water in the butter. An uneven salt distribution may also cause a migration of moisture in the butter.

Mealy or grainy. Attributed to oiling-off of the butterfat at some stage of the butter-making process, improper melting of frozen cream, or improper neutralization of sour cream. The oiled-off fat, upon being cooled, crystallizes into small particles that cannot be worked into a smooth texture.

Ragged boring. Attributable to certain types of dry feeds, such as when dry feeds are not offset by succulent feeds. It is caused by a combination of the factors generally associated with crumbliness and stickiness, particularly when the melting point of the continuous (nonglobular) fat phase of butter is unusually high. Although this condition is related to crumbliness and stickiness, it differs in appearance, as the butter tends to roll on the trier. It may be minimized by procedures that permit the fat in the cream to crystallize at relatively high temperatures, and by rapid chilling of the fat after the butter granules have formed.

Short. Attributable to a predominance of high-melting-point fats with relatively small fat globules and a comparatively low curd content of the butter. Certain types of manufacturing processes, where partial melting of the fat takes place and normal granules are not produced, usually result in a short- and brittle-bodied butter. Too rapid cooling to too low temperature may also be a factor.

Sticky. Associated with dry feeds and late lactation period and a predominance of high-melting-point fats. This defect may result from not having the correct proportion of liquid and solid fat in the butter, as well as the proper proportion of large and small crystals of fat. The condition may be accentuated by too rapid cooling, cooling of the cream to too low a temperature, or overworking the butter.

Weak. Attributable to churning cream that has not been cooled to a low enough temperature, or not held long enough at a low temperature following pasteurization to properly firm the granules. Churning at a too high temperature and incorporating too much air into the butter may also be a cause.

Color

The natural color of butter varies according to seasonal and sectional conditions. The color of butter is considered defective when it is uneven or lacks uniformity within the same churning or package.

The following is a listing of certain color defects and their probable causes:

Mottled. Attributable to insufficient working of the butter, resulting in an uneven distribution of salt and moisture. Diffusion of the moisture toward the undissolved salt, or areas of high salt concentration, causes the irregular color spots. Churning at a too high temperature, resulting in soft granules that do not have sufficient resistance to stand the necessary amount of working, may also cause a mottled condition.

Specks. Attributable to incorporation of small particles of coagulated casein or coloring. White specks present may be small particles of curd formed during heating of improperly neutralized sour cream or as a result of partial coagulation caused by sweet-curdling organisms during pasteurization. The addition of a coarse-bodied starter may also be a contributing factor. Yellow specks may result from the use of butter color that has precipitated because of age or freezing.

Streaked. Attributable to insufficient working of the butter, faulty mechanical condition of the churn causing uneven working of the butter, and addition of remnants from previous churning.

Wavy. Attributable to insufficient working, resulting in an uneven distribution of the water and salt in the butter; may also be caused by faulty mechanical condition of the churn and addition of remnants from previous churnings.

Salt

In grading butter, the factor of salt is considered from the standpoint of the degree of salt taste, and whether it is completely dissolved. A range in the salt content or salty taste of butter is permitted without considering it a defect. This range provides for the various market preferences for salt taste in butter. Uniformity of salt content between churnings from the same factory is desirable.

The following are salt defects and their probable causes:

Sharp salt. Attributable to the use of too much salt or lack of sufficient working to obtain thorough distribution of salt and water.

Gritty salt. Attributable to the use of too much salt or insufficient working of the butter.

Butter Specifications

The specifications for the U.S. grades of butter are as follows:

1. U.S. Grade AA or U.S. 93 Score butter conforms to the following:
- (a) It possesses a fine and highly pleasing butter flavor.
- (b) It may possess a slight feed and a definite cooked (fine) flavor.
- (c) It is made from sweet cream of low natural acid to which a culture (starter) may or may not have been added.
- (d) The permitted total disratings in body, color, and salt characteristics are limited to one-half (½).

For detailed specifications and classification of flavor characteristics, see Table 4.1. For body, color, and salt characteristics, and disratings, see Table 4.2.

2. U.S. Grade A or U.S. 92 Score butter conforms to the following:
- (a) It possesses a pleasing and desirable butter flavor.
- (b) It may possess any of the following flavors to a slight degree: aged, bitter, coarse-acid, flat, smothered, and storage.
- (c) It may possess feed and cooked (coarse) flavors to a definite degree.
- (d) The permitted total disratings in body, color, and salt characteristics are limited to one-half (½), except that when the flavor classification is AA, a disrating total of one (1) is permitted.

For detailed specifications and classification of flavor characteristics, see Table 4.1, and for body, color, and salt characteristics, and disratings, see Table 4.2.

3. U.S. Grade B or U.S. 90 Score butter conforms to the following:
- (a) It possesses a fairly pleasing butter flavor.
- (b) It may possess any of the following flavors to a slight degree: lipase, malty, musty, neutralizer, scorched, utensil, weed, whey, and woody.
- (c) It may possess any of the following flavors to a definite degree: aged, bitter, coarse-acid, smoth-

Table 4.1 Classification of Flavor Characteristics

Identified Flavors[a]	AA	A	B	C
	Flavor Classification			
Feed	S*	D*	P*	—
Cooked (fine)	D	—	—	—
Aged	—	S	D	—
Bitter	—	S	D	—
Coarse-acid	—	S	D	—
Flat	—	S	—	—
Smothered	—	S	D	—
Storage	—	S	D	—
Cooked (coarse)	—	D	—	—
Lipase	—	—	S	D
Malty	—	—	S	D
Musty	—	—	S	D
Neutralizer	—	—	S	D
Scorched	—	—	S	D
Utensil	—	—	S	D
Weed	—	—	S	D
Whey	—	—	S	D
Woody	—	—	S	D
Old cream	—	—	D	—
Barny	—	—	—	S
Metallic	—	—	—	S
Sour	—	—	—	S
Wild onion or garlic	—	—	—	S
Yeasty	—	—	—	S
Stale	—	—	—	D

[a]S—Slight; D—Definite; P—Pronounced.

*When more than one flavor is discernible in a sample of butter, the flavor classification of the sample shall be established on the basis of the flavor that carries the lowest classification.

Table 4.2 Characteristics and Disratings in Body, Color, and Salt

BODY			
	Disratings		
Characteristics[a]	S	D	P
Crumbly	½	1	—
Gummy	½	1	—
Leaky	½	1	2
Mealy or grainy	½	1	—
Short	½	1	—
Weak	½	1	—
Sticky	½	1	—
Ragged boring	1	2	
COLOR			
	Disratings		
Characteristics	S	D	P
Wavy	½	1	—
Mottled	1	2	—
Streaked	1	2	—
Color Specks	1	2	—
SALT			
	Disratings		
Characteristics	S	D	P
Sharp	½	1	—
Gritty	1	2	—

[a]S—Slight; D—Definite; P—Pronounced.

ered, storage, and old cream; feed flavor to a pronounced degree.

(d) The permitted total disratings in body, color, and salt characteristics are limited to one-half (½), except that when the flavor classification is AA, a disrating total of one and one-half (1½) is permitted. When the flavor classification is A, a disrating total of one (1) is permitted.

For detailed specifications and classification of flavor characteristics, see Table 4.1, and for body, color, and salt characteristics, and disratings, see Table 4.2.

4. U.S. Grade C or U.S. 89 Score butter conforms to the following:
   (a) It may possess any of the following flavors to a slight degree: barny, sour, wild onion or garlic, and yeasty.
   (b) It may possess any of the following flavors to a definite degree: lipase, malty, musty, neutralizer, scorched, stale, utensil, weed, whey, and woody.
   (c) The permitted total disratings in body, color, and salt characteristics are limited to one (1), except that when the flavor classification is A, a disrating total of one and one-half (1½) is permitted.

For detailed specifications and classifications of flavor characteristics, see Table 4.1, and for body, color, and salt characteristics, and disratings, see Table 4.2.

5. General:
   (a) Butter of all U.S. grades shall be free of foreign materials and visible mold.
   (b) Butter possessing a flavor rating of AA or A and workmanship disratings in excess of one and one-half (1½) shall be given a flavor rating only.
   (c) Butter possessing a flavor rating of B or C and workmanship disrating in excess of one (1) shall be given a flavor rating only.

Butter that fails to meet the requirements for U.S. Grade C or U.S. 89 Score shall not be given a U.S. grade.

Butter, when tested, that does not comply with the provisions of the Federal Food, Drug, and Cosmetic Act, including minimum milk fat requirements of 80.0 percent, shall not be assigned a U.S. grade.

Butter produced in a plant found on inspection to be using unsatisfactory manufacturing practices, equipment, or facilities, or to be operating under unsanitary plant conditions, shall not be assigned a U.S. grade.

INTERNET RESOURCES

www.ams.usda.gov/dairy

U.S. Department of Agriculture Agricultural Marketing Service (USDA AMS). Dairy programs, services, and resources; current prices, trends, and grading.

www.idfa.org

International Dairy Foods Association (IDFA). The latest in dairy-related news topics, including industry facts and regulations and food safety.

CHAPTER 5

Eggs

The egg is one of the most nutritious and versatile of human foods. As chickens now produce eggs in abundance, this source of food has become extremely important throughout the world, nutritionally as well as economically. Descriptions of the various qualities of individual eggs should assist in understanding egg quality.

The modern trend in production is toward large, highly specialized flocks of chickens. The high-quality egg produced under this system lends itself very well to handling and processing by automatic equipment. In fact, some in-line systems are designed to carry eggs from the hen house to the carton in one continuous operation.

Egg quality and grading procedures are the "what" and "how" of the job. Coupled with reasonably good judgment, practice, and guidance, graders can determine rapidly the proper classification of shell eggs according to official standards of quality.

STRUCTURE, COMPOSITION, AND FORMATION OF THE EGG

Physical Structure

An average chicken egg weighs about 57 g or 2 oz. The parts of an egg are the yolk, the white, the shell membranes, and the shell.

Yolk (ovum). The yolk consists of the latebra, germinal disc, concentric rings of yolk material, and the vitelline membrane (a colorless membrane), which surrounds and contains the yolk. The yolk constitutes approximately 31 percent of the total weight of the egg.

White (albumen). The white consists of several layers, which together constitute about 58 percent of the weight of the egg. The chalaziferous layer immediately surrounds the yolk. This is a very firm but very thin layer of white, which makes up 3 percent of the total white. The inner thin layer surrounds the chalaziferous layer and constitutes about 17 percent of the white. The firm or thick layer of white provides an envelope or jacket that holds the inner thin white and the yolk. It adheres to the shell membrane at each end of the egg. Approximately 57 percent of the white is firm white. The outer thin layer lies just inside the shell membranes, except where the thick white is attached to the shell, and accounts for about 23 percent of the total white.

Shell Membranes. The shell membranes are tough and fibrous and are composed chiefly of protein, similar in nature to that in hair and feathers.

Shell. The shell is composed of three layers and constitutes approximately 11 percent of the egg. The egg, as laid, normally has no air cell. The air cell forms as the egg cools, usually in the large end of the egg, and develops between the shell membranes. The air cell is formed as a result of the different rates of contraction between the shell and its contents.

Composition

The egg is a very good source of high-quality protein and of certain minerals and vitamins. The chemical composition of the egg, including the shell, is summarized in Table 5.1.

Yolk. The important yolk proteins are ovovitellin (about three-fourths of the yolk protein) and ovolivetin. The fatty substances of the yolk are mostly glycerides (true fat), ovolecithin, and cholesterol. Yolk pigments come from green plants and yellow corn that the birds eat. The yolk contains practically all the known vitamins except vitamin C. The higher concentration of the solids of the yolk causes the yolk to increase in size and become less viscous because of the inflow of water from the white as the egg ages. The yolk contains iron, phosphorus, sulfur, copper, potassium, sodium, magnesium, calcium, chlorine, and manganese, all of which are essential elements.

White. The protein of egg is complete; it contains all the indispensable amino acids in well-balanced proportions. The white also contains some water-soluble B vitamins, especially riboflavin. The latter gives the greenish tint to the white.

Formation

Formation of the Yolk

Each yolk within the ovary starts as a single cell (female reproductive cell or germ) with the vitelline membrane

Table 5.1 Chemical Composition of the Whole Egg

Item	One pound	One average large AA egg
Calories	658.0	86.60
Protein	52.1 Grams	6.86 Grams
Fat	46.4 Grams	6.11 Grams
Carbohydrate	3.6 Grams	0.47 Grams
Calcium	218.0 Milligrams	28.71 Milligrams
Phosphorous	828.0 Milligrams	109.00 Milligrams
Iron	9.3 Milligrams	1.22 Milligrams
Sodium	493.0 Milligrams	64.92 Milligrams
Potassium	521.0 Milligrams	68.61 Milligrams
Vitamin A	4,760.0 Intrntnl. units	626.80 Intrintnl. units
Thiamine	0.4 Milligrams	0.05 Milligrams
Riboflavin	1.2 Milligrams	0.16 Milligrams
Niacin	0.2 Milligrams	0.03 Milligrams
Ascorbic acid	0.0 Milligrams	0.00 Milligrams
Magnesium	44.0 Milligrams	5.79 Milligrams
Total fat	45.9 Grams	6.04 Grams
Saturated fat	15.0 Grams	1.98 Grams
Mono unsaturated	20.0 Grams	2.63 Grams
Polyunsaturated	3.0 Grams	0.39 Grams
Cholesterol	2,200.0 Milligrams	289.71 Milligrams

Developed from: Agriculture Handbook No. 8, U.S. Department of Agriculture; Home and Garden Bulletin No. 72, *Nutritive Value of Foods*, U.S. Department of Agriculture.

around it. The yolk develops slowly at first by the gradual addition of yolk fluid. The yolk matures as more yolk fluid is added. The germ stays at the surface of the yolk, leaving a tube-like structure, the latebra, extending to the center of the yolk. Occasionally, reddish brown, brown, tan, or white spots, commonly known as "meat spots," may be found in the egg.

Formation of the White

The white contains ovomucin, secreted by the magnum as fibers or strands, which make the white thick. The quality of the white is largely dependent on the amount of ovomucin secreted by this part of the oviduct.

Formation of the Shell Membranes

The shell membranes are added as the partly formed egg enters the isthmus. The membranes are a closely knit lace-like nitrogenous compound of a substance similar to that present in the chicken's toenails.

Formation of the Shell

Calcium carbonate makes up about 94 percent of the dry shell. A hen may use as much as 47 percent of her skeletal calcium for eggshell formation. Two layers of the shell are formed in the uterus. Pigment, if any, is laid down in the spongy layer of the shell and is derived from the blood. The entire time from ovulation to laying is usually slightly more than 24 hours.

Abnormalities

Double-yoked eggs result when two yolks are released about the same time or when one yolk is lost into the body cavity for a day and then picked up by the funnel when the next day's yolk is released.

Yolkless eggs are usually formed around a bit of tissue that is sloughed off the ovary or oviduct. This tissue stimulates the secreting glands of the oviduct, and a yolkless egg results. The abnormality of an egg within an egg is due to reversal of direction of the egg by the wall of the oviduct. One day's egg is added to the next day's egg, and shell is formed around both. A rupture of one or more small blood vessels in the yolk follicle causes blood spots at the time of ovulation.

Meat spots have been demonstrated to be either blood spots that have changed color, due to chemical action, or tissue sloughed off from the reproductive organs of the hen. Soft-shelled eggs generally occur when an egg is prematurely laid, and insufficient time in the uterus prevents the deposit of the shell.

Dietary deficiencies, heredity, or disease may cause thin-shelled eggs.

Classy- and chalky-shelled eggs are caused by malfunctions of the uterus of the laying bird. Classy eggs are less porous and will not hatch but may retain their quality.

Off-colored yolks are due to substances in feed that cause off-color. Off-flavored eggs may be due to certain feed flavors.

GRADING

Grading generally involves the sorting of products according to quality, size, weight, and other factors that determine the relative value of the product. Egg grading is the grouping of eggs into lots having similar characteristics as to quality and weight. The grading for quality of shell eggs is the classifying of the individual egg according to established standards.

U.S. standards for the quality of individual shell eggs have been developed on the basis of interior quality factors such as condition of the white and yolk and the size of the air cell, and the exterior quality factors of cleanliness and soundness of the shell. These standards apply to eggs of the domesticated chicken that are in the shell.

Eggs are also classified according to weight (or size), expressed in ounces per dozen. Although eggs are not sold according to exact weight, they are grouped within relatively narrow weight ranges or weight classes, the minimum weight per unit being specified.

Advantages of Grading

Grading aids orderly marketing by reducing waste, confusion, and uncertainty with respect to quality values. The egg production pattern and the marketing system in the United States are such that interstate trading and shipment occur constantly and in large volumes. This situation creates a need for uniform standards throughout the United States so that marketing may be facilitated and the efficiency of distribution increased.

Officials of the USDA and state and industry leaders encourage the use of uniform standards and grades for eggs. Most of the eggs reaching the consumer today are graded and marked according to U.S. standards and grades.

The primary advantage in using official standards and grades for eggs is that they furnish an acceptable common language in trading and marketing the product, thus making possible:

1. Impartial official grading that eliminates the need for personal inspection of the eggs by sellers, buyers, and other interested people
2. Pooling of lots of comparable quality
3. Development of improved quality at the producer level through "buying on grade" programs
4. Market price reporting in terms understood by all interested parties

5. Negotiation of loans on generally accepted quality specifications
6. A basis for settling disputes involving quality
7. A basis for paying damage claims
8. A standard on which advertising may be based
9. A uniform basis for establishing brand names
10. The establishment of buying guides for consumers

SHELL COLOR

Shell color does not affect the quality of the egg and is not a factor in U.S. standards and grades. Eggs are usually sorted for color and sold as either "whites" or "browns." Eggs that are sorted as to color and packed separately sell better than when sold as "mixed colors."

For many years, consumers in some areas of the country have preferred white eggs, believing, perhaps, that the quality is better than that of brown eggs. In other areas, consumers have preferred brown eggs, believing they have greater food value. These opinions do not have any basis in fact, but it is recognized that brown eggs are more difficult to classify as to interior quality than white eggs. It is also more difficult to detect small blood and/or meat spots in brown eggs. Research reports and random sample laying tests show that the incidence of meat spots is significantly higher in brown eggs than in white eggs.

GENERAL STANDARDS

Standards of quality have been developed as a means of classifying individual eggs according to various groups of conditions and characteristics that experience and research have shown to be wanted by producers, dealers, and consumers. The term *standardization* implies uniformity, and uniformity in interpretation will result if the same standard is used and is applied accurately in all instances.

Standards of quality apply to individual eggs; grades apply to lots of eggs such as dozens, 30-dozen cases, and carloads. As egg quality is unstable and grading procedures are largely subjective, it is necessary to provide tolerances in grades for small percentages of eggs of a quality lower than that constituting the major part of the grade. The tolerances are provided to allow for errors in judgment, differences in interpretation, and normal deterioration in quality from the time of grading until the eggs are sold to the consumer.

Grades differ from standards in that they provide tolerances for individual eggs within a lot to be of lower quality than the grade name indicates. To produce an acceptable product at reasonable prices, tolerances must be within the capabilities of the industry. Without tolerances, it would not be possible to produce carton eggs at prices acceptable to consumers.

GENERAL QUALITY FACTORS

Speed and accuracy in grading should be accompanied by constant careful handling of the product. When eggs are placed in a carton or filler, they should be placed carefully, not dropped. An egg should always be packed small end down. Cartons of eggs should be placed on a conveyor belt carefully, not dropped. Eggs should not be placed in dirty or torn fillers and flats nor in packing materials giving off foreign odors. Eggs received in cases or packing material giving off foreign odors should not be graded unless the egg content is carefully checked for flavor. Shell eggs held in official plants should be placed under refrigeration of 60° F or lower promptly after packaging. Officially identified shell eggs with an internal temperature of 70° F or higher when shipped from the official plant should be transported at a temperature of 60° F or less.

Every reasonable precaution should be exercised to prevent the "sweating" of eggs (when there is condensation of moisture on the shell) in order to avoid smearing and staining the shell. Eggs taken from a very cool environment should be tempered in the candling room, or other room, with as moderate a temperature as necessary before candling to avoid "sweating" when the eggs are candled.

In judging egg quality it is helpful to break down a classification into steps, considering separately the various quality factors—shell, air cell, yolk condition, and condition of the white. Quality may be defined as the inherent properties of a product that determine its degree of excellence. Those conditions and characteristics that consumers want and are willing to pay for are, in a broad sense, factors of quality. The quality of an egg is determined by comparing a number of factors. The relative merit of one factor alone may determine the quality score of the egg, inasmuch as the final quality score can be no higher than the lowest score given to any one of the quality factors. Quality factors for eggs may be divided into two general groups: exterior and interior.

Exterior quality factors are apparent from direct external observation. Interior quality factors involve the contents of the egg as they appear before a candling light or when the eggs are broken out and measured by the Haugh unit method (measurement used in determining the albumen quality) plus visual examination of the yolk.

Classification of Exterior Quality

The external factors of the egg—shape, soundness, and cleanliness of the shell—can be determined without using a candling light (backlighting the egg), but soundness of shell should be verified by candling. The method or place where this is accomplished may vary with the type of candling operation used.

In hand-candling operations, the examination for shell cleanliness and the removal of leakers or dented checks and misshapen eggs can be accomplished by using a case light. In flash-candling operations, the segregation of eggs according to these shell factors is quite often the responsibility of a person who scans the eggs for exterior factors prior to or immediately following the mass scanning operation. This should be done in a well-lighted area.

EXTERIOR QUALITY FACTORS

Shell Shape and Texture

The normal egg has an oval shape, with one end larger than the other, tapering toward the smaller end. These ends of an egg are commonly called the large end (air-cell end) and the small end. Investigators measured both strength and appearance of many eggs to develop the "ideal" egg shape. The shape of an egg can be considerably different from the ideal but may still be considered practically normal. Eggs that are unusual in shape may have ridges, rough areas, or thin spots.

Abnormal shells may result from improper nutrition, disease, or the physical condition of the hen. Sometimes a shell is cracked while the egg is still in the body of the hen. An additional deposit of shell repairs these eggs, which are commonly referred to as "body checks," over the cracked area, generally resulting in a ridged area.

Shells with thin areas and some other types of defects are usually weaker than normal shells, and the danger of breakage en route to the consumer lowers the utility value of the egg. Eggs of abnormal shape also lack consumer appeal.

The specifications of the U.S. standards provide degrees of variation:

Practically normal—A shell that approximates the usual shape and is sound and free from thin spots. Ridges and rough areas that do not materially affect the shape and strength of the shell are permitted (AA or A quality).

Abnormal—A shell that may be somewhat unusual or decidedly misshapen or faulty in soundness or strength or that may show pronounced ridges or thin spots (B quality).

Soundness of Shell

The shell of an egg may be sound, checked or cracked, or leaking. The following are definitions of these shell factors:

Sound—an egg whose shell is unbroken.

Check—an individual egg that has a broken shell or crack in the shell, but its shell membranes are intact and its contents do not leak.

Leaker—an individual egg that has a crack or break in the shell and shell membranes to the extent that the egg contents are exuding or free to exude through the shell.

Checks are an unavoidable problem in the marketing of eggs because eggs cannot be assembled, graded, packed, transported, and merchandized without some breakage. Such eggs will not keep well or stand even moderately rough handling, and they should be diverted to immediate use. Eggs with checks may range from plainly visible dented checks that are removed during the grading process, to eggs with very fine, hairlike checks (blind checks) that often escape detection because they cannot be seen. Many of these checks become detectable as time passes (due primarily to contraction caused by cooling); however, the eggs have usually moved into marketing channels and may be at the retail level within one to three days after being laid. Blind checks are the most common and frequently the most difficult to detect in rapid candling, being discernible only before the candling light or by "belling."

"Belling" is the practice of gently tapping two eggs together to assist in the detection of "blind checks" by sound. Candlers follow this practice by candling the eggs to verify and complete the findings arrived at by sound. With the use of automatic equipment, the belling procedure generally is not used in examining the eggs for checks. It is necessary to remove leakers and dented checks from the lot carefully to avoid causing further damage to them and to avoid dripping liquid from the leakers onto clean eggs, onto the packaging material, or into the mechanism of the candling equipment. This is necessary for good housekeeping and the appearance of the packaged product and to keep the mechanisms of automatic weighing equipment in proper adjustment.

Shell Cleanliness

In segregating eggs for shell cleanliness, the grader should make a preliminary examination of the general appearance of the layer of eggs to be candled at the time the covering flat and surrounding filler are removed. Eggs with only very small specks, stains, or cage marks may be considered clean if such small specks, stains, or cage marks are not of sufficient number or intensity to detract appreciably from the appearance of the eggs (see "U.S. Standards for Quality of Individual Shell Eggs," page 76). While the eggs are still in standing position (in cup flats), the grader should remove and candle the eggs with stained or dirty shells. The remaining eggs that appear clean from a top view should then be pushed gently over on their sides, and again the eggs with stained or dirty shells should be removed and candled. These two operations will remove all dirty or stained eggs that are noticed at first glance.

In machine-flash candling, the examination for cleanliness is most often done immediately following the washing operation, or after the mass scanning for interior quality is performed. This operation should be in a well-lighted area, and there should be sufficient lighting directly over the eggs for ease of examination.

Classification of Shell Cleanliness

Freedom from stains and foreign material on the shell must be considered in assigning a quality designation to an individual egg. The following terms are descriptive of shell cleanliness:

Clean—A shell that is free from foreign material and from stains or discolorations that are readily visible. An egg may be considered clean if it has only very small specks, stains, or cage marks, and if such specks, stains, or cage marks are not of sufficient number or intensity to detract from the generally clean appearance of the egg. Eggs that show traces of processing oil on the shell are considered clean unless otherwise soiled.

Dirty—A shell that is unbroken and that has dirt or foreign material adhering to its surface, that has prominent stains, or that has moderate stains covering more than $^1/_{32}$ of the shell surface if localized, or $^1/_{16}$ of the shell surface if scattered.

CLASSIFICATION OF INTERIOR QUALITY

Even under the most favorable conditions, egg quality is relatively unstable, as the interior quality of the egg deteriorates from the time it is laid until it is consumed. Sometimes quality changes render eggs useless for food before they reach consumers. However, when eggs are properly cared for, the quality decline can be minimized. Quality decline is illustrated graphically in Figure 5.1.

In grading eggs and, more specifically, in classifying them according to internal quality, the grader is merely try-

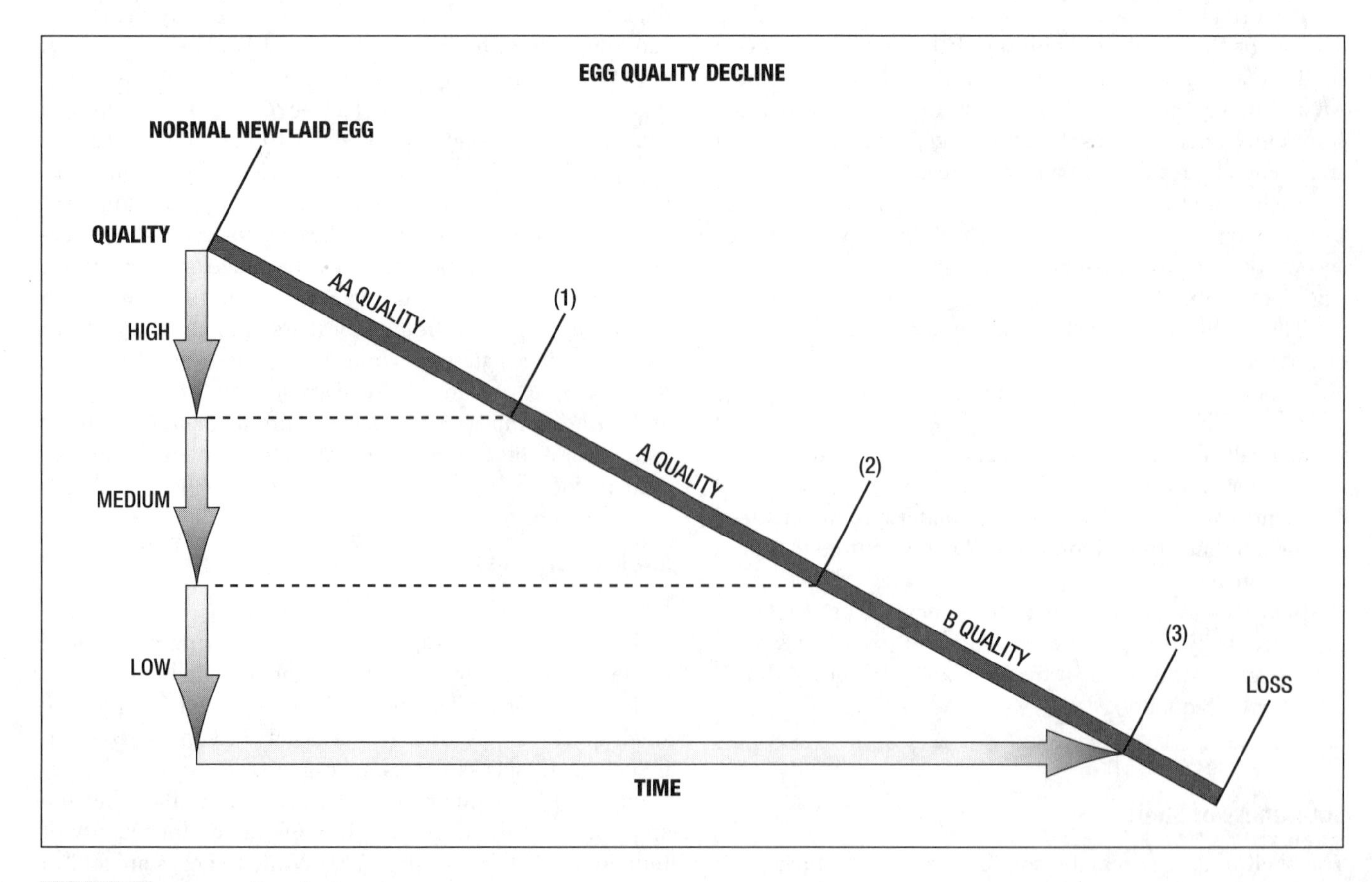

Figure 5.1 Egg quality decline.

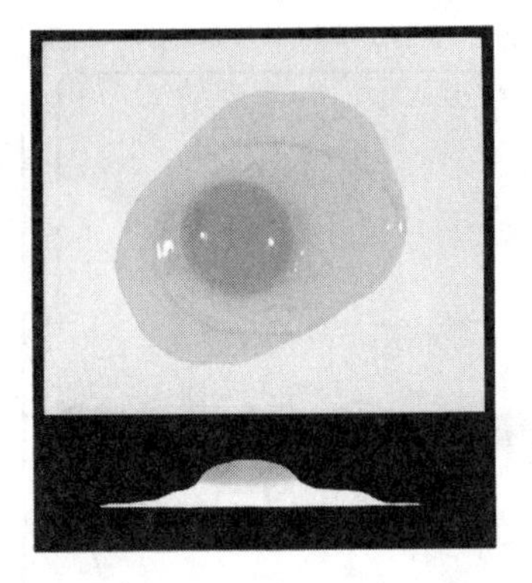

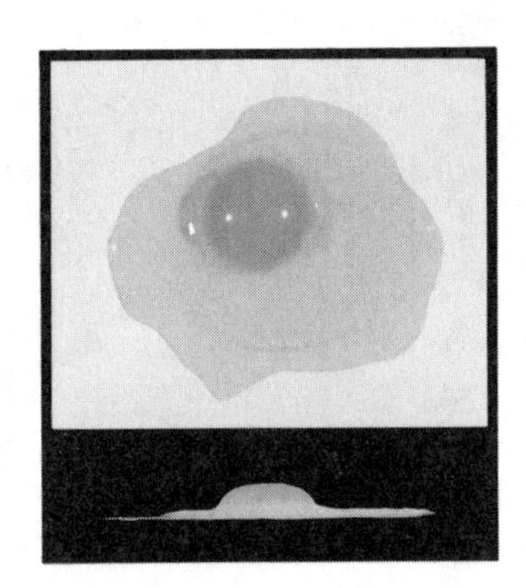

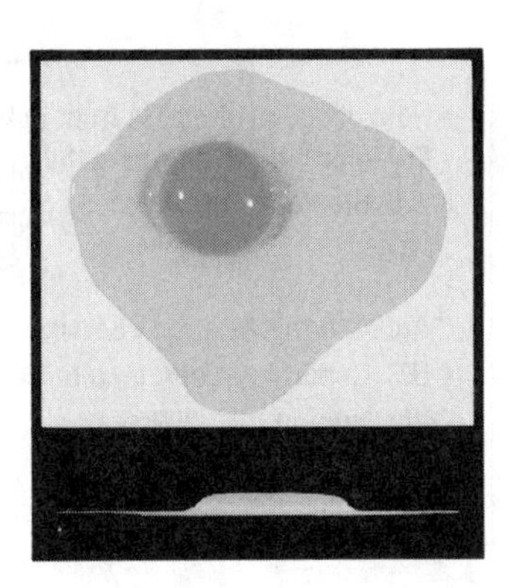

Figure 5.2 The pictures in this chart show the interior quality of eggs that meet the specifications of the U.S. Quality Standards with respect to the condition of the white and yolk. Quality factors dealing with the shell, air cell, and defects are not included. Left to right: Grade AA quality, Grade A quality, and Grade B quality. (Courtesy of the U.S. Department of Agriculture.)

ing to group the eggs according to where each is located on the "quality hill." On the basis of internal quality, edible eggs are divided into three groups, as shown in Figure 5.1. All eggs whose candled characteristics fall between the top line and point 1 on the chart are in the highest quality class, or AA; those between points 1 and 2 are in the next quality class, or A; and those between points 2 and 3 are in the B quality class. Those at and below point X are inedible or "loss" eggs. Pleas see definition on page 76.

Good judgment in determining white and yolk condition can be developed and maintained by having graders break open an egg occasionally. Their estimate of the candled quality should be checked with the broken-out appearance as compared with the chart for scoring broken-out appearance (Figure 5.2).

Interior Quality Factors

Air Cell

As already stated, when an egg is first laid it has no air cell at all or only a small one. Its temperature is about 105° F, and as the egg cools to room temperature the liquids contract more than the shell does. As a result of this contraction, the inner shell membrane separates from the outer membrane to form the air space. Further increase in the size of the air cell beyond that resulting from contraction is due to evaporation of water from the egg. The rapidity with which this takes place is due to many factors, such as age, shell texture, temperature, and humidity. The air cell is normally at the large end of the egg and is one of the first factors observed in candling. The air cell is the easiest quality factor to evaluate, as it can be judged objectively by a simple measuring device—the air cell gauge (Figure 5.3). In candling, the air cell is considered by many as a relatively unimportant quality factor for determining the broken-out quality of an egg. However, the air cell is one of the factors of the U.S. standards and can therefore be the determining factor in classifying the individual egg as to quality. Depth is the only quality factor considered with the air cell. Movement is not considered a quality factor, and the air cell may show unlimited movement and be free or bubbly in all qualities (AA, A, B).

The size of the air cells permitted in the various qualities is as follows:

QUALITY	DEPTH
AA	⅛ in.
A	3/16 in.
B	No limit

The depth of the air cell is measured at the point of greatest distance between the top of the cell and an imaginary plane passing through the egg at the lower edge of the air cell where it touches the shell.

The following terms are descriptive of the air cell:

Depth of air cell (air space between shell membranes, normally in the large end of the egg)—The depth of the air cell is the distance from its top to its bottom when the egg is held with the air cell upward.

Free air cell—An air cell that moves freely toward the uppermost part in the egg as the egg is rotated slowly.

Bubbly air cell—A ruptured air cell resulting in one or more small separate air bubbles, usually floating beneath the main air cell.

Yolk

The appearance of the yolk as the egg is twirled in candling is one of the best indicators of the interior quality of shell eggs. The characteristics of the yolk are determined by

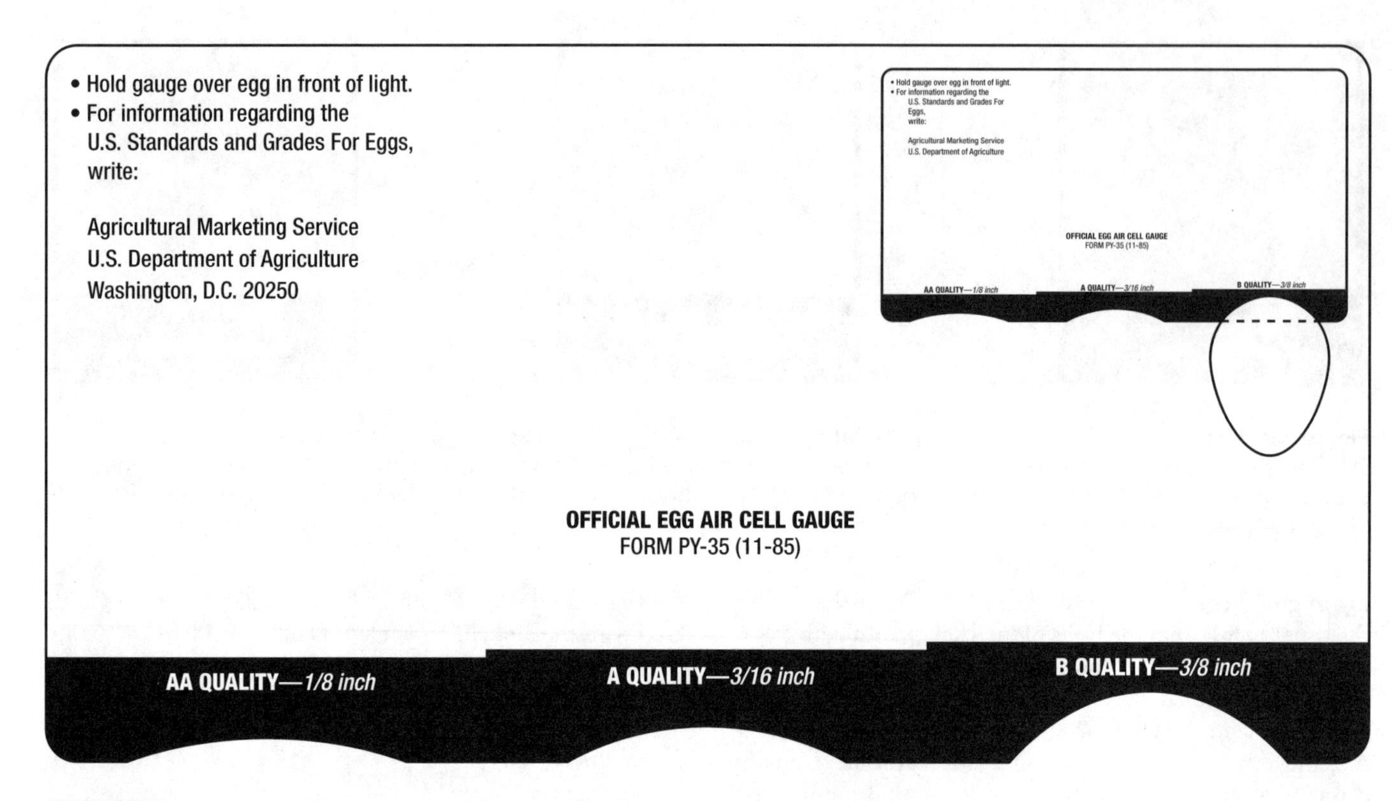

Figure 5.3 Official egg air gauge (form PY-35 (11-85)).

the shadow it casts on the shell before the candling light. The appearance of the yolk is dependent on the condition of the white. However, there are three factors about the yolk itself that are considered in judging egg quality by the yolk:

1. Distinctness of yolk shadow outline
2. Size and shape of yolk
3. Defects and germ development

Distinctness of Yolk Shadow Outline

The distinctness of the yolk outline or shadow outline is governed by three factors:

1. The thickness and consistency of the white. The thicker the white, the less distinct the outline appears, because the yolk is prevented from moving close to the shell.
2. Condition of the yolk. This condition is determined by the presence of blemishes that show up before the candling light as dark shadows on the yolk, or the absence of these blemishes; and the presence or absence of an off-colored yolk appearance that shows as a grayish or greenish shadow.
3. Color of the yolk. It is difficult to determine the color of the yolk before the candling light except off-color. However, extremes in yolk color may influence the candler's judgment of the egg quality. An extremely deep-colored yolk, under some conditions, would cast a darker shadow before the candling light than would a lighter yolk.

By concentrating on the yolk outline instead of the depth of the yolk shadow, therefore, the grader will minimize the influence of yolk color on quality determinations. The color of the yolk and the firmness of the white are two interacting influences affecting the distinctness of the yolk shadow outline. Therefore, a grader cannot be reasonably certain which is the more important factor in any specific case.

The principle of judging distinctness of the outline rather than the depth of darkness of the shadow can be illustrated by holding a ball close to a wall so that its shadow falls on the wall, and then holding it a little farther away from the wall. At the greater distance, the outline of the shadow is less distinct.

The terms used to define the three degrees of distinctness of yolk shadow outline in the U.S. standards for quality of shell eggs are:

Outline slightly defined—Yolk outline that is indistinctly indicated and appears to blend into the surrounding white as the egg is twirled (AA quality).
Outline fairly well defined—Yolk outline that is discernible but not clearly outlined as the egg is twirled (A quality).
Outline plainly visible—Yolk outline that is clearly visible as a dark shadow when the egg is twirled (B quality).

Size and Shape of Yolk

The yolk in a new-laid egg is round and firm. As the yolk ages, it absorbs water from the white. This increases its size and causes it to stretch and weaken the vitelline membrane and to assume a somewhat flattened shape on top and an "out-of-round" shape generally, resembling a balloon partially filled with water. Yolk size and shape are mentioned only in the lowest quality classification for eggs, "B quality," where these factors become apparent.

Defects and Germ Development

Relatively little is known about the exact causes of most yolk defects other than those due to germ development. Some of the causes that have been advanced are irregular deposits of light and dark yolk, blemishes resulting from rubbing, and the development of accumulations or clusters of the fat and oil in droplets. The relative viscosity of the white has a direct bearing on the accurate determination of defects on the yolk before the candling light.

Unless yolk defects are very prominent, detecting them is difficult, particularly when the egg has a thick white. Germ development is visible before the candling light and can generally be detected as a circular dark area near the center of the yolk shadow. If blood is visible, the egg must be rejected as inedible.

The terms used to describe yolk defects are:

Practically free from defects—Yolk that shows no germ development but may show other very slight defects on its surface (AA and A quality).
Serious defects—Yolk that shows well-developed spots or areas and other serious defects, such as olive yolks, that do not render the egg inedible (B quality).
Clearly visible germ development—Development of the germ spot on the yolk of a fertile egg that has progressed to the point where it is plainly visible as a definite circular area or spot with no blood in evidence (B quality).
Blood due to germ development—Blood caused by development of the germ in a fertile egg to a point where it is visible as definite lines or as a blood ring. Such an egg is classified as inedible.

White

Nearly all new-laid eggs contain four layers of white—chalaziferous, inner thin, thick, and outer thin. Largely, the relative proportions of the thick and outer thin layers of white govern the appearance of the egg before the candling light. The white and yolk are very closely associated, and any discussion of either one, of necessity, involves the other. However, there are two important considerations about the white that are included in standards of quality: condition or viscosity and clarity.

The condition of the white is determined by candling the intensity of the yolk shadow and the freedom of movement of the yolk as the egg is twirled before the candling light. These factors are related to the viscosity of the white. Thick whites permit only limited movement of the yolk and an indistinct shadow results.

The reverse is true of thin whites that permit free movement of the yolk, and a distinct shadow results. The grader must judge, from the behavior of the yolk, how the white will appear when the egg is broken out.

The following terms describe the white:

Clear—White that is free from discoloration and from any foreign bodies floating in it. Prominent chalazas, a thin layer of dense albumen, should not be confused with foreign bodies such as spots or blood clots (AA, A quality).
Firm—White that is sufficiently thick or viscous to prevent the yolk outline from being more than slightly defined or indistinctly indicated when the egg is twirled. With respect to a broken-out egg, a firm white has a Haugh unit value of 72 or higher when measured at a temperature between 45° and 60° F (AA quality).
Reasonably firm—White that is somewhat less thick or viscous than a firm white. A reasonably firm white permits the yolk to approach the shell more closely, which results in a fairly well defined yolk outline when the egg is twirled. With respect to a broken-out egg, a reasonably firm white has a Haugh unit value of 60 to 72 when measured at a temperature between 45° and 60° F (A quality).
Weak and watery—White that is weak, thin, and generally lacking in viscosity. A weak and watery white permits the yolk to approach the shell closely, thus causing the yolk outline to appear plainly visible and dark when the egg is twirled. With respect to a broken-out egg, a weak and watery white has a Haugh unit value lower than 60 when measured at a temperature between 45° and 60° F (B quality).
Blood spots or meat spots—Small blood spots or meat spots (aggregating not more than 1/8 in. in diameter)

may be classified as B quality. If a spot is larger, or there is diffusion of blood into the white surrounding a blood spot, the egg is classified as "loss." Blood spots cannot be due to germ development. They may be on the yolk or in the white. Meat spots may be blood spots that have lost their characteristic red color or tissue from the reproductive organs.

Bloody white—An egg that has blood diffused through the white. Such a condition may be present in new-laid eggs. Eggs with bloody whites are classed as loss.

LOSS EGGS

A freshly laid egg is usually free of bacteria on the inside and is well protected from bacteria by the shell, shell membranes, and several chemical substances in the egg white. If subjected to warm temperatures, moisture, or both, bacteria are able to penetrate the egg and overcome the egg's defense. When bacteria grow inside the egg, they may form by-products, cause the contents of the egg to decompose, or both. These conditions result in the characteristic colors, appearance, or odors from which the "rots" take their name.

The U.S. standards of quality also define certain eggs as "loss." A loss is an egg that is inedible, a leaker, cooked, frozen, contaminated, or containing bloody whites, large blood spots, large unsightly meat spots, or other foreign material.

Inedible eggs are described in the U.S. standards to include black rots, white rots, sour eggs, eggs with green whites, musty eggs, and moldy eggs. In addition to the inedible eggs already described, eggs showing severe shell damage and the presence of large blood spots or diffused blood in the white are classified as loss. Leakers are classified as loss.

Eggs not classified as loss but as "no grade" include eggs of possible edible qualities that have been contaminated by smoke, chemicals, or other foreign material that has seriously affected the character, appearance, or flavor of the eggs.

High concentrations of fish oil or garlic fed to hens impart their flavor to the eggs. Eggs exposed to foreign odors after they have been laid may give off these odors. Eggs stored near kerosene, carbolic acid, mold, must, fruits, and vegetables, for example, readily absorb odors from these products.

U.S. STANDARDS, GRADES, AND WEIGHT CLASSES FOR SHELL EGGS

The U.S. standards, grades, and weight classes for individual shell eggs are applicable only to eggs of the domesticated chicken that are in the shell.

U.S. Standards for Quality of Individual Shell Eggs

(Based on candled appearance.) The standards described here are summarized in Table 5.2.

AA quality—The shell must be clean, unbroken, and practically normal. The air cell must not exceed 1/8 in. in depth, may show unlimited movement, and may be free or bubbly. The white must be clear and firm so that the yolk is only slightly defined when the egg is twirled before the candling light. The yolk must be practically free from apparent defects.

A quality—The shell must be clean, unbroken, and practically normal. The air cell must not exceed 3/16 in. in depth, may show unlimited movement, and may be free or bubbly. The white must be clear and at least reasonably firm so that the yolk outline is only fairly well defined when the egg is twirled before the candling light. The yolk must be practically free from apparent defects.

B quality—The shell must be unbroken, may be abnormal, and may have slightly stained areas. Moderately stained areas are permitted if they do not cover more than 1/32 of the shell surface if localized, or 1/16 of the shell surface if scattered. Eggs having shells with prominent stains or adhering dirt are not permitted. The air cell may be more than 3/16 in. in depth, may show unlimited movement, and may be free or bubbly. The white may be weak and watery so that the yolk outline is plainly visible when the egg is twirled before the candling light. The yolk may appear dark, enlarged, and flattened and may show clearly visible germ development but no blood due to such development. It may show other serious defects that do not render the egg inedible. Small blood spots or meat spots (aggregating not more than 1/8 in. in diameter) may be present.

The terms *dirty*, *check*, and *leaker* are defined as:

Dirty—An individual egg that has an unbroken shell with adhering dirt or foreign material, prominent stains, or moderate stains covering more than 1/32 of the shell surface if localized, or 1/16 of the shell surface if scattered.

Check—An individual egg that has a broken shell or crack in the shell but whose shell membranes are intact and whose contents do not leak.

Leaker—An individual egg that has a crack or break in the shell and shell membranes to the extent that the egg contents are exuding or free to exude through the shell.

Table 5.2 Summary of U.S. Standards for Quality of Individual Shell Eggs

	Specifications for each quality factor		
Quality Factor	**AA Quality**	**A Quality**	**B Quality**
Shell	Clean. Unbroken. Practically normal.	Clean. Unbroken. Practically normal.	Clean to slightly stained.[a] Unbroken. Abnormal.
Air	1/8 inch or less in depth.	3/16 inch or less in depth.	Over 3/16 inch in depth.
Cell	Unlimited movement and free or bubbly. Clear.	Unlimited movement and free or bubbly. Clear.	Unlimited movement and free or bubbly. Weak and watery.
White	Firm. Outline—slightly defined.	Reasonably firm. Outline—fairly well defined.	Small blood and meat spots present.[b] Outline—plainly visible.
Yolk	Practically free from defects.	Practically free from defects.	Enlarged and flattened. Clearly visible germ development but no blood. Other serious defects.

[a]Moderately stained areas permitted (1/32 of surface if localized, or 1/16 if scattered).
[b]If they are small (aggregating not more than 1/8 inch in diameter).

For eggs with dirty or broken shells, the standards of quality provide two additional qualities. These are:

Dirty	*Check*
Unbroken. Adhering dirt or foreign material, prominent stains, moderate stained areas in excess of B quality	Broken or cracked shell but membranes intact, not leak.ing.[c]

[c]Leaker has broken or cracked shell and membranes, and contents leaking or free to leak.

Note: The C quality classification for individual shell eggs was eliminated in 1981. The percentage of C quality eggs found in the total egg production had decreased to an insignificant point, about 1 percent of eggs.

U.S. CONSUMER GRADES AND WEIGHT CLASSES FOR SHELL EGGS

Consumer Grades

The grading regulations for shell eggs provide for "origin" and "destination" consumer grades. "Origin grading" is defined as a grading made on a lot of eggs at a plant where the eggs are graded and packed.

Table 5.3 gives a summary of the consumer grades, and Table 5.4 gives the tolerance for individual cases within a lot.

U.S. consumer grade AA (at origin)—shall consist of eggs that are at least 87 percent AA quality. The maximum tolerance of 13 percent that may be below AA quality may consist of A or B quality in any combination, except that within the tolerance for B quality not more than 1 percent may be B quality due to air cells of more than 3/8 in., blood spots (aggregating not more than 1/8 in. in diameter), or serious yolk defects. Not more than 5 percent (7 percent for Jumbo size) checks are permitted and not more than 0.50 percent leakers, dirties, or loss (due to meat or blood spots) in any combination, except that such loss may not exceed 0.30 percent. Other types of loss are not permitted.

U.S. consumer grade AA (destination)—shall consist of eggs that are at least 72 percent AA quality. The remaining tolerance of 28 percent shall consist of at least 10 percent A quality and the rest shall be B quality, except that within the tolerance for B quality not more than 1 percent may be B quality due to air cells of more than 3/8 in., blood spots (aggregating not more than 1/8 in. in diameter), or serious yolk defects. Not more than 7 percent (9 percent for Jumbo size) checks are permitted, and not more than 1 percent leakers, dirties, or loss (due to meat or blood spots) in any combination, except that such loss may not exceed 0.30 percent. Other types of loss are not permitted.

U.S. consumer grade A (at origin)—shall consist of eggs that are at least 87 percent A quality or better. Within the

Table 5.3 Summary of U.S. Consumer Grades for Shell Eggs

U.S. consumer grade (origin)	Quality required[a]	Tolerance permitted[b]	
		Percent	Quality
Grade AA	87 percent AA	Up to 13	A or B[5]
		Not over 5	Checks[6]
Grade A	87 percent A or better	Up to 13	B[5]
		Not over 5	Checks[6]
Grade B	90 percent B or better	Not over 10	Checks
U.S. consumer grade (destination)	**Quality required[1]**	**Tolerance permitted[c]**	
		Percent	**Quality**
Grade AA	72 percent AA	Up to 28[d]	A or B[e]
		Not over 7	Checks[f]
Grade A	82 percent A or better	Up to 18	B[e]
		Not over 7	Checks[f]
Grade B	90 percent B or better	Not over 10	Checks

[a]In lots of two or more cases, see Table 5.4 of this section for tolerances for an individual case within a lot.

[b]For the U.S. Consumer grades (at origin), a tolerance of 0.50 percent leakers, dirties, or loss (due to meat or blood spots) in any combination is permitted, except that such loss may not exceed 0.30 percent. Other types of loss are not permitted.

[c]For the U.S. Consumer grades (destination), a tolerance of 1 percent leakers, dirties, or loss (due to meat or blood spots) in any combination is permitted, except that such loss may not exceed 0.30 percent. Other types of loss are not permitted.

[d]For U.S. Grade AA at destination, at least 10 percent must be A quality or better.

[e]For U.S. Grade AA and A at origin and destination within the tolerances permitted for B quality, not more than 1 percent may be B quality due to air cells over ⅜ inch, blood spots (aggregating not more than ⅛ inch in diameter), or serious yolk defects.

[f]For U.S. Grades AA and A Jumbo size eggs, the tolerance for checks at origin and destination is 7 percent and 9 percent, respectively.

Table 5.4 Tolerance for Individual Case Within a Lot

U.S. consumer grade	Case quality	Origin	Destination
		Percent	
Grade AA	AA (min)	77	62
	A or B	13	28
	Check (max)	10	10
Grade A	A (min)	77	72
	B	13	18
	Check (max)	10	10
Grade B	B (min)	80	80
	Check (max)	20	20

maximum tolerance of 13 percent that may be below A quality, not more than 1 percent may be B quality due to air cells of more than ⅜ in., blood spots (aggregating not more than ⅛ in. in diameter), or serious yolk defects. Not more than 5 percent (7 percent for Jumbo size) checks are permitted, and not more than 0.50 percent leakers, dirties, or loss (due to meat or blood spots) in any combination, except that such loss may not exceed 0.30 percent. Other types of loss are not permitted.

U.S. consumer grade A (destination)—shall consist of eggs that are at least 82 percent A quality or better. Within the maximum tolerance of 18 percent that may be below A quality, not more than 1 percent may be B quality due to air cells of more than ⅜ in., blood spots (aggregating not more than ⅛ in. in diameter), or serious yolk defects. Not more than 7 percent (9 percent for Jumbo size) checks are permitted, and not more than 1 percent leakers, dirties, or loss (due to meat or blood spots) in any combination, except that such loss may not exceed 0.30 percent. Other types of loss are not permitted.

U.S. consumer grade B (at origin)—shall consist of eggs that are at least 90 percent B quality or better; not more than 10 percent may be checks and not more than 0.50 percent leakers, dirties, or loss (due to meat or blood spots) in any combination, except that such loss may not exceed 0.30 percent. Other types of loss are not permitted.

U.S. consumer grade B (destination)—shall consist of eggs that are at least 90 percent B quality or better; not more than 10 percent may be checks and not more than 1 percent leakers, dirties, or loss (due to meat or blood spots) in any combination, except that such loss may not exceed 0.30 percent. Other types of loss are not permitted.

Additional tolerances—in lots of two or more cases:

For grade AA—No individual case may exceed 10 percent less AA quality eggs than the minimum permitted for the lot average.

For grade A—No individual case may exceed 10 percent less A quality eggs than the minimum permitted for the lot average.

For grade B—No individual case may exceed 10 percent less B quality eggs than the minimum permitted for the lot average.

For grades AA, A, and B—No lot shall be rejected or downgraded because of the quality of a single egg except for loss other than blood or meat spots.

Note: The fresh fancy quality control program was eliminated in 1981 because it was used very little. The grade A quality control program was also eliminated then because it was not being used.

1. For lots of two or more cases, see Table 5.4 of this section for tolerances for an individual case within a lot.
2. For the U.S. consumer grades (at origin), a tolerance of 0.50 percent leakers, dirties, or loss (due to meat or blood spots) in any combination is permitted, except that such loss may not exceed 0.30 percent. Other types of loss are not permitted.
3. For the U.S. consumer grades (destination), a tolerance of 1 percent leakers, dirties, or loss (due to meat or blood spots) in any combination is permitted, except that such loss may not exceed 0.30 percent. Other types of loss are not permitted.
4. For U.S. Grade AA at destination, at least 10 percent must be A quality or better.
5. For U.S. Grade AA and A at origin and destination within the tolerances permitted for B quality, not more than 1 percent may be B quality owing to air cells of more than 3/8 in., blood spots (aggregating not more than 1/8 in. in diameter), or serious yolk defects.
6. For U.S. Grades AA and A Jumbo-size eggs, the tolerances for checks at origin and destination are 7 percent and 9 percent, respectively.

Table 5.5 **Weight Classes of U.S. Consumer Grades for Shell Eggs**

Size or weight class	Minimum net weight per dozen	Minimum net weight per 30 dozen	Minimum weight for individual eggs at rate per dozen
	Ounces	*Pounds*	*Ounces*
Jumbo	30	56	29
Extra large	27	50-1/2	26
Large	24	45	23
Medium	21	39-1/2	20
Small	18	34	17
Peewee	15	28	

Weight Classes

The weight classes for U.S. consumer grades for shell eggs shall be as indicated in Table 5.5 and shall apply to all consumer grades.

A lot-average tolerance of 3.3 percent for individual eggs in the next lower weight class is permitted as long as no individual case within the lot exceeds 5 percent.

U.S. WHOLESALE GRADES AND WEIGHT CLASSES FOR SHELL EGGS

Wholesale Grades

Table 5.6 gives a summary of the wholesale grades listed here.

U.S. Specials—percentage AA quality shall consist of eggs of which at least 20 percent are AA quality; the actual percentage of AA quality eggs shall be stated in the grade name. Within the maximum of 80 percent that may be below AA quality, not more than 7.5 percent may be B quality, dirties, or checks in any combination, and not more than 2 percent may be loss.

U.S. Extras—percentage A quality shall consist of eggs of which at least 20 percent are not less than A quality; the actual total percentage of A quality and

Table 5.6 **Summary of U.S. Wholesale Grades for Shell Eggs**

Wholesale grade designation	Maximum tolerance permitted (lot average): AA quality	A quality or better	B quality or better	Minimum percentage of eggs of specific qualities required[a]: dirties and checks	B quality Dirties and checks	Loss
					Percent	
U.S. Specials—% AA Quality[b]	20	Balance	None except for tolerance	7.5	—	2
U.S. Extras—% A Quality[b]	—	20	Balance	—	11.7	3
U.S. Standards—% B Quality[b]	—	—	84.3	—	11.7	4

[a]Substitution of eggs possessing higher qualities for those possessing lower specified qualities is permitted.

[b]The actual total percentage must be stated in the grade name.

better shall be stated in the grade name. Within the maximum of 80 percent that may be below A quality, not more than 11.7 percent may be dirties or checks in any combination, and not more than 3 percent may be loss.

Note: Three wholesale grades—U.S. Trades, U.S. Dirties, and U.S. Checks—were eliminated in 1981 because they had not been used for years.

Weight Classes

The weight classes for the U.S. wholesale grades for shell eggs shall be as indicated in Table 5.7.

U.S. NEST-RUN GRADE AND WEIGHT CLASSES FOR SHELL EGGS

Nest-Run Grade

Table 5.8 summarizes the nest-run grade described here.

U.S. Nest Run—percentage AA quality shall consist of eggs of current production of which at least 20 percent are AA quality; the actual percentage of AA quality eggs shall be stated in the grade name. Within the maximum of 15 percent that may be below A quality, not more than 10 percent may be B quality for shell shape, for interior quality (including meat or blood spots), or due to rusty or blackish-

Table 5.7 Weight Classes for U.S. Wholesale Grades for Shell Eggs

	For 30 dozen eggs		Weights for individual eggs at rate per dozen	
Weight classes	**Average net weight on a lot[a]**	**Minimum net weight individual case[b] basis**	**Minimum weight**	**Weight variation tolerance for not more than 10 percent, by count, of individual eggs**
	At least—			
Extra large	50-½ pounds	50 pounds	26 ounces	Under 26 but not under 24 ounces
Large	45 pounds	44 pounds	23 ounces	Under 23 but not under 21 ounces
Medium	39-½ pounds	39 pounds	20 ounces	Under 20 but not under 18 ounces
Small	34 pounds	None	None	None

[a]Lot means any quantity of 30 dozen or more eggs.

[b]Case means standard 30 dozen egg case as used in commercial practice in the U.S.

Table 5.8 Summary of U.S. Nest-Run Grade for Shell Eggs

	Minimum percentage of quality required (lot average)[a]		Maximum percentage tolerance permitted (15% lot average)[a]			
Net Run grade description[b]	AA Quality[c]	A Quality or better[d]	B Quality for shell shape, interior quality (including blood and meat spots), or cage marks[e] and blood stains	Checks	Loss	Adhering dirt or foreign material ½ inch or larger in diameter
U.S. Nest Run— % AA Quality[f]	**20**	85	10	6	3	5

[a]Substitution of eggs of higher qualities for lower specified qualities is permitted.

Stains (other than rusty or blackish appearing cage marks or blood stains), and adhering dirt and foreign material on the shell less than ½ inch in diameter shall not be considered as quality factors in determining the grade designation.

[b]No case may contain less than 10 percent AA quality.

[c]No case may contain less than 75 percent A quality and AA quality eggs in any combination.

[d]Case marks which are rusty or blackish in appearance shall be considered as quality factors. Marks which are slightly gray in appearance are not considered as quality factors.

[e]The actual total percentage must be stated in the grade name.

appearing cage marks or blood stains; not more than 5 percent may have adhering dirt or foreign material on the shell ½ in. or larger in diameter; not more than 6 percent may be checks; and not more than 3 percent may be loss. Marks that are slightly gray in appearance and adhering dirt or foreign material on the shell less than ½ in. in diameter are not considered quality factors. The eggs shall be officially graded for all other quality factors. No case may contain less than 75 percent A quality and AA quality eggs in any combination.

Weight Classes

The weight classes for the U.S. nest-run grade for shell eggs shall be as indicated in Table 5.9.

Note: U.S. procurement grades and weight classes for shell eggs were eliminated in 1981 because they were obsolete. Procurement grade 1 had not been used for a number of years. This grade was practically identical to U.S. consumer grade A; thus, the consumer grade standard could readily be used in place of the procurement standard. The U.S. Department of Defense was the principal user of procurement grade 1.

Table 5.9 Weight Classes for U.S. Nest-Run Grade for Shell Eggs

Weight classes	Minimum average net weight on lot basis 30-dozen cases
	Pounds
Class XL	51
Class 1	48
Class 2	45
Class 3	42
Class 4	39

No individual sample case may vary more than 2 pounds (plus or minus) from the lot average.

EGG STANDARDS

Fresh Eggs

A fresh egg is one that has been stored 29 days or less. Eggs that are held in storage for more than 29 days are called "storage eggs." These must be held at a temperature of 29° to 31° F, with a humidity of 90 to 92.

Processed Eggs

No U.S. grade standards are available for processed eggs. Standards of identity have been established for frozen eggs, whites and yolks; dried eggs, whites and yolks; and liquid eggs, whites and yolks. All processed egg products are produced and packed under continuous federal inspection.

Whole Eggs

A whole egg is the entire egg, without the shell, cracked, pasteurized, and packaged in 4-lb, 5-lb, 10-lb, or 30-lb containers. These are pasteurized at 140° F for 3½ minutes. A 30 lb container is the equivalent of approximately 23 dozen eggs. Egg whites and egg yolks are packed in the same manner.

Frozen Eggs

Frozen eggs are available as mixed whole eggs, whites, yolks, or those with salt or sugar added. The freezing of eggs produces a toughened product upon thawing; the addition of sugar or glycerin prevents the toughening. A 30 lb container is the equivalent of 60 dozen yolks or 35 dozen whites.

Dried Eggs

Eggs are dried by either a roller process or a spray process. They are also treated with enzymes to remove glucose. Freeze-dried eggs are also available. One pound of dried eggs is the equivalent of 32 large eggs.

Egg Substitutes

Egg substitutes are available fresh, frozen, and dried. They are the equivalent of whole eggs, but contain no cholesterol, for individuals who may wish to reduce the amount of cholesterol in their diets.

HOW TO BUY SHELL EGGS

Specify the grade and size. U.S. Grade AA is best for visual appearance, as in fried eggs. U.S. Grade A is suitable for hard-cooked and soft-cooked eggs; Medium or Large size is best for those uses. U.S. Grade A eggs should be used for cracked egg purposes, omelets, or scrambled eggs. Size is not as important in this case. U.S. Grade B eggs are suitable for general-purpose cooking or baking. Eggs should be delivered within two to three days of grading and kept at a temperature of less than 42° F in an area away from foods with strong odors. U.S. Grade AA eggs will hold that grade for 10 days at 42° F.

An egg specification purchase form should contain the following items:

U.S. Grade:	AA
Type:	Fresh, shell
Size:	Large
Color:	White
Package size:	30 dozen/case net weight not under 45 lb.
Price:	Priced by the case

INTERNET RESOURCES

www.aeb.org

American Egg Board. This site is a resource for egg safety, egg nutrition, egg recipes, and a glossary of terms.

www.ams.usda.gov/howtobuy/eggs.htm

How to Buy Eggs, by the U.S. Department of Agriculture Agricultural Marketing Service (USDA AMS).

CHAPTER

6

Poultry

STANDARDS AND GRADES

The difference between standards of quality and the grades assigned to poultry is sometimes misunderstood. The standards of quality cover the various factors that determine the grade. These factors, such as fat covering, fleshing, exposed flesh, discoloration, and so forth, are used collectively to determine the grade of the bird. The U.S. consumer grades for poultry are used at the retail level. The U.S. consumer grades are U.S. Grade A, U.S. Grade B, and U.S. Grade C. The U.S. procurement grades are designed primarily for institutional use. These grades are U.S. Procurement Grade 1 and U.S. Procurement Grade 2. The procurement grades place more emphasis on meat yield than on appearance.

OFFICIAL IDENTIFICATION BY GRADERS LICENSED BY THE U.S. DEPARTMENT OF AGRICULTURE

Anyone having a financial interest in a lot of processed poultry may make application to the U.S. Department of Agriculture (USDA) to have an official grade designation placed on the lot. This service, which is available throughout the country, is operated on a self-supporting basis. A nominal fee is charged that covers the time and travel expense of the grader, plus the cost of administering the program. The USDA enters into cooperative agreements with state departments of agriculture, making it possible to license qualified state employees to grade and certify the quality of poultry.

Many processors utilize full-time resident graders in their plants. This enables a plant to apply the U.S. grade mark to each individual package, or each individual bird.

The military and federal, state, county, and city institutions, as well as other large-scale buyers such as steamship lines, independent and chain stores, and private hospitals, make use of the grading service by specifying U.S. grades in their contracts for poultry products. Commercial firms often use the U.S. standards and grades as a basis for establishing specifications for their own products.

CLASSES

Some states provide a voluntary grading and inspection program. Such programs generally follow the U.S. standards and grades in whole or in part. Producers, as well as processors, may use the standards of quality as a basis for sorting or selecting birds for market.

Kind refers to the different species of poultry, such as chickens, turkeys, ducks, geese, guineas, and pigeons. Groups that are essentially of the same physical characteristics, such as fryers or hens, divide the kinds of poultry into classes. These physical characteristics are associated with age and sex.

The kinds and classes of live, dressed, and ready-to-cook poultry listed in *U.S. Classes, Standards, and Grades for Poultry* (revised 9/5/2002) are in general use in all segments of the poultry industry. The uniform application of the U.S. grading is the classifying and sorting of poultry and poultry products by the USDA classes, standards, and products according to various groups of conditions and grades for poultry.

Besides information about the classes, standards, and grades for poultry, this manual also covers the voluntary poultry grading program and the grading of poultry for compliance with special purchase requirements.

Ready-to-cook poultry carcasses and parts, as well as certain poultry products, may be graded for quality according to official standards and grades.

Ready-to-cook poultry may be certified as meeting the special requirements of a buyer's purchase contract or specification. Certificates or other documents may also be issued relative to the product graded.

These grading services aid in the nationwide marketing of poultry and poultry products. They change from time to time as the types of products available change. Today, less than 25 percent of all chickens and less than 20 percent of all turkeys are marketed as whole birds. The rest reach the market in value-added forms, such as cut-up parts, boneless and/or skinless cuts, and further processed products. These products include a growing variety of freshly prepared ready-to-cook, as well as precooked, refrigerated poultry entrées. This array of poultry and poultry products can usually be found year-round either as fresh or frozen products.

The development of grade standards and the identification and certification of class, quality, quantity, and condition of agricultural products is authorized by the Agricultural Marketing Act (AMA) of 1946. Regulations to implement the AMA were developed in cooperation with state agriculture officials, producers, processors, and consumers.

The use of grading is voluntary, but anyone using it must pay for it. Those who do apply for service must also provide the space, equipment, lighting, or other facilities that are needed by the grader or are required by the regulations. In addition, when grading is performed in a plant, the USDA must approve the plant and its facilities.

Regulations also provide penalties for the misuse of grading, such as a false representation that a poultry has been officially graded. Products labeled with the terms *Prime*, *Choice*, and *Select*, which indicate superior or top quality, must be equivalent to U.S. Grade A.

The letters "U.S." or "USDA" may be used with a poultry grademark only if an authorized grader has graded the poultry. The letter grades "A," "B," and "C" may be used without the grademark only if the product has been graded by a USDA grader and meets the requirements of the appropriate letter grade. The words "Prepared from," or similar wording, may be used with the U.S. grademark only if the product was produced from graded poultry.

THE POULTRY DIVISION

The voluntary poultry grading program is administered by the Poultry Division of the USDA's Agricultural Marketing Service. The division's national office oversees regional and state offices that supervise graders stationed across the country.

A grader assigned to a specific processing plant on a full-time or part-time basis usually performs resident grading. If more than one grader is assigned to a plant, one of them will be designated as the grader-in-charge.

Most plants process large volumes of poultry and are assisted by resident graders. The authorized grading personnel grade the poultry according to quality. The resident graders perform final check grading and certification.

Resident grading costs include an amount equal to the salary of the resident grader, plus a charge based on the volume of product handled in the plant to cover supervisory and administrative costs. The cost per pound of poultry for this service is generally little more than that incurred by plants employing their own graders.

Most grading services provided to resident plants are also available on a temporary basis, that is, on an as-needed basis. Temporary plant fees are based on the time needed to do the work.

Fee grading is performed when an applicant requests grading of a particular lot or carload of poultry, often for compliance with special contract requirements. Requests for this type of service are usually made on an irregular basis, and the charges are based on the time needed to do the work. Most fee grading work is done at locations other than processing plants, such as at distribution facilities or destination points.

Cooperative agreements between the USDA and various state agencies set forth procedures for the collection and disbursement of fees.

Under a State Trust Fund agreement, arrangements for service are made between the state and the individual firms, with the concurrence of the USDA. Fees are collected by the state. The USDA is reimbursed by the state for federal supervision of the program.

Under a Federal Trust Fund agreement, applications for service are made between the USDA and the individual firms. The USDA collects the fees. The USDA pays the state for the costs it incurs in performing grading work and supervision.

POULTRY ELIGIBLE FOR GRADING

The kinds of poultry that may be graded include, but are not limited to, chickens, turkeys, ducks, geese, pigeons, and guineas. The form may be as a ready-to-cook carcass or part or as a further processed product. Descriptions of the kinds are given in the following sections.

Chickens

Rock Cornish Game Hen or Cornish Game Hen

A Rock Cornish game hen or Cornish game hen is a young, immature chicken (usually 5–7 weeks of age), weighing not more than 2 lb ready-to-cook weight, which was dressed from a Cornish chicken, or the progeny of a Cornish chicken crossed with another breed of chicken.

Broiler or Fryer

A broiler or fryer is a young chicken (usually 9–12 weeks of age), of either sex, that is tender-meated, with soft, pliable, smooth-textured skin and flexible breastbone cartilage.

Roaster

A roaster is a young chicken (usually 3–5 months of age), of either sex, that is tender-meated, with soft, pliable, smooth-textured skin, and breastbone cartilage that may be somewhat less flexible than that of a broiler or fryer.

Capon

A capon is a surgically unsexed male chicken (usually under 8 months of age) that is tender-meated, with soft, pliable, smooth-textured skin.

Stag

A stag is a male chicken (usually under 10 months of age), with coarse skin, somewhat toughened and darkened flesh, and considerable hardening of the breastbone cartilage. Stags show a condition of fleshing and a degree of maturity intermediate between those of a roaster and a cock or rooster.

Hen, Stewing Chicken or Fowl

A hen, or stewing chicken or fowl, is a mature female chicken usually more than 10 months of age), with meat less tender than that of a roaster, and nonflexible breastbone tip.

Cock or Rooster

A cock or rooster is a mature male chicken with coarse skin, toughened and darkened meat, and hardened breastbone tip.

Turkeys

Fryer-Roaster Turkey

A fryer-roaster turkey is a young, immature turkey (usually under 16 weeks of age), of either sex that is tender-meated with soft, pliable, smooth-textured skin and flexible breastbone cartilage.

Young Hen Turkey

A young hen turkey is a young female turkey (usually 5–7 months of age) that is tender-meated, with soft, pliable, smooth-textured skin, and breastbone cartilage that is somewhat less flexible than that in a fryer-roaster turkey.

Young Tom Turkey

A young tom turkey is a young male turkey (usually 5–7 months of age) that is tender-meated, with soft, pliable, smooth-textured skin, and breastbone cartilage that is somewhat less flexible than that of a fryer-roaster turkey.

Yearling Hen Turkey

A yearling hen turkey is a fully matured female turkey (usually under 15 months of age) that is reasonably tender-meated, with reasonably smooth-textured skin.

Yearling Tom Turkey

A yearling tom turkey is a fully matured male turkey (usually under 15 months of age) that is reasonably tender-meated, with reasonably smooth-textured skin.

Mature Turkey or Old Turkey (Hen or Tom)

A mature or old turkey is an old turkey of either sex (usually in excess of 15 months of age), with a coarse skin and toughened flesh.

(For labeling purposes, the designation of sex within the class name is optional, and the three classes of young turkeys may be grouped and designated as "young turkeys.")

Ducks

Broiler Duckling or Fryer Duckling

A broiler duckling or fryer duckling is a young duck (usually under 8 weeks of age), of either sex, that is tender-meated and has a soft bill and soft windpipe.

Roaster Duckling

A roaster duckling is a young duck (usually under 16 weeks of age), of either sex, that is tender-meated and has a bill that is not completely hardened and a windpipe that is easily dented.

Mature Duck or Old Duck

A mature duck or an old duck is a duck (usually over 6 months of age), of either sex, with toughened flesh, hardened bill, and hardened windpipe.

Geese

Young Goose

A young goose may be of either sex, is tender-meated, and has a windpipe that is easily dented.

Mature Goose or Old Goose

A mature goose or an old goose may be of either sex and has toughened flesh and a hardened windpipe.

Guineas

Young Guinea

A young guinea may be of either sex, is tender-meated, and has a flexible breastbone cartilage.

Mature Guinea or Old Guinea

A mature guinea or an old guinea may be of either sex, has toughened flesh and a hardened breastbone.

Pigeons

Squab

A squab is a young, immature pigeon of either sex and is extra-tender-meated.

Pigeon

A pigeon is a mature pigeon of either sex, with coarse skin and toughened flesh.

POULTRY INSPECTION

All poultry that is graded must first be inspected. *Inspection* refers to the wholesomeness of poultry and its fitness for food. It is not concerned with quality or grade. The inspection mark means that the poultry has passed examination by a qualified USDA veterinarian or inspector during slaughter and/or processing. A USDA veterinarian supervises all slaughter plants. Plants that apply for inspection service and are accepted are known as official establishments.

Under the Poultry Products Inspection Act of 1957, the USDA provides mandatory federal inspection of poultry and poultry products shipped in interstate or foreign commerce. The Wholesome Poultry Products Act of 1968 strengthened that law and opened the way for vastly improved state poultry inspection systems.

All poultry slaughtered for human food destined for sale in commerce must be processed, handled, packaged, and labeled in accordance with the Act and its regulations. The government pays for mandatory federal inspection service, except for overtime and holiday work. States with inspection systems equal to the federal program conduct inspections in plants shipping only within that state. In states without such programs, intrastate plants are also under federal inspection. Voluntary inspection service for squab and game birds is provided under the authority of the AMA, and the applicant pays the cost.

Administration of both mandatory and voluntary federal inspection is the responsibility of USDA's Food Safety and Inspection Service.

After poultry has passed inspection and is ready to be graded, the grader must be sure that it is still in a wholesome condition to be graded. Ready-to-cook poultry that is off-condition (slimy or slippery condition of the skin, putrid or sour odor) is ineligible for grading. In addition, ready-to-cook poultry showing other not-acceptable conditions cannot be graded. If the grading is done at the processing plant, the poultry must be sent back for reworking. If the grading is performed elsewhere, the number of carcasses or parts found with these conditions would be recorded, and the lot would be ineligible for grading. These conditions include protruding pinfeathers, bruises requiring trimming, lungs incompletely removed, parts of the trachea present, and feathers or extraneous material of any type inside or outside the carcass.

MARKS ON CONSUMER CONTAINERS

The inspection mark (Figure 6.1) is required on consumer containers and shipping containers of poultry inspected under the Poultry Products Inspection Act.

Figure 6.1 Inspection mark on consumer containers. (Courtesy of the U.S. Department of Agriculture.)

Figure 6.2 Grademark on consumer containers. (Courtesy of the U.S. Department of Agriculture.)

Figure 6.4 Grademark on shipping containers. (Courtesy of the U.S. Department of Agriculture.)

Figure 6.3 Grademark on consumer containers. (Courtesy of the U.S. Department of Agriculture.)

Figure 6.5 Sample grading stamp on shipping containers. (Courtesy of the U.S. Department of Agriculture.)

The grademark must be printed with either light-colored letters on a dark field (Figure 6.2) or dark-colored letters on a light field (Figure 6.3).

Any wing tag, insert label, or other label that bears the inspection mark, the grademark, or both, must also show either the plant number or the firm name and address. When both marks are shown on a tag, they must appear on the same side of the tag, or they may appear on both sides of the tag. Wing tags of shield design may not be used for showing the inspection mark only, but may be used to show both marks or the grademark only. Wing tags bearing the grademark must show the class of the product.

MARKS ON SHIPPING CONTAINERS

The grademark (Figure 6.4) is used on a shipping container to designate a consumer grade. The date the product was graded shall be used in conjunction with the grademark when applied by stamping. Preprinted grademarks do not require a date.

The sample-graded stamp (Figure 6.5) is used when grade certification is made in conjunction with other factors, such as test weighing and temperature certification. This stamp may be used only when a certificate is actually issued and shall show the certificate number in the stamp. It is not used for condition certification of nongraded product for Puerto Rico.

The contract compliance stamp (Figure 6.6) is used when identifying a product that has been produced according to approved specifications and is not to be applied until the product is determined to be in compliance with all contract requirements. Either the certificate number or the date the product was certified may be used in the stamp, but the certificate number is preferred.

The provisional stamp (Figure 6.7) is used on poultry graded for military contracts for troop issue and/or resale when purchased under military "orders for subsistence." It is

Figure 6.6 Contract compliance stamp on shipping containers. (Courtesy of the U.S. Department of Agriculture.)

also to be used on Veterans Administration products when purchased on Defense Procurement Supply Center purchase contracts. The certificate number is in this stamp. It is to be applied at the time the product is determined to be in compliance with all contract requirements.

The lot stamp (Figure 6.8) may be used to identify product that has been graded according to special purchase requirements, such as product for commodity purchase programs placed in a freezer until sufficient quantities accumulate for subsequent sale and delivery. It may also be used to identify further processed items produced according to approved specifications. The consecutive day of the year when the product was graded is in this stamp.

USDA
FOR MILITARY
167895
PROVISIONAL
P-7399

Figure 6.7 Provisional stamp on shipping containers. (Courtesy of the U.S. Department of Agriculture.)

USDA
LOT
255
P-963

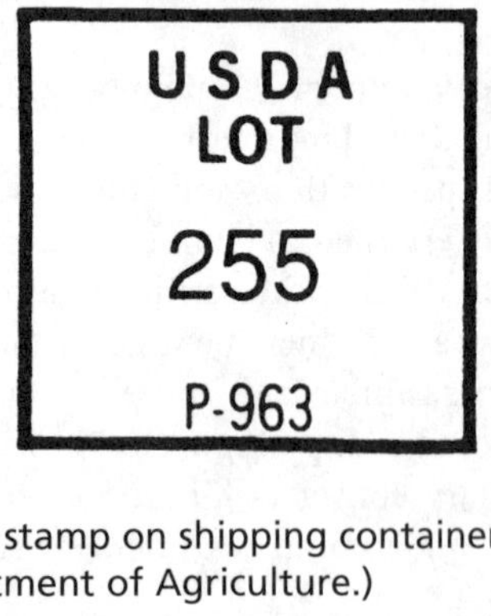

Figure 6.8 Lot stamp on shipping containers. (Courtesy of the U.S. Department of Agriculture.)

Figure 6.9 Officially certified mark on shipping containers. (Courtesy of the U.S. Department of Agriculture.)

The officially certified mark (Figure 6.9) is used on poultry sampled for temperature, condition, weight, or other certifications where the grade is not referenced on the certificate or when the product is not identified with an official grade. This stamp may be used only when a certificate is actually issued and shall show the certificate number in the stamp. It is used for condition certification of nongraded product for Puerto Rico.

Grading According to Quality Standards and Grades

Quality refers to the inherent properties of a product that determine its relative degree of excellence or value. Experience and research have identified certain properties in poultry that are desired by producers, processors, and consumers. Some of these properties are a good proportion of meat to bone, adequate skin covering, absence of pinfeathers, and freedom from discoloration. Standards of quality enumerate the factors that affect these properties and apply to individual ready-to-cook poultry carcasses, parts, and products. There are no grade standards for giblets, detached necks and tails, wing tips, and skin. For carcasses and parts, the factors include conformation, fleshing, fat covering, exposed flesh, and discoloration. For poultry products, factors besides those already mentioned include presence of tendons and blood clots, as well as product-specific factors.

Grades apply to lots of poultry of the same kind and class, each of which conforms to the requirements for the grade standard. The U.S. consumer grades for poultry are U.S. Grades A, B, and C.

Grading involves evaluating poultry in terms of the standards to determine the grade. A given lot of poultry may contain a small percentage of a quality lower than the grade specified because some defects are permitted. This is an unavoidable necessity due to today's production-type processing methods.

Poultry grade standards have changed over the years to reflect developments in poultry production, processing, and marketing. Standards for ready-to-cook poultry were added to the regulations in 1950, roasts were added in 1965, parts and certain other products were added in 1969, and all provisions for grading live and dressed poultry were eliminated in 1976. As the need arises and meaningful quality factors are established for other products, additional grade standards will be developed.

Uses of Standards and Grades

Standards and grades are used extensively throughout the marketing system. Grade standards furnish an acceptable common language that buyers and sellers use in trading and market news reporters use when communicating about market prices and supplies.

Poultry firms that base the price paid to the grower on grade yields will grade the product to USDA standards.
Processors request grading service so that they may use the official grademark on each bird or package.
The Department of Defense, airlines, grocery stores, fast-food restaurants, hospitals, schools, and other large-scale food buyers specify USDA-graded poultry products in their purchasing contracts.
Commercial firms use the standards and grades as a basis for their own product specifications, for advertising, and for establishing brand names.
Firms participating in the USDA's Export Enhancement Program must use the USDA's grading services.
Both buyers and sellers may use the grades as the basis for negotiating loans, settling disputes involving quality, or paying damage claims.

EXAMINING CARCASSES AND PARTS TO DETERMINE QUALITY

In today's high-volume processing operations, plant employees specifically authorized by the USDA to perform this work initially grade poultry. Subsequently, the product is check-graded by a USDA grader to determine compliance with applicable grade standards.

Poultry being graded should be at approximately reading distance from the grader's eye. Graders who require reading glasses should use them, especially when grading white-feathered birds where minute pinfeathers, hairs, and small wet feathers are hard to see.

Each carcass or part, including those used in preparing a poultry food product bearing the grademark, must be graded and identified in an unfrozen state on an individual basis. The class must also be determined, inasmuch as the quality factors of fat, fleshing, and conformations vary with the age and sex of the bird.

Most carcasses and parts are graded on the processing line or out of the cooler. A whole carcass can be examined by grasping the hocks of the bird in one hand with the breast up. With the bird in this position, the grader can easily observe the breast, wings, and legs, and by a mere twist of the wrist, the back of the bird can be turned into position to be observed. When parts are graded, the side nearest the grader should be examined first, then the parts are turned around so the opposite side can be examined.

The usual procedure is to decide on the fleshing and fat-covering factors first. This makes it easier to decide on the other factors. The intensity, aggregate area, location, and number of defects must also be evaluated. In each quality level there are maximum defects permitted, some varying with the weight of the ready-to-cook carcass.

The final quality rating is based on the factor with the lowest rating. Thus, if the requirements for A quality are met in all factors except one, and this factor is B quality, the final grade designation would be "B." The carcass or part is then placed in the proper grade location by the grader or by an electronically controlled computer system.

When the resident grader check-grades the product, the acceptable quality level (AQL) procedures used are based on the percentage of defective undergrade birds or parts present. If the AQL results show that there are excessive undergrades, the grader will work with plant management to identify and correct the problem.

Quality Factors for Carcasses and Parts

The following factors must be considered in determining the quality of an individual ready-to-cook carcass or part.

Conformation. The structure or shape of the bird may affect the distribution and amount of meat, and certain defects detract from its appearance. Some of the defects that should be noted are breasts that are dented, crooked, knobby, or V-shaped; backs that are crooked or hunched; legs and wings that are deformed; and bodies that are definitely wedge-shaped.

Fleshing. The drumsticks, thighs, and breast carry the bulk of the meat. There is, however, a definite correlation between the covering of the flesh over the back and the amount of flesh on the rest of the carcass. Females almost invariably carry more flesh over the back and generally have a more rounded appearance to the breast, thighs, and legs.

The common defects in fleshing are breasts that are V-shaped or concave, rather than full and rounded; breasts that are full near the wishbone, but taper sharply to the rear; legs and drumsticks that are thin; and backs that have insufficient flesh to cover the vertebrae and hip bones.

Fat covering. Fat in poultry is judged entirely by accumulation under the skin. This is true even for chicken parts. Accumulations occur first around the feather follicles in the heavy feather tracts.

Poorly fattened birds may have some accumulation of fat in the skin along the heavy feather tracts on the breast. Then, accumulations can be noted at the juncture of the wishbone and keel and where the thigh skin joins the breast skin. At the same time, accumulations can be noted around the feather follicles between the heavy feather tracts and over the back and hips. Well-finished older birds have sufficient fat in these areas and over the drumsticks and thighs, so that the flesh is difficult to see. Fowl that have stopped laying have a tendency to take on excessive fat in the abdominal area. Younger birds generally have less fat under the skin between the heavy feather tracts on the breast and over the drumsticks and thighs than mature birds.

Pinfeathers. Processors try to eliminate the problem of pinfeathers, particularly those just coming through the skin, by moving poultry to slaughter after feathering cycles are over. There is, for instance, a very short period within which the slaughtering of ducklings must be done. With other classes, the period is longer and attention is given primarily to noting whether the bulk of the pins have sufficient brush on them to facilitate picking.

The types of pinfeathers—protruding and nonprotruding—are considered in grading. Protruding pinfeathers have broken through the skin and may or may not have formed a brush. Nonprotruding pinfeathers are evident but have not pushed through the outer layer of skin.

Before a quality designation can be assigned, ready-to-cook poultry must be free of protruding pinfeathers that are visible to a grader during examination of the carcass at normal operating speeds. However, a carcass may be considered as being free from protruding pinfeathers if it has a generally clean appearance (especially on the breast) and if not more than an occasional protruding pinfeather is in evidence during a more careful examination of the carcass. Vestigial feathers; hair on chickens, turkeys, guineas, and pigeons; and down on ducks and geese must also be considered.

Exposed flesh, cuts, tears, and broken bones. Exposed flesh can result from cuts, tears, missing skin, or broken or disjointed bones. It detracts from the appearance of the carcasses and parts and permits the flesh to dry out during cooking, thus lowering the eating quality. The number and extent of such defects permitted depend on their location, whether on the breast or legs or elsewhere.

Skin discoloration and flesh blemishes and bruises. Most poultry are packaged in material that reduces exposure to air. However, in time the skin will still dry out and become discolored. If drying has occurred, the size of the areas is taken into consideration as a discoloration. Firms marketing poultry in areas that prefer a yellow finish on chicken will scald at lower temperatures (124°–126° F), thus leaving the cuticle (bloom) on the chicken and enhancing its yellow appearance. For the fast-food market wanting batter/breaded fried products or customers preferring a white appearance on poultry, firms scald poultry at higher temperatures (132°–136° F) to remove the cuticle. This enhances the bonding of the batter/breading to the chicken skin, but extra precautions must be taken to reduce the exposure of the product to air. Bruises in the flesh or skin are permitted only to the extent that there is no coagulation or clotting (discernible clumps of red cells). Small clots in the skin or on the surface of the flesh may be cut to allow them to leach out in the chilling process. Such cuts would be taken into consideration in determining the quality. Blue or green bruises must be removed before grading. Excessive grade loss because of bruises should be called to the attention of management.

Some breeds of older turkeys may have a condition called "blue back" that causes a discoloration over the back and wings. Occasionally, chickens may have a bluish or bluish green color in the body lining. These discolorations detract from the appearance and are included in the total aggregate areas permitted for discoloration.

Freezing defects. Discoloration and dehydration of poultry skin during storage is commonly called "freezer burn." This defect detracts from the appearance and, in the case of either moderate or severe freezer burn, lowers the quality of the poultry. Other freezing defects are the darkening of the meat due to slow freezing and seepage of moisture from a defrosting product, resulting in clear, pinkish, or reddish layers of ice.

CUTTING POULTRY PARTS

The USDA standards of quality apply to poultry parts cut in the manner described in this section. These descriptions were originally developed when parts were cut from a carcass by hand. Today, many processors disjoint whole carcasses by machine. Machine-cut parts may also be graded as long as they are not misshapen and have nearly the same appearance as they had prior to cutting from the carcass. Under certain conditions, parts cut in other ways may also be officially identified when properly labeled. Skin or fat not normally associated with a part may not be included unless stated on the label.

Halves are prepared by making a full-length back and breast split of the carcass so as to produce approximately equal right and left sides. Portions of the backbone must remain on both halves. The cut may be no more than 1/4 in. from the sternum (breastbone). Quarters consist of the entire eviscerated poultry carcass that has been cut into four equal parts, excluding the neck.

"Breast quarter" consists of half a breast with the wing and a portion of the back attached.

"Breast quarter without wing" consists of a front quarter of a poultry carcass from which the wing has been removed.

"Leg quarter" consists of a thigh and drumstick, with a portion of the back attached. It may also include attached abdominal fat and up to two ribs. A leg with a complete or entire rear back portion attached may also be grade-identified if certain criteria are met.

Breasts are separated from the back at the shoulder joint and by a cut running backward and downward from that point along the junction of the vertebral and sternal ribs. The ribs may be removed from the breasts, and the breasts may be cut along the breastbone to make two approximately equal halves; or the wishbone portion, as described later in this section, may be removed before cutting the remainder along the breastbone to make three parts. Pieces cut in this manner may be substituted for lighter or heavier pieces for exact-weight-making purposes, and the package may contain two or more of such parts without affecting the appropriateness of the labeling as "chicken breasts." Neck skin is not included with the breasts, except that "turkey breasts" may include neck skin up to the whisker.

"Breasts with ribs" are separated from the back at the junction of the vertebral ribs and back.

Breasts with ribs may be cut along the breastbone to make two approximately equal halves; or the wishbone portion may be removed before cutting the remainder along the breastbone to make three parts. Pieces cut in this manner may be substituted for lighter or heavier pieces for exact-weight-making purposes, and the package may contain two or more of such parts without affecting the appropriateness of the labeling as "breasts with ribs." Neck skin is not included, except that "turkey breasts with ribs" may include neck skin up to the whisker.

"Split breasts with back portion" or breast halves with back portion do not require centering of the cut to make two approximately equal halves if labeled "split breast."

"Boneless-skinless breasts" shall be prepared from breasts cut as described earlier. After deboning, they shall be free of tendons, cartilage, bone pieces, blood clots, discoloration, and undue muscle mutilation.

"Boneless-skinless breast with rib meat" consists of a breast cut as described earlier with rib meat attached.

"Wishbones" (pulley bones) with covering muscle and skin tissue are severed from the breast approximately halfway between the end of the wishbone (hypocledium) and from point of breastbone (cranial process of the sternal crest) to a point where the wishbone joins the shoulder. Neck skin is not included.

Legs include the whole leg, that is, the thigh and the drumstick, whether jointed or disjointed. Back skin is not included.

"Leg with pelvic bone" consists of a poultry leg with adhering meat and skin and pelvic bone. The patella (kneebone) may be included on either the drumstick or the thigh.

Thighs are disjointed at the hip joint and may include the pelvic meat but shall not include the pelvic bones. Back skin is not included.

"Thigh with back portion" consists of a poultry thigh with back portion attached. Thighs may also include abdominal meat (flank meat), but shall not include rib bones.

Drumsticks are separated from the thigh by a cut through the knee joint (femorotibial and patellar joint) and from the hock joint (tarsal joint).

Wings include the entire wing with all muscle and skin tissue intact, except that the wing tip may be removed.

"Wing drummette" consists of the humerus (first portion) of a wing with adhering skin and meat attached.

Backs include the pelvic bones and all the vertebrae posterior to the shoulder joint. *Note:* When labeled "back rear portion," the product is permitted to carry the Grade C identification when applicable grade criteria are met. The meat may not be peeled from the pelvic bones. The vertebral ribs and/or scapula may be removed or included. Skin shall be substantially intact. Grade standards may be applied when necks are packed with backs if certain procedures are followed.

STANDARDS FOR QUALITY OF READY-TO-COOK INDIVIDUAL CARCASSES AND PARTS

Carcasses and Parts—"A" Quality

Conformation. The carcass or part is free of deformities that detract from its appearance or that affect the normal distribution of flesh. Slight deformities, such as slightly curved or dented breastbones and slightly curved backs, may be present.

Fleshing. The carcass has a well-developed covering of flesh, considering the kind, class, and part.

1. The breast is moderately long and deep and has sufficient flesh to give it a rounded appearance, with the flesh carrying well up to the crest of the breastbone along its entire length.
2. The leg is well fleshed and moderately thick and wide at the knee and hip joint area and has a well-rounded, plump appearance, with the flesh carrying well down toward the hock and upward to the hip joint area.
3. The drumstick is well fleshed and moderately thick and wide at the knee joint and has a well-rounded, plump appearance, with the flesh carrying well down toward the hock.
4. The thigh is well to moderately fleshed.
5. The wing is well to moderately fleshed.

Fat covering. The carcass or part, considering the kind, class, and part, has a well-developed layer of fat in the skin.

The fat is well distributed so that there is a noticeable amount of fat in the skin in the areas between the heavy feather tracts.

Defeathering. The carcass or part has a clean appearance, especially on the breast. The carcass or part is free of pinfeathers, diminutive feathers, and hairs that are visible to the grader.

Exposed flesh. Parts are free of exposed flesh resulting from cuts, tears, and missing skin (other than slight trimming on the edge). The carcass is free of these defects on the breast and legs. Elsewhere, the carcass may have cuts or tears that do not expand or significantly expose flesh, provided the aggregate length of all such cuts and tears does not exceed 3/4 in. for poultry weighing up to 2 lb; 1 1/2 in. for poultry weighing over 2 lb, but not more than 6 lb; 2 in. for poultry weighing over 6 lb, but not more than 16 lb; and 3 in. for poultry weighing over 16 lb. The carcass may have exposed flesh elsewhere other than on the breast and legs due to slight cuts, tears, and areas of missing skin, provided the aggregate area of all exposed flesh does not exceed an area equivalent to the area of a circle of the diameter specified.

Disjointed and broken bones and missing parts. The carcass is free of broken bones and has not more than one disjointed bone. The wing tips may be removed at the joint, and in the case of ducks and geese, the parts of the wing beyond the second joint may be removed, if removed at the joint and both wings are so treated. The tail may be removed at the base. Cartilage separated from the breastbone is not considered as a disjointed or broken bone.

Discoloration of the skin and flesh. The carcass or part is practically free of such defects. Discoloration due to bruising shall be free of clots (discernible clumps of red or dark cells). Evidence of incomplete bleeding, such as more than an occasional slightly reddened feather follicle, is not permitted. Flesh bruises and discoloration of the skin, such as "blue back," are not permitted on the breast or legs of the carcass or on these individual parts, and only lightly shaded discolorations are permitted elsewhere.

The total areas affected by flesh bruises, skin bruises, and discoloration, such as "blue back," singly or in any combination, shall not exceed one-half of the total aggregate area of permitted discoloration. The aggregate area of all discoloration for a part shall not exceed an area equivalent to the area of a circle 1/4 in. in diameter for poultry weighing up to 6 lb and 1/2 in. in diameter for poultry weighing over 6 lb. The aggregate area of all discoloration for a carcass shall not exceed an area equivalent to the area of a circle of the diameter specified.

Freezing defects. With respect to consumer packaged poultry, parts, or specified poultry food products, the carcass, part, or specified poultry food product is practically free from defects that result from handling or occur during freezing or storage. The following defects are permitted if they, alone or in combination, detract only very slightly from the appearance of the carcass, part, or specified poultry food product:

1. Slight darkening over the back and drumsticks, provided the frozen bird or part has a generally bright appearance.
2. Occasional pockmarks due to drying of the inner layer of skin (derma) (however, none may exceed the area of a circle 1/8 in. in diameter for poultry weighing 6 lb or less and 1/4 in. in diameter for poultry weighing over 6 lb).
3. Occasional small areas showing a thin layer of clear or pinkish-colored ice.

Backs. A-quality backs shall meet all applicable provisions pertaining to parts and shall include the meat contained on the ilium (oyster), pelvic meat and skin, and vertebral ribs and scapula with meat and skin.

Carcasses and Parts—"B" Quality

Conformation. The carcass or part may have moderate deformities, such as a dented, curved, or crooked breast; crooked back; or misshapen legs or wings, which do not materially affect the distribution of flesh or the appearance of the carcass or part.

Fleshing. The carcass has a moderate covering of flesh, considering the kind, class, and part.

1. The breast has a substantial covering of flesh, with the flesh carrying up to the crest of the breastbone sufficiently to prevent a thin appearance.
2. The leg is fairly thick and wide at the knee and hip joint area and has sufficient flesh to prevent a thin appearance.
3. The drumstick has a sufficient amount of flesh to prevent a thin appearance, with the flesh carrying fairly well down toward the hock.
4. The thigh has a sufficient amount of flesh to prevent a thin appearance.
5. The wing has a sufficient amount of flesh to prevent a thin appearance.

Fat covering. The carcass or part has sufficient fat in the skin to prevent a distinct appearance of the flesh through the skin, especially on the breast and legs.

Defeathering. The carcass or part may have a few nonprotruding pinfeathers or vestigial feathers that are scattered sufficiently so as not to appear numerous. Not more than an occasional protruding pinfeather or diminutive feather shall be in evidence under a careful examination.

Exposed flesh. A carcass may have exposed flesh, provided that no part on the carcass has more than one-third of the flesh exposed and the meat yield of any such part on the carcass is not appreciably affected. A part may have no more than one-third of the flesh normally covered by skin exposed. A moderate amount of meat may be trimmed around the edges of a part to remove defects.

Amount of disjointed and broken bones and missing parts. The carcass may have two disjointed bones, or one disjointed bone and one nonprotruding broken bone, but is free of broken bones. Parts of the wing beyond the second joint may be removed at a joint. The tail may be removed at the base. The back may be trimmed in an area not wider than the base of the tail and extending from the tail to the area halfway between the base of the tail and the hip joint.

Discoloration of the skin and flesh. The carcass or part is free of serious defects. Discoloration due to bruising shall be free of clots (discernible clumps of red or dark cells). Evidence of incomplete bleeding shall be no more than very slight. Moderate areas of discoloration due to bruises in the skin or flesh and moderately shaded discoloration of the skin, such as "blue back," are permitted, but the total areas affected by such discoloration, singly or in any combination, may not exceed one-half of the total aggregate area of permitted discoloration. The aggregate area of all discoloration for a part shall not exceed an area equivalent to the area of a circle having a diameter of ½ in. for poultry weighing up to 2 lb; 1 in. for poultry weighing over 2 lb, but not more than 6 lb; and 1½ in. for poultry weighing over 6 lb. The aggregate area of all discoloration for a carcass shall not exceed an area equivalent to the area of a circle of the diameter specified.

Freezing defects. With respect to consumer packaged poultry, parts, or specified poultry food products, the carcass, part, or specified poultry food product may have moderate defects that result from handling or occur during freezing or storage. The skin and flesh shall have a sound appearance but may lack brightness. The carcass or part may have a few pockmarks due to drying of the inner layer of skin (derma). However, no single area of overlapping pockmarks may exceed that of a circle ½ in. in diameter. Moderate areas showing layers of clear pinkish- or reddish-colored ice are permitted.

Backs. B-quality backs shall meet all applicable provisions pertaining to parts, and shall include either the meat contained on the ilium (oyster) and meat and skin from the pelvic bones or the vertebral ribs and scapula with meat and skin.

Carcasses and Parts—"C" Quality

1. A part that does not meet the requirements for A or B quality may be of C quality if the flesh is substantially intact.
2. A carcass that does not meet the requirements for A or B quality may be C quality. Both wings may be removed or neatly trimmed. Trimming of the breast and legs is permitted, but not to the extent that the normal meat yield is materially affected. The back may be trimmed in an area not wider than the base of the tail and extending from the tail to the area between the hip joints.

C-quality backs shall include all the meat and skin from the pelvic bones, except that the meat contained in the ilium (oyster) may be removed. The vertebral ribs and scapula with meat and skin and the backbone located anterior (forward) of the ilia bones may also be removed (front half of back).

STANDARDS FOR QUALITY OF SPECIFIED POULTRY FOOD PRODUCTS

Poultry Roast—"A" Quality

This standard applies to raw poultry products labeled in accordance with the poultry inspection regulations as ready-to-cook "roasts" or similar descriptive terminology.

1. The deboned poultry meat used in the preparation of the product shall be from young poultry.
2. Bones, tendons, cartilage, blood clots, and discoloration shall be removed from the meat.
3. All pinfeathers, bruises, hair, discoloration, and blemishes shall be removed from the skin, and where necessary, excess fat shall be removed from the skin covering the crop area or other areas.
4. Seventy-five percent or more of the outer surface of the product shall be covered with skin, whether attached to the meat or used as a wrap. The skin shall not appreciably overlap at any point. The combined weight of the skin and fat used to cover the outer surface and that used as a binder shall not exceed 15 percent of the total net weight of the product.
5. The product shall be fabricated in such a manner that it can be sliced after cooking, and each slice can be served with minimum separation.
6. Seasoning or flavor enhancers, if used, shall be uniformly distributed.
7. Product shall be fabricated or tied in such a manner that it will retain its shape after defrosting and cooking.
8. Packaging shall be neat and attractive.
9. Product shall be practically free of weepage after packaging and/or freezing, and, if frozen, shall have a bright, desirable color.
10. Product packaged in an oven-ready container shall meet all the requirements of paragraphs (1) through

(9) of this section, except that with respect to skin covering, the exposed surface of the roast need not be covered with skin. If skin is used to cover the exposed surface, it may be whole or emulsified. In addition, for roasts packaged in oven-ready containers, comminuted (mechanically deboned) meat may be substituted in part for skin, but may not exceed 8 percent of the total weight of the product.

Boneless Poultry Breast and Thigh—"A" Quality

These standards apply to raw poultry products labeled as ready-to-cook boneless poultry breasts or thighs, or as ready-to-cook boneless poultry breast fillets or thigh fillets, or with words of similar import.

1. The breast or thigh shall be cut as specified.
2. The bone or bones shall be removed in a neat manner without undue mutilation of adjacent muscle.
3. With skin, the breast or thigh shall meet A-quality requirements for ready-to-cook poultry parts for defeathering, exposed flesh, and discoloration, and shall be free of tendons and cartilage.
4. Skinless breasts or thighs shall be free of tendons, cartilage, blood clots, and discoloration. Minor abrasions due to preparation techniques are permitted.
5. Slight trimming is permitted around the edges and inner muscle surface of the part.

U.S. CONSUMER GRADES FOR READY-TO-COOK POULTRY AND SPECIFIED POULTRY FOOD PRODUCTS

Regulations currently provide for both consumer and procurement grades that are based on the U.S. standards; however, consumer grades are used almost exclusively.

U.S. Grade A. A lot of ready-to-cook poultry, parts, or poultry food products consisting of one or more ready-to-cook carcasses or parts, or individual units of poultry food products of the same kind and class, each of which conforms to the requirements for A quality, may be designated as U.S. Grade A.

U.S. Grade B. A lot of ready-to-cook poultry or parts consisting of one or more ready-to-cook carcasses or parts of the same kind and class, each of which conforms to the requirements for B quality or better, may be designated as U.S. Grade B.

U.S. Grade C. A lot of ready-to-cook poultry or parts consisting of one or more ready-to-cook carcasses or parts of the same kind and class, each of which conforms to the requirements for C quality or better, may be designated as U.S. Grade C.

U.S. PROCUREMENT GRADES FOR READY-TO-COOK POULTRY

U.S. Procurement Grade 1. Any lot of ready-to-cook poultry composed of one or more carcasses or parts of the same kind and class may be designated and identified as U.S. Procurement Grade 1 when:

1. 90 percent or more of the carcasses or parts in such lot meet the requirements of A quality, with the following exceptions:
   (a) Fat covering and conformation may be as described in this manual for B quality.
   (b) Trimming of skin and flesh to remove defects is permitted to the extent that not more than one-third of the flesh is exposed on a separated part or on any part on a carcass, and the meat yield of a separated part or any part on a carcass is not appreciably affected.
   (c) Discoloration of the skin and flesh may be as described in this manual for B quality.
   (d) One or both drumsticks on a carcass may be removed if the part is severed at the joint.
   (e) The back on a carcass may be trimmed in an area not wider than the base of the tail and extending to the area between the hip joints.
   (f) The wings or parts of wings on a carcass may be removed if severed at a joint.
2. The rest of the carcasses or parts meet the same requirements, except that they may have only a moderate covering of flesh.

U.S. Procurement Grade 2. Any lot of ready-to-cook poultry composed of one or more carcasses or parts of the same kind and class that fails to meet the requirements of U.S. Procurement Grade 1 may be designated and identified as U.S. Procurement Grade 2, provided that:

1. Trimming of flesh from a separated part or from any part on the carcass does not exceed 10 percent of the meat.
2. Portions of a carcass weighing not less than one-half of the whole carcass may be included, if the portion approximates in percentage the meat-to-bone yield of the whole carcass.

GRADING ACCORDING TO SPECIAL REQUIREMENTS

The Poultry Division of the USDA provides services in addition to the grading of poultry according to the U.S. standards and grades. Basically, these services involve determining the compliance of poultry products with the purchase or processing contracts of institutional food buyers. This type of food acceptance or certification work has been growing.

More and more value-added poultry products are available that look less and less like whole carcasses and parts. Buyers often cannot identify what they have bought as easily as they could when only ice-packed whole birds were available.

The foodservice industry continues to expand, especially fast-food-type establishments. Although each company has its own menu identity, all units within a company have uniform food purchase requirements.

Government purchasers such as the military, hospitals, and schools must utilize competitive bidding to help control their food purchasing and processing costs.

USDA grading specialists can help food purchasers prepare explicit product specifications and can certify that purchases comply with these specifications. This helps to assure purchasers that they are getting what they order, facilitates competitive bidding, can result in overall higher-quality food, permits long-range meal planning, and eliminates controversies between the buyer and seller over the compliance of products.

Depending on the purchase specification, poultry and poultry products can be examined to determine compliance with the requirements for:

- Quality of raw materials, including grade of poultry used
- Product formulation and percentage of ingredients
- Batter breading pickup and coverage
- Product weight, dimension, temperature, freedom from defects
- Condition of product packaging, such as freedom from cuts, tears, holes
- Metal detection
- Condition of shipping containers and transportation vehicles

The number of samples examined and the number of defects allowed in a sample or production lot are outlined in the USDA Poultry Division Acceptable Quality Levels (AQL) sample plans.

Grading Live Poultry

Modern poultry industry practices make the grading of live and dressed poultry unnecessary, and all provisions for such grading were deleted from the regulations in 1976.

Rabbits

Fryer or Young Rabbit

A fryer or young rabbit is a young, domestic rabbit carcass, weighing not less than $1\frac{1}{2}$ lb, and rarely more than $3\frac{1}{2}$ lb, processed from a rabbit usually less than 12 weeks of age. The flesh of a fryer or young rabbit is tender and fine-grained and of a bright, pearly pink color.

Roaster or Mature Rabbit

A roaster or mature rabbit is a mature or old domestic rabbit carcass of any weight, but usually over 4 lb, processed from a rabbit usually 8 months of age or older. The flesh of a roaster or mature rabbit is firmer and coarser-grained, the muscle fiber is slightly darker in color and less tender, and the fat may be more creamy in color than that of a fryer or young rabbit.

Carcasses found to be unsound, unwholesome, or unfit for food shall not be included in any of the quality designations specified.

"A" Quality. To be of A quality, the carcass:

1. Is short, thick, well-founded, and full-fleshed.
2. Has a broad back, broad hips, broad, deep-fleshed shoulders, and firm muscle texture.
3. Has a fair quantity of interior fat in the crotch and over the inner walls of the carcass, and a moderate amount of interior fat around the kidneys.
4. Is free of evidence of incomplete bleeding, such as more than occasional slight coagulation in a vein; is free from any evidence of reddening of the flesh due to fluid in the connective tissues.
5. Is free from all foreign material (including, but not being limited to, hair, dirt, and bone particles) and from crushed bones caused by removing the head or the feet.
6. Is free from broken bones, flesh bruises, defects, and deformities; ends of leg bones may be broken due to removing the feet.

"B" Quality. To be of B quality, the carcass:

1. Is short, thick, fairly well-founded, and fairly well fleshed.
2. Has a fairly broad back, fairly broad hips, fairly broad and deep-fleshed shoulders, and fairly firm muscle texture.
3. Has at least a small amount of interior fat in the crotch and over the inner walls of the carcass, with a small amount of interior fat around the kidneys.

4. Is free of evidence of incomplete bleeding, such as more than an occasional slight coagulation in a vein; is free from any evidence of reddening of the flesh due to fluid in the connective tissues.
5. Is free from all foreign material (including, but not being limited to, hair, dirt, and bone particles) and from crushed bones caused by removing the head or feet.
6. Is free from broken bones, flesh bruises, defects, and deformities; ends of leg bones may be broken due to removing the feet.

"C" Quality. A carcass that does not meet the requirements of A or B quality may be of C quality, and such carcass:

1. May be long, rangy, and fairly well fleshed.
2. May have thin, narrow back and hips, and soft, flabby muscle texture.
3. May show very little evidence of exterior fat.
4. May show very slight evidence of reddening of the flesh due to blood in the connective tissues.
5. Is free from all foreign material (including, but not being limited to, hair, dirt, and bone particles) and from crushed bones caused by removing the head or feet.
6. May have moderate bruises of the flesh, moderate defects, and moderate deformities; may have not more than one broken bone in addition to broken ends of leg bones due to removal of the feet; and may have a small portion of the carcass removed because of serious bruises; discoloration due to bruising in the flesh shall be free of clots (discernible clumps of dark or red cells).

INTERNET RESOURCES

www.ams.usda.gov/poultry/pdfs/PYGRDMANUAL.pdf
USDA Poultry Grading Manual

www.fsis.usda.gov/OA/pubs/ingrade.htm

USDA Food Safety Inspection Service. The inspection and grading of meat and poultry are two separate programs within the U.S. Department of Agriculture (USDA). Inspection for wholesomeness is mandatory, but grading for quality is voluntary; the service is requested and paid for by poultry producers and processors.

CHAPTER

7

Fish

Where seafood is concerned, most consumers prefer to dine out rather than prepare seafood at home. This has caused foodservice professionals to expand their knowledge of seafood products that are available for sale. According to the available figures, U.S. consumers spent about $55.3 billion for fishery products in 2001, including $38.2 billion in sales in restaurants and institutions. In the United States, consumption of fishery products was 14.8 lb of seafood per person in 2000. In 1960, consumption was about 10.3 lb per person, and in 1980 consumption rose to 12.5 lb. Fresh and frozen fish accounts for 5.7 lb, and fresh and frozen shellfish consumption was 4.6 lb per capita in 2001. Consumption of canned fishery products was 4.2 lb per person in 2001. Canned tuna made up 3.1 lb of that quantity.

The grading and inspection of fresh fish are entirely voluntary, although all processing and packing plants are inspected for proper sanitation and handling procedures. There are two types of inspection marks, federal inspection and quality.

PACKED UNDER FEDERAL INSPECTION

The designation "Packed under Federal Inspection" (PUFI) (Figure 7.1) is a product guarantee that:

- The seafood has been properly labeled.
- The seafood is pure, wholesome, and safe.
- The seafood has been processed and packaged under sanitary conditions.
- The product has not been graded as to a specific quality level.

Seafood so designated is of an acceptable quality as determined by federal inspectors. This standard applies to whole or dressed fish, fresh or frozen, of any species fit for human consumption and processed and maintained using good manufacturing practices. The Food and Drug Administration (FDA) regulates seafood labeling as to the product, net weight, and country of origin.

Figure 7.1 Packed under Federal Inspection seal

QUALITY

The grading and inspection of fresh fish, which are currently voluntary, are conducted by the U.S. Department of Commerce. The department provides this service to processors that use the grade or inspection shield on their products (Figure 7.2), for a fee. This mark indicates the processors that choose to have their plants under continuous federal inspection. These packers display both the PUFI seal and a grading stamp. At the present time, the grades (A, B, and C) signify that:

1. The product is clean, safe, and wholesome.
2. The product is of a specified quality, indicated by the appropriate grade designation.
3. The product was produced in an acceptable establishment, with proper equipment, and in an appropriate processing environment as required by food-control authorities.
4. The product was processed under supervision by federal food inspectors and packed by sanitary food handlers.
5. The product is truthfully and accurately labeled as to the common or usual name, optional ingredients, or quantity.

Figure 7.2 U.S. Grade mark. (Courtesy of the U.S. Department of Agriculture.)

U.S. grade designations are as follows:

Grade A—top or best quality (can change season to season). Uniform in size and free from defects. Good condition and good flavor for the species.
Grade B—good quality, but has less uniformity of size; some blemishes or defects.
Grade C—fairly good quality; little uniformity and many defects.

Grade B and Grade C are usually marketed without any grade designations on the label.

PRODUCT DESCRIPTION

Vertebrate fish, or finfish, are characterized by backbone and fins and are of two types: lean and fat. Fat fish, such as mackerel, salmon, and swordfish, are best prepared by broiling or baking. Lean types, such as cod, flounder, and haddock, are best for frying.

There are approximately 200 varieties of fish marketed in the United States. The most popular are classified here in categories of freshwater fish and saltwater fish.

Look for bright, shiny fish, with no loose scales; bright, bulging, clean eyes; red inside the gills; firm flesh that bounces back when pressed; no strong odors; no slime. Because fish deteriorates so quickly, it is advisable to buy fresh fish, but only if it is to be used within 48 hours. All the most popular varieties of fish are available frozen.

MARKET FORMS OF FISH

Fresh fish is available in several market forms, ranging from the natural state to a processed end product. The price varies according to availability, amount of labor involved in processing, and market needs. The market forms of fish include the following:

Round or whole fish. As it comes from the water. It is dressed before cooking; edible portion—45 percent.
Drawn. Eviscerated (with the entrails removed). Entrails cause rapid deterioration and spoilage; drawn fish can be stored for a longer period; edible portion—48 percent.

Dressed. Eviscerated, head on, and completely cleaned; used for stuffing and usually served as a single portion; edible portion—65 percent.
Steaks. Cross sections of one fish; cut from a dressed fish with the skin off; edible portion—85 percent.
Headed and gutted. Head, tail, fins, and entrails removed.
Fillet. One-half the fish (backbone to belly) removed from the head, tail, fins, and skin; ready to cook; 100 percent edible; premium product.
Fish sticks. Pieces of fish meat of uniform size, usually 3 in. by 1 in.

CATEGORIES OF CUTS AND SHAPES OF PREPARED FISH

Aberdeen cut. Rhombus shape, cut from block, sides squared or tapered.
American cut. Tapered or beveled edges of fish portions or fillets.
Bits. Also known as bites, nuggets, cubes—small pieces less than ½–1 oz, each in square, round, irregular shapes of fish from blocks.
Cakes. Rounded, flat cakes of minced or ground fish, usually breaded.
Chunks. Large pieces, cross section slices, similar to steaks.
Diamond cut. French cut—same as Aberdeen cut.
Fillet. Boneless piece of fish cut lengthwise from backbone. With/without skin. Butterflied fillets are two sides of fish held together by skin and flesh of back.
Natural cut. Cut from block.
Portion. Usually a square or rectangle, cut from block, 1½–6 oz, breaded or unbreaded, raw/precooked. "Grated fish portions" is a term used for portions made from mechanically separated fish flesh.
Squares. Same as portions.
Steaks. Cross section slices of dressed fish, with backbone portion in center, ½–1 in. thick.
Sticks. Rectangles of fish cut from block, usually 2 oz each, breaded; also cut from mixed block.
Tail. One side of tail of fish, or portion that resembles fish tail. Boneless, usually breaded or battered, raw or precooked, from 3½–8 oz. Entire tail, bone-in, also breaded, and sold as tail.
Tidbits. Same as bits.

Note: Many companies have their own identifying names for portions and cuts, such as Club, Imperial, Dover, and so on.

FROZEN FISH

WHITE-FLESHED OCEAN FISH

Cod (Scrod)	Pollock
Flounder (Fluke, Sole)	Sea Trout
Greenland	Snapper (Red)
Haddock (Scrod)	Sole (Grey, Lemon, Dover)
Halibut	Turbot
Ocean Catfish	Whiting (Hake)
Ocean Perch	Whiting, Cape

These fish are usually available in the following forms:

Fillets. Raw, breaded, or in batter, IQF (individually quick frozen); precooked, breaded, or in batter, IQF; raw, unbreaded, skinless and boned, or skin on and boned; IQF or IQF glazed, bulk (cello wrap)
Steaks. Usually raw, unbreaded, with or without skin, IQF (cod, halibut, pollock)
Whole. Usually raw, unbreaded, with or without skin, IQF (sole, snapper, flounder, whiting, cape whiting)
Bites. Very small cubes, raw, breaded, IQF
Portions. Raw or precooked, breaded or unbreaded, IQF, square or rectangular, or shaped
Sticks. Usually 1 by 3 in., raw or precooked, breaded, IQF, 1–2 oz.

OTHER POPULAR SPECIES

Freshwater Catfish	Shad (and with roe)
Mackerel (Spanish, King)	Smelt
Mullet	Trout (Rainbow, Brook, Speckled, Golden)
Salmon (Sockeye, Red, Coho)	Yellow Perch

These fish are usually in these forms:

Fillets. Raw, breaded, or in batter, IQF; precooked, breaded, or in batter, IQF; raw, unbreaded, skinless and boned, or skin on and boned; IQF or IQF glazed, bulk (cello wrap); butterflied and breaded; stuffed.

Steaks. Raw, unbreaded, ½–1 in. thick (catfish, salmon).

Whole. With skin, with or without head, with or without tail, raw; or breaded, or unbreaded, dressed, boned, stuffed, butterflied (catfish, trout, smelt, shad, perch).

Portions. Raw or cooked, breaded, or unbreaded; square, or rectangular, stuffed or shaped.

Freshwater

Blue Gill. Also known as blue perch, blue sunfish, coppernose, gold perch, chainside; peak season in summer months; available whole or as fillets.

Brook Trout (*Salvelinus fontinalis*). Caught in local streams from May to October.

Buffalo Fish (*Ictiobus cyprinella*) (bigmouth, gourdhead, or red mouth). Caught in the Mississippi River and its tributaries, mostly in Louisiana, from February to August; Chicago, Gulf States, and New York markets.

Carp (*Cyprinus carpio*). Also known as German carp; caught in the Great Lakes and the Mississippi River; sold in California (February through July); and in Chicago (December and February through June, with April being the biggest month); available in fillets, but mostly sold fresh; Chicago and New York markets.

Catfish (*Ictalurus*, *Ameiurus*, and *Leptops olivaris* are all known as bullheads or catfish). Also known as blue channel fish, yellow catfish, and fiddler or spotted fish; caught along all coasts, Florida lakes, and the Mississippi River, March through October; available headless, round, skinned, and dressed; Chicago and Gulf States markets.

Chub (*Leucichthys zenithicus*). Also known as blackfin, bloater, bluefin, and tullibee; caught in Lake Huron, Lake Michigan, and Lake Superior in June through September, most are sold as smoked fish; only the blackfin is used as a fresh fish.

Eel (*Anguilla bostoniensis*). Also known as anguilla, capitone, sand boy, and shoestring; can be bought all year.

Frog's Legs. Bullfrogs and common frogs found in the South, grass frogs mostly from the Mississippi River and its tributaries; available all year, but most are caught from April through October; the white-meated bullfrog is the best grade.

Lake Herring (*Leucichthys artedi*). Also known as blueback and cisco of Lake Erie, caught in Lake Huron, Lake Michigan, Lake Ontario, and Lake Superior; available drawn and round, fresh and frozen, salted and smoked; best lake herring is ½ lb. Lake Erie; New York and Chicago markets.

Lake Trout (*Cristivomer namaycush*). Also known as gray trout, Great Lakes trout, Mackinaw trout, and salmon trout; caught in Lake Huron, Lake Michigan, Lake Superior, and Lake Ontario, mostly from May to October; available filleted, but most are sold drawn and fresh; Chicago market.

Pickerel Pike (*Esox lucius* (common pike), *Esox masquinongy* (muskellunge), *Esox reticulatus*, and *Esox vermiculatus* (pickerels). Also known as grass pike of Lake Erie, jack pike of Canada, and lake pickerel. Caught mostly in the northern Canadian lakes, especially Lake of the Woods; can be bought all year, but the peak is in June; available dressed, round, and filleted; fresh and frozen; Chicago market.

Pike Perch. Includes blue pike, sauger, and yellow pike perch; blue pike also known as blue pickerel, blues, grass pike, Great Northern pike, jack salmon, lake pickerel, and pike perch; saugers also known as sand pike; yellow pike also known as dore, pike perch, salmon jack, wall-eyed pike, and yellows; blue pike caught mostly in Lake Erie, with some from Lake Ontario; saugers caught in Lake Erie; yellow pike caught in Canada, with a few from Lake Erie, Lake Huron, Saginaw Bay, and other lakes; available year-round, with the blue pike peak from October to April, sauger peak from October to February, and yellow pike peak from April to June; blue pike available round and as fresh and frozen fillets; saugers available round and as fresh and frozen fillets; yellow pike available dressed and filleted, but most are sold fresh; Chicago and New York markets.

Rock Bass. Also known as goggle-eye bass, red eye, red-eye perch, sunfish; peak season is the winter months; available filleted, whole.

Sheepshead (*Aplodinotus grunniens*). Also known as white perch, croaker, freshwater drum, gaspergou, and gray bass. Caught in Lake Erie, Lake Huron, and Lake Ontario and in the Mississippi Valley to Texas and Louisiana; peak in April through June; available as fillets, but most are sold round; fish from the South are the best; Chicago and Gulf States markets.

Smelts (*Osmerus mordax*). Lake Michigan supplies most of the freshwater smelts, with the peak in mid-March; Chicago and Great Lakes markets.

Sturgeon (*Acipenser oxyrhynchus*). Also known as green, lake, sea, and common sturgeon; lake sturgeon is caught in large rivers and the Great Lakes, and common sturgeon is found in the North; in season all year; available brined, canned, fresh, frozen, and smoked.

Terrapin. Found in Minnesota, Missouri, Wisconsin, Iowa, Michigan, New York, Delaware, New Jersey, Maryland, Virginia, and Florida; in season from June through October; available live; Gulf States market.

White Bass. Also known as silver lake bass and striped lake bass; season in winter months; available filleted or whole.

Whitefish (*Coregonus clupeaformis*). Also known as Lake Champlain shad; caught in the Great Lakes and Canadian lakes, mostly from May through August; available drawn, dressed, filleted; the best, in order of desirability, are from Lake Superior, Lake Michigan, Lake Erie, and Lake Huron; Chicago and New York markets.

Yellow Perch (*Perca flavescens*). Also known as red fin; jumbos or English perch (very large perch from Canada), Lake Erie perch (large sizes), and lake perch (small sizes); caught in the Great Lakes and coastal streams and lakes from Maine to North Carolina; in season all year, but mostly from April through November; available as fillets and rounds, fresh and frozen; the best are Lake Erie yellow perch, which are of two grades: pound net, which is the best, and gill net, which is inferior.

Saltwater

Abalone (*Haliotis cracherodi*). Also known as aurora, black, green, grand ear shell, pink, rainbow, red, and rough abalone; found along the Pacific coast all year; available canned, dried, fresh, and salted.

Barracuda (*Sphyraena barracude*). Also known as gauchanche; found along the Atlantic coast from the Carolinas to Florida and the West Indies; can be bought all year.

Black Sea Bass. Also known as channel bass and sea wolf; peak season is during the summer months; available whole and as fillets.

Black Fish. Also known as bowfin, cottonfish, dog fish, grindle, lawyer, and speckled cat; can be bought all year; available whole and as fillets.

INTERNET RESOURCES

www.nfi.org

National Fisheries Institute. Represents companies in the fish and seafood industry. Includes a Glossary, Species Guide, Top Ten Seafoods, Recipes and Nutrition, and Industry Links.

CHAPTER

8

Meat

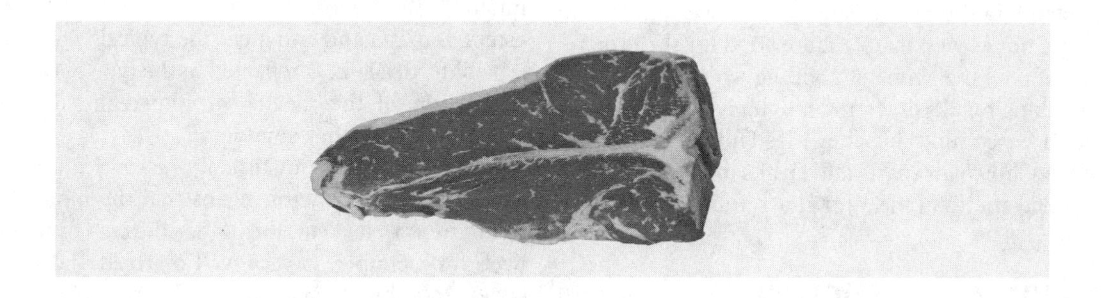

Meat has been the basic item in the human diet since prehistoric times. Historians say that the earliest people started the development of more advanced cultures because of their quest for meat. It was the desire and need for meat that caused the development of the first snare traps, which later led to the use of traps to capture animals for their fur. This sequence is encountered in the economic development of every region of the world.

It was also the quest for meat that initiated the development of the first spears and arrows. It was the desire for a stable supply of meat that led to abandoning nomadic meat hunting in favor of meat raising and farming and to the eventual creation of stable cultures.

Civilizations have been built on meat agriculture, and nations have been destroyed when their cattle succumbed to disease. Indeed, there is no chapter of history that has not been affected by the need for, the desire for, and the uses of meat.

Do we really need meat in our diet? The answer to that question is no, if based on pure physical need. The nutrients we get from meat are all available from other sources. The protein that is so vital to the sustenance of life is available from the plant world, from indirect meat sources such as milk and eggs, and from the marine world of fish and crustaceans. But from a psychological point of view, meat is certainly a necessity. The aroma of a steak on the broiler, the succulence of a potted roast, the sight of a pig on the spit, the sound of a stew simmering are so enticing to us as almost to suggest an intuitive desire for meat. The economy of a nation and the well-being of its people are reflected by the supply and demand status of the meat industry. When a nation prospers, the prices of the better cuts of meat rise sharply as people can afford, and therefore demand, the expensive cuts.

Although we can get our nutrients from other sources, the minerals and proteins we need are most readily available from meat.

MEAT IN THE FOODSERVICE BUDGET

There is no single item in the foodservice budget that is as large or as important as meat. In food purchases alone, meat may range from 30 percent of total food expenditures for an institution to 70 percent of food expenditures for a steak restaurant. Meat purchases may range from as much as 20 percent of all operating expenditures for an institutional foodservice to 40 percent for a commercial establishment. There is no way to overemphasize the importance of the proper purchasing and receiving of meat.

The purchase of the proper cut of meat for the use intended is of paramount importance. As a general rule for cooking proteins, one should buy the higher grades of meat for the dry methods of cooking (broiling, roasting), as such cuts have adequate marbling (fat) to retain the moist, juicy texture expected when they are eaten. The lower grades of meat, which are not as well marbled, are good for the moist-heat methods of cooking (braising, boiling, stewing), as the slow, moist cooking breaks down the tough meat fibers to an acceptable level for comfortable chewing. Thus, it is wise to purchase USDA Choice for roasts and steaks, but wasteful to use USDA Choice grades of meat for stews and pot roasts.

COMPOSITION AND STRUCTURE

The meat we eat is the muscle of the animal. Muscles are of two types with two respective functions. The involuntary muscles are the smooth muscles of the blood vessels and digestive tract, which do their jobs automatically and virtually continuously. These muscles are lacking in interspersed fat and are therefore very tough. The voluntary muscles are composed of many tiny fibers joined together by connective tissues in a fashion that is analogous to the way wire cables are used in building bridges. This type of formation is called striated. Muscles are composed of cells of sizes and shapes that vary according to the feed, age, sex, and type of animal. The connective tissue cells are joined together in bundles by two types of connective tissue. White connective tissue, called collagen, is found in all the muscles of the animal; it is dominant in the more tender cuts of meat. Yellow connective tissue, called elastin, is predominant in the muscles that carry the heaviest load of constant strain and work. The white connective tissue breaks up and becomes moist in cooking, whereas the yellow connective tissue requires mechanical means of tenderizing such as pounding, cubing, or grinding. The younger and less exercised an animal is, the less yellow connective tissue it will have and the more tender its meat will be.

Fat and Marbling

Of all the factors that go into the measurement of the quality of meat, fat is the most important. The amount, the distribution, and the condition of an animal's fat are essential ingredients in the mix of cost, toughness or tenderness, and palatability of meats. The total amount of fat may range from as little as 5 percent on a very lean animal to as much as 35 percent on a very fat animal. The fat on the outside of an animal's muscles is called cover, or fat cap. The fat that lies between the bundles of muscle fibers is called marbling. The amount of marbling in the meat and between the connective tissues is a very important consideration, in that it creates the texture desired in cooked meat.

Meat in the United States is usually well marbled, because the animals are fed specialized grain feeds that convert rapidly to fat for a period preceding slaughter. This meat is in marked contrast to South American or Spanish meat, which is usually raised only on grazing grasses and is therefore less fatty and much tougher to chew. Fat accounts for much of the flavor of meat, because a fat subcomponent, esters, imparts and enhances the typical "meat" flavors. The type of feed, again, is reflected in the type of fat and the eventual flavor of the meat; thus the regional preference for "corn-fed" beef, for example.

The more fat on an animal, the less lean, usable product can be sold. This factor means that the more fat an animal has in ratio to its lean, the higher the cost of that lean. Prime beef, for example, is very well marbled and has a thick fat cover and thus costs much more than Choice, Select, or Standard. The positive nutritive value of meat is in the lean tissue and not in the fat. The food value of a pound of lean is the same whether it is from costly Prime or from inexpensive Commercial grade beef. As mentioned earlier, the means of cooking has much to do with the toughness, and therefore the smart operator will buy according to need. Prime meat should be purchased only for the "fine dinner house" business of expensive steak and dinner operations where the finest quality is called for and paid for by a clientele that can afford it.

Pigments, Extracts, and Water

The color of meat depends on the animal's age at slaughter. The older an animal is, the more red pigment (myoglobin) is found in the meat. The range of color varies with the type of animal, as well as the use of particular muscles. Heavily exercised muscles, such as the muscles of the legs, and the involuntary muscles tend to be darker.

The connective tissue of meat contains extracts that contribute to the flavor of the meat. These extracts are water soluble; thus, the meat that is cooked by the moist-heat method is less flavorful than meat cooked by dry-heat methods. The stock made of boiled meats is especially flavorful because it contains so much of these extracts. The amount of water in meat increases as the percentage of fat decreases. Because meat may contain more than 60 percent water, the longer meat is cooked and the higher the temperature at which it is cooked, the greater the water loss, and therefore the more "shrinkage" or loss of weight that occurs. Because fat is desirable, meat that is firm, as opposed to soft, is of preferred quality.

Aging Meat

Aged meat is meat that has had the fibers broken down through material processes. Meat contains many fermenting agents, called enzymes. These enzymes break down the cell structures of the lean and the connective tissue and so make the meat easier to chew. By storing meat at a controlled temperature and humidity, for a controlled period of time, the aging or ripening process will increase the tenderness and flavor of the meat. Generally, it is not economical to age meat more than two weeks, as water loss and discoloration may cause a loss of 15–20 percent of the meat.

Beef may be aged as much as 45 days if held at 34° F and 88 percent humidity, with at least 10 cubic feet per minute (cfm) air circulation. Lamb and mutton may be aged up to 10 days, but pork and veal are never aged.

All meat should be held at a temperature between 34° F and 40° F for at least four days after slaughter. This period of time will allow the enzymes to combat the rigor mortis, or postslaughter hardening, to a point where the meat has softened and is no longer considered "green."

REFRIGERATION

Fresh meat is best stored at a temperature of 32°–34° F and a humidity of no less than 80 percent in a well-ventilated refrigerator. The meat should be allowed to "breathe," which means that it should be covered with material no denser than a loose-net treated cheesecloth. The higher the temperature and the longer the fresh storage of meat, the greater the amount of weight that will be lost through evaporation, and the greater the amount of meat that will have to be trimmed off, and therefore lost. A very important part of meat purchasing is keeping loss due to storage time to a minimum. Because of this potential for loss, it is often the wisest purchasing decision to specify delivery of fabricated meat in the frozen state.

FREEZING

When meat is frozen, the water in the meat expands. If meat is frozen slowly, then the water in the meat will expand enough to form ice crystals, and the result will be broken cell walls in the meat. For this reason, it is important that meat be frozen very quickly in order to prevent formation of destructive ice crystals. Meat should be frozen at temperatures of at least –25° F in an atmosphere with air circulating at 40 cfm. The center of the meat should be at 0° F in a maximum of four hours. Do not accept frozen meat from a purveyor unless you are certain that this procedure has been followed.

When meat is frozen, a certain amount of evaporation takes place. This drying of the surface is commonly known as "freezer burn." Freezer burn may be prevented by wrapping the meat so that it is as airtight as possible before freezing. Many meat purveyors freeze their meats in cryovac, or vacuum (totally air-free), bags. Some of these bags can be used to cook the meat in, saving the operator steps in handling and cutting down on pot washing.

TEMPERING

Tempering is the process by which you defrost foods that are frozen. Frozen meat must be tempered under refrigeration only. An ideal medium for the growth of harmful bacterial will exist if meat is tempered at room temperature (or any temperature between 50° F and 140° F). It generally takes about 24 hours to completely defrost an 8 lb roast of 34° F to 40° F; 72 hours for a 24 lb roast. For health safety reasons, never put any kind of meat in a sink of standing water or thaw it at room temperature for any reason. Meat may be roasted or potted from the frozen state only if it is done at a carefully controlled low temperature. Many tests have shown that the difference in yield of roasts cooked at 250° F, as opposed to 350° F, may be as great as 10 percent.

MEAT BUYING KNOW-HOW

Meat buying is a complicated matter. It takes much experience in all phases of meat production, from slaughter to stew, to know all that is desirable for the wise purchasing and usage of meats. This chapter presents the basic facts about meat so that the buyer will know what is behind the terminology. It must be emphasized that the rules set forth later in this chapter for the writing and the use of specifications must be strictly adhered to, or else substantial money and/or quality loss will be suffered. The specifications that appear later in this chapter should be used with the pictures of the various meat cuts and with a ruler or meat-measuring device. Only after long experience can one be certain that a visual check, without using these aids, will be accurate; few people have that ability. There is a large variety of ways in which meat can be cut, and "let the buyer beware" applies more to meat than to any other product used in the foodservice industry.

INSPECTIONS

There are many labels, stamps, and codes used in the meat industry. The ones to be trusted most are those of the U.S. Department of Agriculture (USDA). These federal certifications are standardized throughout the country. Some state and local governments maintain their own inspection programs. These programs vary widely from state to state and area to area and, consequently, generally cannot be relied on to be of the quality of the USDA programs. The military inspection programs are of the same quality as the USDA programs and are interchangeable so far as military meat purchasing is concerned.

Religious inspections, such as kosher or kashruth, provide certification that the meat has met certain religious codes in regard to the method of slaughter and butchering. These religious certifications carry no guarantee of quality.

Federal Inspection

A label certifying that the meat has met certain criteria of wholesomeness must appear on all meat and meat products that are offered in interstate commerce. This label on the carcass of the meat, or on the label attached to further processed meat or meat product, has met the following standards: The carcass has been thoroughly examined for disease, and any unwholesome part has been removed and destroyed. The meat has been handled and prepared in a sanitary fashion in a meat plant that meets rigid federal standards of sanitation. No known harmful substances have been added to the meat. The federal stamp has been affixed. The number in the circle identifies the plant where the meat was prepared. Labels attached to further processed meats must meet certain federal criteria and must be approved before they can be used. The information on the label must include the circular inspection stamp, the name and address of the plant, the net weight, and a list of ingredients in the order of their dominance. If meat is packed in a container for shipping, the container must be labeled. The Domestic Meat Label must be affixed to the outside of the container.

Imported Meat

Imported meats that meet USDA standards are stamped at the port of entry.

Exemption from inspection: Although the cost of maintaining an on-premises USDA inspector is less than $.50 per animal, there are some companies engaged in interstate commerce that are allowed an exemption from the requirement. These are the hotels or restaurants that raise their own meat. However, their plants must undergo periodic inspection, and their exemption is revocable.

Other Inspection Programs

The state and local programs for inspection are varied in their standards. Some areas have no inspection. Before a buyer decides to purchase meat that has been inspected by state or local inspectors, it would be wise to make a personal comparison between those standards and their enforcement and the federal standards.

The kosher stamp is not a substitute for federal or state wholesomeness inspection. This is an additional label that indicates religious wholesomeness only.

The Military Inspection Program closely relates to the USDA Inspection Program. Meat inspected in this program is for military service only.

GRADING

The USDA maintains a voluntary meat-grading service. The federal grades are applied to meat as a means of giving the consumer a clear-cut guide as to the quality of the meat he or she buys. The grading standards are the same throughout the country. Because the grading standards are complicated, it is important that the meat buyer understand the meaning of the various grades and how they are determined. With this understanding, the buyer will be able to choose the grade of meat that is appropriate for the use intended.

Although the Federal Grading Service grades much of the meat in the United States, there are some meat packers who choose to apply their own grades instead of subscribing to federal grading. The buyer should beware of such grading. The USDA grades are carefully worked out and consistent throughout the land. The professional meat buyer buys the measured, standardized, reliable federal grades of meat and nothing else.

There are two types of federal grading. The first type, with which most people are familiar, is quality grading. Quality grading classifies meat according to a number of factors that contribute to how well the meat "eats" and "tastes." The quality grade is determined by an examination of the kind (class) of animal, the sex, the shape (conformation), the amount of exterior fat (finish), the amount of interior fat (marbling), and the firmness of the lean and the fat. The specifics vary, depending on the class of animal. These specifics of grading quality are explained in the following section, in which specifications are given for each separate class.

The second type of grading is cutability grading or yield grading. The yield grade tells the buyer how much lean he or she can expect to obtain from the cut carcass.

Effective January 24, 1989, new standards took effect for the grading of meat. The changes affected the amount of grain that must be fed to cattle of a particular age. Essentially, this means that cattlemen will be able to save money by feeding their livestock less grain without taking the risk of having the meat from these grass-fed cattle fall into lower grades. Before the grading change, approximately 4 percent of cattle were graded Prime; 50 percent, Choice; 35 percent, Good; and the rest below Good. Under the new grading standards, the percentage are said to be 4 percent, Prime; 60 percent, Choice; 25 percent, Select; and the rest Standard and lower grades, which include Commercial, Utility, Cutter, and Canner.

CLASS

The classes (kinds) of meat and the grades for which they are eligible are listed in the following sections. Stamps appear in red on all the major cuts of meat. Because the stamping device is rolled down the uncut carcass, there are some interior cuts of meat that cannot bear the stamp.

Beef Classes

Steer—a male animal that has had the male hormones removed at a very young age.
Heifer—a female animal that has never borne a calf.
Cow—a female animal that has borne one or more calves.
Bull—a male animal that has matured with male hormones intact.
Bullock—a young bull.
Stag—a male animal that is heavier and fatter than a bull. Male hormones are removed after a certain stage of maturity is reached.

When buying beef, the buyer will not know whether he or she is buying steer, heifer, or cow except by examination of the pelvic area. A steer has rough fat in the cod area, a very small pelvic cavity, and a small "pizzle eye." A heifer has smooth cod fat and a larger pelvic cavity than the steer. A cow has a very large pelvic cavity and, usually being an old animal, its bones are hard and white as opposed to the softer, pink bones of a young animal. Bulls and stags are graded on a different set of standards than the steer, heifer, and cow. A bull will have very heavy muscle development at the rounds, neck, and shoulder; a large "pizzle eye"; and coarse, dark meat. The stag will be less heavy and coarse than the bull, but not as light or as finely fleshed as a steer.

Beef Grades

STEER, HEIFER, BULLOCK

USDA GRADES—Prime, Choice, Select, Standard, Commercial, Utility, Cutter, Canner

COW

USDA GRADES—Choice, Select, Standard, Commercial, Utility, Cutter, Canner

BULL, STAG

USDA GRADES—Choice, Select, Commercial, Utility, Cutter, Canner

VEAL AND CALF

a bovine of an age up to three months is classified as veal. Veal has a very light pink, actually almost gray colored, lean meat. The lean and fat are very soft, smooth, and flexible. The bones are very red, and the rib bones are very narrow. A calf is a bovine that is older than three months. The calf has a pink to reddish-pink lean, which is soft and pliable to a degree not as great as veal, but is not nearly as firm as beef. The bones are pink and starting to harden. The rib bones are wider on a calf than on veal.

USDA GRADES—Prime, Choice, Select, Standard, Utility, Cull

Ovine

Lamb is classified according to age. The very young, under weaning age, lamb is often called hothouse lamb or genuine spring lamb. A lamb between two and five months of age may be called spring lamb or, sometimes, milk lamb or milk-fed lamb. Lamb has soft, porous bones, very light pink lean, and creamy white to pink fat, which is soft and pliable. Perhaps the foremost indicator is the break joint of the foreleg. Such joints break into four well-defined ridges, which are smooth, soft, and blood red.

Yearling mutton has flesh that is light red to red; fat that is white and fairly firm; and graining in the flesh, which is fine, but it does not have as little grain as lamb. The break joint in yearling mutton has four ridges, as it does with lamb, but the edges are not smooth and the bone is hard and white.

Mutton has flesh that is medium red to dark red with a grain that is similar to that of calf or young steer. The fat is quite white and has a brittle consistency. The foreleg of mutton will not break as it does with lamb and yearling mutton. Instead, a hard, shiny, smooth knuckle with two prominent ridges forms. Whereas sex has no bearing on the classification of lamb and yearling mutton, mutton is classified according to sex.

MUTTON

USDA GRADES—Choice, Good, Utility, Cull

Ewe—a female animal.

Wether—a male animal that has had its male hormones removed at a young age.

Buck—a mature male that has not been sexually altered.

LAMB, YEARLING, MUTTON

USDA GRADES—-Prime, Choice, Good, Utility, Cull

Pork

The cuts of pork meat that have the greatest monetary value in the marketplace are from the back or hindquarter of the animal. The female animal usually has a heavier and therefore more valuable hindquarter than the male animal. When a male animal has its male hormones removed at an early age, its muscular development is then much like that of the female, with the hindquarters being heavy.

PORK (HOGS)

Barrow—a male animal that has had the male hormones removed when very young

Gilt—a female animal that has never borne pigs

Sow—a female animal that has borne pigs

Boar—a male animal that has matured with male hormones intact

Stag—a male animal that has had its male hormones removed after reaching maturity

BARROWS, GILTS

USDA GRADES—-U.S. No. 1, U.S. No. 2, U.S. No. 3, U.S. No. 4, U.S. Utility.

SOWS

USDA GRADES—U.S. No. 1, U.S. No. 2, U.S. No. 3, Medium, Cull

Hogs are essentially graded in three categories. U.S. No. 1, U.S. No. 2, and U.S. No. 3 are the Choice grades. The differences between these grades have to do with the ratio of lean to fat and the yield of the loin, the ham, the picnic, and the Boston butt. The U.S. No. 4 (barrows, gilts) and U.S. Medium (sow) grades are the next level down in the grading and the bottom of the consumer grades. U.S. Utility (barrow, gilt) and U.S. Cull are not marketed for retail consumption. Sows are graded on a different set of standards than barrows and gilts. Some 70 percent of the hog production is further processed before it reaches the retail marketplace. Hogs are raised for a variety of purposes. In the old days, they were referred to as Roasting (10 to 30 pound pigs), Shipping (large sows for barrel packing), Meat-Type (fairly lean for bacon and hams), and Fat-Type (hard and skinned hams). Today, hogs are marketed within a very narrow range of weights and carcass lengths. General well-defined grading of pork cuts on the basis of quality is not yet done, and pork is marketed mostly on the basis of weight and lean-to-fat ratio. When better control is possible over muscular development, amount of finish, evenness of marbling, and uniformity and texture of the lean, then accurate quality grading may be possible. For buying carcass or packer-style (split carcass) hogs, the following are guidelines to acceptable quality.

Cuts—muscles should not have more than a moderate amount of intramuscular fat.

Bones—should be porous with cartilage present. Brittle or flinty bones are not acceptable.

Lean—should range from light pink to light red, with a smooth, fine texture similar to that of calf.

Fat—should be firm, creamy white, and fairly evenly distributed.

Skin—thin, smooth, pliable.

MEAT CONFORMATION

Conformation has nothing to do with the taste quality of an animal. The term *conformation* refers to the animal's shape. The more meat in the higher-priced regions of the body, the better the conformation. An animal with excellent conformation will have little meat in such areas as the neck and the shanks, but will have thick, plump loins, ribs, rounds, and chucks.

FINISH

Finish is the fat that covers the animal carcass and that is distributed inside the animal and between the muscles. As mentioned earlier, finish is an important determinant in grading all meats. White fat is desirable and is the result of careful breeding and grain feeding. Heavy, rough fat is the result of too fast, too much, or too long feedlot feeding. Good fat is creamy white in appearance and occurs in increasing amounts according to age. Thus, bovine finish will be least on veal, more on calves, and heaviest on beef animals. Lambs are less well finished than yearling muttons; yearling muttons have less finish than muttons. Bovine and ovine fats have a firm-to-brittle texture, whereas hogs have a soft fat. Very soft hog fat is an indication of slop feeding and is quite undesirable, as it is difficult to butcher a soft hog.

MARBLING

The amount of finish that appears in the flesh itself is called marbling. The more numerous the flecks of fat in the meat, the more flavorful, juicy, and tender the meat will be. That is why marbling is such an important factor in palatability and a prime determinant of grade. It should be pointed out again that the positive food value is in the lean, and that an animal with a lesser finish, or a yellow finish, or a lower grade will offer just as much nutritive value per pound as the "higher-quality" meats.

QUALITY

Quality in meat is that combination of factors that contribute to the typical excellent taste of a particular animal. This is called *palatability*. Besides finish and marbling, the factors for quality grading include the color of the flesh and the age of the animal. In grading veal, calf, ovine, and hogs, the grader determines the grade on the whole carcass and without benefit of a cut surface of the lean. Finish is perhaps the most important of the grading factors. The grader depends not on the amount of external finish, but rather on the fat streaking in the ribs (feathering) and the fat streaking in the inside flank muscles. In beef cattle, the carcass is cut at the 12th rib, and an evaluation can be made of the degree of ossification (hardening) of the bones and the firmness and marbling of the rib eye muscle.

The age or maturity of the animal has much to do with the quality. Age refers to the physiological age of the animal as indicated by the color and texture of the lean, condition of the bone, and hardening or ossification of the cartilage, rather than the actual age of the animal. The physiological and actual age of an animal will be similar only in well-bred, properly raised animals.

There is a substantial overlap in the different quality factors, and therefore the lack of some factors may be compensated for by the presence of others.

CUTABILITY OR YIELD GRADING

Above and beyond the grading for the quality of meat, there is a grading available for beef, lamb, and mutton that identifies the yield of trimmed retail cuts that will be achieved in cutting. These yield grades are Nos. 1, 2, 3, 4, and 5 and refer, in declining order, to the amount of yield. The method used for determining the yield grade is discussed in the beef section. Because one objective of a meat-purchasing program is to obtain the most value for the purchasing dollar, it is a wise meat buyer who understands and uses yield grades in meat purchasing.

CUT IT, OR BUY IT CUT?

The decision regarding whether to buy carcasses, sides, or quarters and butcher them in-house, or to buy fabricated cuts of meat, is not a difficult one to make. The procedures involved in making the decision are simple, but the time consumed may be substantial. The information needed to make such a decision is as follows:

1. Do you have a butcher on staff? Not every operation has a butcher; a person who cuts meat is not necessarily a butcher, as butchers must go through lengthy training and apprenticeship. If a butcher does not know exactly what he or she is doing, then a large amount of the meat dollar may be wasted as a result of cutting meat in a way that provides too few of the expensive cuts and too many of the cheap cuts. If an operation does not have a good butcher, then it should buy only fabricated and portion-cut meats.
2. Is your operation equipped to butcher meat? Not every operation is. A butcher shop needs hard and mechanical saws, specialized cleavers and knives, special types of tables, proper refrigeration, and receiving facilities.
3. Can you afford a butcher's salary? Consider first that an average butcher's salary is $1,000 per week. If an operation saves an average of $.40 per pound of meat through its own butchering, then the operation must use more than 2,500 lb of meat per week to break even with buying fabricated or portion-controlled meat cuts. Because most butchers spend 20 percent of their time (8 hours per week) in setup and cleanup, this means they must be able to butcher almost 70 lb of meat per hour. If the operation uses a lot of roasts, this will be possible, but if the operation uses mostly steaks, chops, ground meat, and diced meat, this level of productivity may well not be possible. In addition, an institution that uses 2,500 lb of butchered meat per week must feed an average of more than 700 people per meal.
4. What is the cost per pound of meat butchered in-house? The use of a Yield Test Card can help in making this determination. Consider, for example, a hindquarter that has been broken down using the New York cutting method, which yields a good institutional mixture of cuts. The $ ____ per pound for the hindquarter is the actual price paid. The "market value per pound." figures are obtained either from the National Provisioner "*Yellow Sheet*" or from market quotations on the date of fabrication.

   There is a figure called a "fabrication multiplier," which is the ratio of the in-house fabricated cost to the market value. When this figure is known, it may be used to calculate future costs of in-house fabrication as the price of hindquarters changes.
5. What are the other implications of butchering at the operation? The biggest drawback is not being able to use all of the cuts that are obtained by butchering. For in-house butchering to be beneficial, the menu must call for the use of all the cuts obtained and in the amounts obtained. Close coordination of the menu and the butchering plan is needed to make such butchering pay off. Another drawback is that all the cuts obtained by butchering will be of the same grade. It is generally accepted that an operation will need U.S. Choice for broiling and dry roasting, but that U.S. Select is sufficient for pot roasting and stewing.
6. If you buy fabricated meat, how can you cut it the way you want it? The USDA maintains an inspection service that verifies that the packer has prepared the meat to your specifications. This USDA Acceptance Service is available, for a fee, to any operator using substantial quantities of meat, as an aid in the meat-purchasing program.

HOW MUCH TO ORDER?

To know how much meat to order, an operator needs to know how much he or she uses. Knowing how much to use means knowing both how much is used per serving and how many servings are needed. The question of how much meat to use per serving is a matter of operational philosophy and type of institution. Our society places a large psychological value on meat. Nutritional adequacy calls for only 56 g of protein per adult per day. If given their preference, people want as much meat as possible. This fact can be traced to the close correlation between our image of success and the meat we eat. By far, the most popular meat in the United States is beef. In order to make a menu popular, the mix of meat dishes should take this preference into consideration. The means of keeping down the cost of meat on the menu are varied.

1. Balance the menu entrée offerings with popular, less expensive, extender items when offering the more expensive, solid entrées. A clientele that likes a relatively expensive, solid entrée, like sauerbraten, may well choose a less expensive option dish as stew or goulash.
2. Never buy a better grade of meat than needed for the use intended.
3. Buy meat on tight specifications. It is a costly mistake to pay for fat that is not needed or to fail to pay for fat when it is needed.
4. Cook meat at the temperatures at which it should be cooked. Cooking losses increase in proportion to oven temperature. A study has shown cooking losses to be as much as 10 percent greater for meat cooked fast at high temperatures than for meat cooked slowly.
5. Knowing how much meat to order means keeping accurate records.

The first record needed is a file of Cooking Loss Cards. This record has to be computed only once for each menu item, as the loss should always be about the same for the same grade and cut of meat. With the usable, after-cooking, weight as a base, one may calculate the number of portions per cooked pound and then the number of raw pounds needed for *X* portions by determining a multiplier.

The second necessary record is a card that records the number of portions used in the past. Note that the percentage of items used, in the same menu combination, will stay within a narrow percentage range. Although the customer count in an individual operation may vary greatly from day to day, the percentage of the customers who favor a particular entrée will remain very close to the same. Seasons of the year may have some bearing on the ratio but, again, the ratio will remain true within a particular season. Keeping such a record for a few months will be sufficient to start using it as an ordering guide; keeping the record over the years will help identify unpopular menu combinations, changing taste trends, and need for menu change.

The third record needed is some type of inventory/order/receiving record. Consideration should be given to the needs of the institution—that is, volume, refrigeration and freezing capacity, management time, and purchasing plan (butchering vs. fabricated and portion-controlled product). Determining how much to order can be done in several ways.

1. Operations that have a standing menu (restaurants, hotels) or a short-cycle menu for seven or ten days will probably find a par-stock or a minimum-maximum stock inventory system best.
   (a) Par-Stock. In this system, the operator determines a stock level that will cover the institution's needs for the period of the ordering cycle. For purposes of example, we will call it one week. If the maximum weekly need for Item No. 137 Ground Beef is 100 lb, then the par-stock is 100 lb. Each week the ground beef order will be the difference between the on-hand inventory and 100 lb. Thus, if the current inventory is 30 lb, the order will be for 70 lb, and so on.
   (b) Minimum-Maximum Stock. In this method, the operator determines the maximum amount of an item that may be needed in the ordering cycle and the minimum quantity that must be kept in inventory to prevent running out. This system is very good for operations that tend to have a week-to-week fluctuation that does not follow a set pattern. Whenever the inventory falls to the minimum level, an order is placed to bring the stock back up to a maximum. Again, using the 100 lb of ground beef as a maximum inventory, then a minimum inventory of, say, 20 lb, may be determined. Whenever the inventory falls to 20 lb, an automatic reorder of 80 lb is placed.
2. Operations that have a cyclical menu of some length (three to six weeks), or operations that have high volume and/or limited refrigerated storage space, often find it most convenient to order on an as-needed basis. The drawbacks of this system are twofold. The first is that, unlike the par-stock and minimum-maximum stock systems, it requires management to review the stock and menu needs at reg-

ular intervals instead of placing standing orders or leaving the ordering to inexperienced employees. The second drawback is that there is no reserve stock to use in an emergency. The advantages of this system are that no inventory is being carried on the books, and there is no excess stock in the house and therefore no chance of spoilage. Simply stated, the procedure is to estimate, from the production history cards, the number of portions (converted to raw pounds) that are needed, and then order just that quantity. In an inventory/order/receiving record the amount ordered is entered above the diagonal line, and the amount actually received is entered below that line. The ending inventory column provides for a weekly inventory so that any excess stock may be subtracted from the amount needed to order.

HOW MUCH TO PAY?

High quality and high price do not necessarily go hand in hand any more than do low quality and low price. A full discussion of purchasing techniques is presented in Chapter 1, "Purchasing Policies." When buying meat, the buyer must be aware of certain things:

1. It costs the purveyor money to make deliveries. The larger the order and the less frequent the deliveries, the more likely it is that the price will be lower.
2. Excessive "fill-in" orders, unclear specifications, and unreasonable demands put on a purveyor will result in higher prices.
3. If you deal with a purveyor that has an inefficient plant operation, the cost of its inefficiency will be passed on to you either in price, poor trim, or improperly cared-for product.
4. Each grade of meat has a range of tolerance; thus, there is such a thing as "high" Choice and "low" Choice meat. Beware of the purveyor that always claims its meat is high Choice; appreciate the purveyor that tells you when the meat is low, middle range, or high in the grade.
5. Beware of the "special" or the "lowball" price. Chances are that the meat is too fresh or has been either aged too long or aged improperly.
6. Receiving controls must be very tight. Each piece of meat must be weighed and measured against the specifications. A simple ruler is an excellent receiving tool.
7. Storage temperatures must be exact, and storage time as brief as possible. Meat shrinks as a result of evaporation during storage.
8. The buyer and the receiver must have complete loyalty to the operation. A system of checks and balances must be established to ensure a high degree of honesty. Dishonesty, as discussed in Chapter 1, takes many forms and is often difficult to uncover.
9. Meat buying must be conducted either on a bid basis or by contract. Base contract prices should be arrived at by using one of the various USDA market reports or the "*Yellow Sheet*" that is published by the National Provisioner.

HOW MUCH TO CHARGE?

What an operator must charge for its meat entrées is contingent both on what the cost per portion is and on the financial goals of the operation. The method for arriving at the charge is the same for all types of operation, but the goals may be very different. A public establishment such as a hotel or restaurant is in business to make a profit. Here, the price charged will have to contribute to a profit mixture. If the total profit from operations must be, for example, 12 percent, then every item must contribute to the profit to a degree of 12 percent. Some price balancing may be done, but the average gain must be 12 percent over all costs involved. If the food cost of the operation is 40 percent of expenditures, then each item will have to sell at 2.5 times its cost plus 12 percent.

Schools and colleges generally must run foodservice operations on a no-profit–no-loss basis. In this case, the price of food must be the amount that contributes to breaking even. Using the preceding example (40 percent food cost), the food in these institutions would be priced 2.5 times the raw cost.

Some institutions, such as industrial feeding operations and hospital employee cafeterias, operate on company subsidies in order to keep food charges low. The idea is to provide eating facilities for the employees in close proximity to their work. Prices in this situation are based either on the ceiling that top management wants on food prices or on the amount of subsidy.

In the first case, the cost of raw food bears no relationship to the price charged. In the second case, it is necessary to estimate the total operating costs, subtract the subsidy, and charge prices for food that contribute proportionately to that expense figure.

WHAT IS THE COST?

The steps to computing the cost per portion are as follows:

1. Compute the raw ready-to-cash (RTC) weight (meat either purchased processed or see Yield Test Card).
2. Cook, trim, and carve the meat. Subtract the combined weight of the carved portions from the cooked weight.
3. Divide the total raw cost by the total weight of the carved portions.

Example:

100 lb inside round, RTC, at $1.98 = $198.00
76 lb carved weight
$198.00 total cost/76 lb carved weight =
$2.61/lb cost per salable pound
$2.61 oz portion weight
$2.61 / 16 oz =$.163 per oz
6 oz portion = $.978, or $.98, per portion

This calculation should be made for every menu item served. The carved weight percentage (76 percent in the preceding example) can be used to quickly recalculate the cost per portion when the raw product cost changes or portion size is changed. A comparison of the costs per portion on various cuts of meat can quickly reveal the most economical cut of meat to use for any given menu item.

WRITING MEAT SPECIFICATIONS

The are two purposes of written specifications for meat and meat products. The first purpose is to give the operator and its employees knowledge of which product should be used for what purpose. The second purpose is to tell the purveyor exactly what the expected product is. It is not possible to give a quotation on a meat item honestly if the purveyor and the operator do not agree on the exact definition of the meat item. Once written, a copy of the operation's specifications must be sent, for reference, to every purveyor from which the operation buys. The operator must insist that the purveyor's deliveries meet the exact specifications the operator has set. Any variance from these standards is dishonest, costly to the operation, and therefore forbidden.

What Must be Included

A properly written meat specification should include the operation's exact requirements in regard to the following:

1. Class of animal
2. USDA grade and division of grade
3. USDA yield grade
4. Acceptable weight range
5. State of refrigeration
6. Fat limitations

The Institutional Meat Purchase Specifications

The Institutional Meat Purchase Specifications (IMPS) are the official USDA requirements for the inspection, packaging, packing, and delivering of specific meats and meat products, and for the certification of those products by USDA meat graders. The IMPS are the most widely known and accepted meat specifications in our country. Formally, the IMPS are applied with the use of the Meat Acceptance Service of the USDA. In operations too small to use the Meat Acceptance Service, the IMPS can still be used as long as the person receiving the meat knows how to relate the written specifications to the physical object.

The Meat Buyers Guide

The Meat Buyer's Guide's contains the meat specifications agreed upon by the North American Meat Processors Association (NAMP), the largest and most powerful of all wholesale meat purveyor associations. The publication was written as a means of simplifying the IMPS and to give pictorial definition to the most common meat cuts. There is no better set of meat specifications anywhere in the country than the IMPS and Meat Buyer's Guide's (MBG). Because The Meat Buyer's Guide is illustrated with photographs that show the proper dimensions of the cut products, they are the specifications that should be used by all operators. Every meat purveyor knows the MBG specifications and should therefore have no difficulty in delivering products that meet the MBG specifications.

The specifications set forth in the rest of this chapter are both IMPS and MBG specifications. The MBG specifications are the specifications of first choice. Because the MBGs do not include all the IMPS, the IMPS are woven into the text. IMPS and MBGs have replaced the prevalent regional names for the various meat cuts. To help the reader identify cuts, the most common regional names are included, as well as the names used in the Uniform Retail Meat Identity Standards, published by the National Livestock and Meat Board. Most common uses for each cut of meat are listed as a preparation aid to the reader.

INSTITUTIONAL MEAT PURCHASE SPECIFICATIONS GENERAL REQUIREMENTS

Beef Specifications

USDA GRADE—To be specified by purchaser

The purchaser must specify either

1. A quality grade, or
2. A combination of quality grade and yield grade. Yield grades 1 through 5 are applicable to all quality grades.

DIVISION OF QUALITY GRADE—To be specified by purchaser (not applicable to yield grade). If the upper half or lower half of a quality grade is desired, it must be so specified, otherwise the full range of the grade is acceptable.

WEIGHT RANGE—To be specified by purchaser, Range A, B, C, D, or E, or actual weight range in 20/24, and so forth.

STATE OF REFRIGERATION—To be specified by purchaser

1. Chilled
2. Frozen

FAT LIMITATIONS—Carcasses, Sides, or Quarters (not applicable if yield grade is specified). Except when yield grade is specified by thickness of external fat measured at the thinnest point over the rib or loin eye, fat thickness must not exceed that indicated in Table 8.1.

In addition, carcasses, sides, or quarters are not acceptable if, because of uneven distribution of external fat or large deposits of kidney and pelvic fat, they do not meet the maximum permitted thickness of fat over the rib eye.

FAT LIMITATIONS—Wholesale and Fabricated Cuts: To be specified by purchaser (not applicable if yield grade is specified). Except when yield grade is specified, or fat limitations are indicated in detailed specifications, for all wholesale and fabricated beef products the purchaser must specify a maximum average thickness of surface fat: 1¼ in., ¾ in., ½ in., ¼ in.

Table 8.1 Maximum Thickness of Fat at Thinnest Point Over Ribeye

Grade	Weight range A&B	Weight range C, D, & E
U.S. Prime	⅞ in.	1¼ in.
U.S. Choice	⅝ in.	1 in.
U.S. Select	⅜ in.	¾ in.
U.S. Standard	¼ in.	⅜ in.
U.S. Commercial	⅝ in.	⅞ in.
U.S. Utility	¼ in.	½ in.
U.S. Cutter or Canner	⅛ in.	¼ in.

When string tying is required, roasts must be made firm and compact and held intact by individual loops of strong twine uniformly spaced at approximately 2 in. intervals girthwise. In addition, some roasts may require string tying lengthwise. In lieu of string tying, it is permissible to enclose roasts in a stretch netting, provided it complies with the Regulations Governing the Meat Inspection of the USDA. Purchasers may specify that roasts be string tied when this requirement is not specified in the detailed roast item specifications.

MATERIAL—Beef products described must be derived from sound, well-dressed split and quartered beef carcasses or from sound, well-trimmed primal cuts from such carcasses. The beef must be prepared and handled in accordance with good commercial practice and must meet the type, grade, style of cut, weight range, and state of refrigeration specified.

Beef cuts that have been excessively trimmed in order to meet specified weights, or that are substandard according to the specifications for any reason, are excluded. The beef must be of good color normal to the grade and must be practically free of residue remaining from sawing the meat and bones. It should be free of blood clots, scores, odor foreign to strictly fresh beef, mutilations (other than slight), ragged edges, superficial appendages, blemishes, discoloration (e.g., green, black, blue, etc.), deterioration, damage, or mishandling. The spinal cord must be completely removed, and the beef must also be free of bruises and evidence of freezing or defrosting and must be in excellent condition to the time of delivery. Stag and bull beef are not acceptable.

Portion-cut items described herein must be prepared from fresh-chilled carcasses or bone-in cuts that are in excellent condition, and they must be of the applicable kind and of the U.S. grade or selection specified. The meat must show no evidence of off-condition, including, but not restricted to, off-odor, being slightly sticky, gassy, rancid, sour, or discolored, and evidence of defrosting or mishandling. In addition, the portion-cut items must be free of blood clots, scores (other than slight), ragged edges, bruises, and spinal cord, and must be practically free of bone and meat dust. Portion-cut items supplied must be in compliance with the applicable requirements specified herein and with other requirements specified by the purchaser.

PORTION WEIGHT OR THICKNESS OF STEAKS, CHOPS, CUTLETS—Purchasers must specify either the portion weight or thickness desired, but not both. However, in order to control uniformity of portion sizes, the weight range of the trimmed meat from which the portions are to be produced may also be specified.

PORTION WEIGHT—If portion weight is specified, the actual portion weights desired (3 oz, 6 oz, 12 oz, etc.) must be indicated.

WEIGHT TOLERANCES—Unless otherwise specified by the purchaser, depending on the portion weight specified, the following tolerances over and under the weight specified will be permitted:

WEIGHT SPECIFIED	TOLERANCES (OVER AND UNDER)
Less than 6 oz	1/4 oz
6 oz but less than 12 oz	1/2 oz
12 oz but less than 18 oz	3/4 oz
18 oz or more	1 oz

Example: When 8 oz steaks are specified, individual steaks weighing 7 1/2 to 8 1/2 oz are applicable.

THICKNESS—If thickness is specified, the actual thickness desired must be indicated (cubed steaks, ground beef patties, and cubed cutlets excepted).

THICKNESS TOLERANCE—Unless otherwise specified by the purchaser, depending on the thickness specified, the following tolerances over and under the thickness specified will be permitted:

THICKNESS SPECIFIED	TOLERANCES (OVER AND UNDER)
1 in. or less	3/16 in.
More than 1 in.	1/4 in.

Example: When 1 1/4 in. steaks are specified, individual steaks measuring 1 to 1 1/2 in. are acceptable.

WEIGHT RANGE FOR ROASTS—Purchasers must specify the actual weight range (4–6, 8–10, 17–19 lb, etc.) desired. If purchasers want roasts further reduced in size, this must also be specified.

SURFACE FAT STEAKS—Unless otherwise specified by the purchaser, or unless definite fat limitations are indicated in the detailed item specifications, on surfaces where fat is present, the fat must not exceed an average of 1/2 in. in thickness, and the thickness at any point must be not more than 3/4 in.

CHOPS, CUTLETS, AND FILLETS—Unless otherwise specified by the purchaser, surface fat, where present, must not exceed an average of 1/4 in. in thickness, and the thickness at any one point must be not more than 3/8 in.

ROASTS—The purchaser must specify the maximum average surface fat thickness desired as expressed in the following list, unless definite fat limitations are indicated in the detailed item specifications.

MAXIMUM AVERAGE SURFACE FAT THICKNESS (SEAM FAT EXCEPTED)

3/4 in. (1 in. maximum at any one point)

1/2 in. (3/4 in. maximum at any one point)

Defatting must be done by smoothly removing the fat by following the contour of the underlying muscle surface. Beveling of the edges only is not acceptable. In determining the average thickness of surface fat or the thickness of fat at any one point on steaks and roasts that have an evident, natural depression into the lean, only the fat above the portion of the depression that is more than 3/4 inch in width will be considered.

INSPECTION

All meats, prepared meats, meat food products, and meat by-products (as defined in Rules and Regulations of the Department of Agriculture Governing the Grading and Certification of Meats, Prepared Meats, and Meat Products) covered by these specifications must originate from animals that were slaughtered or from product items that were manufactured or processed in establishments regularly operated under the supervision of the Meat and Poultry Inspection Program (MPIP) of the Consumer and Marketing Service (C&MS) of the USDA or under any other system of meat inspection approved by the Consumer and Marketing Service of the USDA.

ORDERING DATA

The purchaser will requisition product items by specifying the item number, name, and the desired options, such as grade or selection, weight range, formula, state of refrigeration, indicated in each specification, and products must be offered for delivery on such basis by the contractor, subject to official examination, acceptance, and certification by USDA meat graders or other designated personnel. The examination, acceptance, and certification of products by the USDA shall be in accordance with USDA Meat Grading instructions. (Copies of Meat Grading Instructions may be obtained from the local Meat Grading Branch Main Station Office.)

CERTIFICATION

In connection with the issuance of meat grade certificates, one or more kinds of official USDA meat grade certificates will be involved, depending on whether the product is for delivery chilled or frozen.

1. Products for Delivery Chilled. When products are to be delivered chilled, an official final certificate will be issued by the responsible USDA meat grader to cover all factors and details of the products.
2. Products for Delivery Frozen. When products are to be delivered frozen, the responsible USDA meat grader will issue an official preliminary certificate, identified as such, to cover all factors and details of the chilled product prior to freezing. The responsible USDA meat grader will issue an official final certificate covering all factors and details of the frozen product prior to loading for delivery.

Disposition of Certificates

The original and up to two extra copies of all preliminary and final certificates are available to the contractor. The purchaser may request the contractor to supply copies of all final certificates.

The contractor shall pay the cost of the examination, acceptance, and certification.

Time Limitation

Products prepared for delivery under a purchase order shall not be offered to USDA meat graders for examination and acceptance more than 72 hours before shipment.

STATE OF REFRIGERATION

The detailed specifications for the various products indicate two different states of refrigeration. These are defined as follows:

1. Chilled. Chilled products are those which after preparation and in accordance with good commercial practice are thoroughly chilled (but not frozen or defrosted) to an internal temperature of not higher than 50° F. They must be held at suitable temperatures (32°–38° F) and must be in excellent condition at the time of delivery.
2. Frozen. Products to be delivered frozen must be promptly and thoroughly frozen in suitable and reasonably uniform temperatures not higher than 0° F. Products thus frozen must be maintained and delivered in a solidly frozen state. The products must show no evidence of defrosting, refreezing, freezer burn, contamination, or mishandling.

When the state of refrigeration is not specified in the purchase order, the product must be maintained and delivered chilled.

PACKAGING AND PACKING

Packaging

1. All carcass meat and wholesale cuts that are normally wrapped in commercial practice must be completely and properly packaged in suitable material (cryovac bags, grease- and moisture-resistant paper, suitable plastic or metal foil covering, stockinettes, etc.) to ensure sanitary delivery.
2. Fabricated and boneless cuts (including units of diced and ground meat); cured, smoked, and dried meat; and edible by-products that are normally wrapped in commercial practice must be separately and closely packaged with suitable grease- and moisture-resistant paper or suitable plastic or metal foil covering, etc.
3. Portion-control products must be suitably packaged in accordance with good commercial practice and, unless otherwise specified in the purchase order, such packages must contain not more than 25 lb net weight.

4. Unless otherwise specified in the purchase order, products such as frankfurters, sliced bacon, sliced dried beef, linked or bulk pork or breakfast sausage, and the like, must be suitably packaged and placed in containers of the kind conventionally used for such products as illustrated in the following:
   (a) Frankfurters and linked sausage—1 lb. retail-type individual packages packed not more than 10 lb per unit in the outer container, or layer packed 1 link deep with parchment or waxed paper separators between layers in a 5–10 lb container.
   (b) Sliced bacon—1 lb retail-type individual packages such as folded or sleeve-type cartons, cello covering, or flat hotel-style packets snugly packed in a substantial outer container not to exceed 50 lb net packed weight.
   (c) Sliced dried beef—Either 1/4 lb, 1/2 lb, or 1 lb retail-type individual packages, snugly packed not more than 10 lb net weight per outer container, or bulk or layer packed in a substantial inside-waxed or plastic or waxed paper–lined container not to exceed 10 lb net packed weight.
   (d) Bulk pork or breakfast sausage—1 lb retail-type individual packages such as cello rolls, plastic bags, waxed paper cups, or folded or sleeve-type cartons, packed not more than 10 lb net weight per container, or in waxed or plastic-coated paper tubs of either 5 or 10 lb net weight.
5. It is the contractor's responsibility to ensure that the products to be frozen are suitably wrapped and packaged in a material that is grease and moisture resistant and that will also prevent freezer deterioration.

Packing

Unless otherwise specified in the purchase order, products customarily packed in shipping containers must not be packed in excess of 100 lb, net weight, except that portion-control products are limited to units weighing not more than 25 lb. Containers must be of a size and shape normally used and adapted to the product being packed. The shipping containers must be made from material that will impart no odor, flavor, or color to the product and must be packed to full capacity without slack filling or overfilling. Containers used for packaging, which meet the packing requirements, may be used as shipping containers. Otherwise, immediate containers must be placed in a master container meeting the packing requirements, except that part or whole shipments of not more than five packaged units may be shipped in their immediate containers. Chilled products must be packed in wire-bound wooden boxes or in fiberboard boxes. Products to be frozen must be packed in fiberboard boxes. Fiberboard used in making these boxes shall be as described in Federal Specification PPP-F-320 and must comply with the requirements given in the following list. Products that are normally bulk packed (spareribs, oxtails, tongues, etc.) and those that are normally moist or subject to dripping moisture must be packed in either wax-resin-impregnated fiberboard boxes or in boxes that are protected by one or both of the following methods:

1. Appropriate moisture-proof plastic or wax coated on the inside (not applicable to wire-bound wood boxes). The quantity of wax applied to the interior must be sufficient to be visible when lightly scraped with the fingernail.
2. Completely lined on the inside (sides, ends, top, and bottom) with suitably waxed kraft or parchment paper or with appropriate moisture-proof plastic liners.

Cured products in pickle brine may be either put in plastic bags and then packed into fiberboard boxes of the type specified for use with products subject to dripping moisture, or they may be packed directly into fiber, wooden, or metal drums. Drum interiors shall be suitably protected by a wax or plastic coating or lined with a plastic bag liner. When used, plastic bags and plastic bag liners must be securely closed. Drums shall have full opening tops with lock rim closures which permit sealing by USDA, and opening and reclosure by the purchaser. Unless otherwise specified in the purchase order, in lieu of the aforementioned shipping containers, new or reconditioned wooden slack barrels or fiber drums may be used for packing full ribs, oven-prepared ribs, roast-ready ribs, strip loins, lamb backs, and similar cuts to be delivered in the fresh state. The wooden slack barrels must be protected from absorption and leakage by an inside lining of plastic or crinkled kraft paper. The fiber drums must be protected from absorption and leakage by a wax or plastic inside coating or by an inside lining of plastic or crinkled kraft paper. After packing, the barrels or drums must be properly headed and covered.

Closure

Fiberboard boxes shall be securely closed with the use of one or more of the following methods:

1. Strapping—Boxes may be strapped with one of the following:
   (a) Flat steel straps, at least ¼ in. in width that are protected with an enameled or rust-resistant coating.

(b) Pressure-sensitive adhesive, filament-reinforced tape at least 0.5 in. in width.
(c) Nonmetallic strapping at least 3/8 in. wide by 0.015 in. thick, or at least 1/4 in. wide by 0.027 in. thick.
(d) Substantial round or twist-tie galvanized steel wire. Containers with a net weight of more than 25 lb must be strapped by placing one strap, wire, or tape, girthwise at the approximate center around the top, sides, and bottom of the container; also by a second strap, wire, or tape centrally located around the top, ends, and bottom of the container at a right angle to the first. For containers with a net weight of 25 lb or less, the second strap may be omitted.

2. Stapling—Staples shall be sufficient in number and shall be properly distributed to ensure a secure closure and to prevent lifting of edges and corners of outer flaps.
3. Gluing—The top and bottom flaps shall be firmly glued together over a sufficient area to ensure a secure closure and to prevent lifting of edges and corners of outer flaps. The USDA maintains the right to have the packing and closure materials determined by appropriate test procedures.

SEALING

When individual products are not stamped, the shipping containers in which the products are packed must be sealed in accordance with USDA meat grading instructions.

Marking

Containers Packed to More than 25 Pounds Net Weight

1. The following markings must be legibly and conspicuously stenciled or printed on one end of the container in letters and numbers not less than 1/2 in. high:
   (a) Upper left-hand corner—The true name of the product and the code identification of these specifications (IMPS), together with the product item number (Roast Ready Ribs, IMPS No. 109; Bologna, IMPS No. 801, etc.).
   (b) Upper right-hand corner—The date of examination and acceptance by the USDA meat grader (month, day, and year).
   (c) Lower left-hand corner—The grade or selection of product (U.S. Prime, U.S. Choice, etc.; or Selection No. 1, Selection No. 2, as applicable).
   (d) Lower right-hand corner—The number of pieces or packages of product in the container and the net weight of the product. (This information may be applied with a felt-tip pen, crayon, or pencil.)
2. The following markings must be stenciled or printed on the top or side of the container in letters and numbers not less than 1/2 in. high:
   (a) The name and address of the contractor.
   (b) The name and address of the supplier if other than the contractor.

   The name and address of the consignee (not applicable to stockpiled products).

In addition to the aforementioned markings, when product is prepared for stockpiling (not prepared under a purchase order), it shall have any deviations from specification requirements and all applicable options such as weight range, formula, portion size, and so on, stenciled or printed on the lower left-hand corner of the same end on which the other markings appear.

When lack of space precludes listing all the information required on the lower left-hand corner of the end, it may be stenciled or printed on the opposite end or on a side panel. The marking material must be flat, waterfast, nonsmearing, and black in color.

Containers Packed to 25 Pounds or Less Net Weight

In lieu of stenciling, such containers may have a printed or typewritten label firmly attached by adhesive material on one end of the container, which legibly and conspicuously bears the markings of the name and address of the contractor.

Examination for Condition of Containers

Definitions and procedures contained in the U.S. Standards for Conditions of Food Containers shall apply for lots containing 50 or more shipping containers or 300 or more primary containers. (Copies of the Standards may be obtained from the Livestock Division, C&MS, USDA, Washington, DC 20250.) In lots that include fewer than 50 shipping containers or fewer than 300 primary containers, the containers shall be examined individually and all defective containers must be replaced or corrected, as applicable. Examination for the condition of the containers shall be made in conjunction with the final examination and acceptance of the product. However, at the option of the contractor, on a product to be delivered frozen, such an examination may also be made in conjunction with the preliminary examination and acceptance.

Condition of Product at Time of Delivery

Refrigerated trucks must be used when necessary to protect products during transport, and these trucks must be clean and free from foreign odors. At destination, all products will be reexamined by the consignee for cleanliness and soundness.

Contractors furnishing products under these specifications are expected to furnish such assistance as may be necessary to expedite the grading, examination, and acceptance of these products. These specifications will be strictly enforced by the using agencies. The consignee may, by purchasing acceptable products in the open market, immediately replace any products that are not delivered or which he or she rejects. In such case, any increase in cost of these products will be charged against the defaulting contractor.

WAIVERS AND AMENDMENTS TO SPECIFICATIONS REQUIREMENTS

Waivers of a few specification requirements may be made, provided:

1. The change can be indicated clearly and precisely.
2. There is agreement between purchaser and contractor on the changes.
3. The purchaser furnishes the USDA meat grader who is to perform the examination and acceptance of the product with a written statement indicating the precise nature of the changes.

The following are examples of waivers that may be made:

1. Substitution of weight ranges for those specified
2. Substitution of grade of meat specified
3. Modification of fat content in ground or diced meat
4. Slight variations in trim or style of cutting
5. Slight variations in sausage formulas

Changes involving extensive rephrasing of specification requirements must be considered as amendments to the specification and may be placed in effect only after such changes have been submitted and approved by the Standardization Branch, Livestock Division, C&MS, United States Department of Agriculture.

INSTITUTION INSPECTION

Final acceptance of all products will be by the consignee at the point of delivery. Products that are not appropriately identified with the "USDA Accepted As Specified" stamp will be rejected. Products that are appropriately identified with that stamp but that have other obvious, major deviations from specification requirements also will be rejected. Products appropriately identified with the "USDA Accepted As Specified" stamp but that, in the opinion of the consignee, have minor deviations from the specification requirements that do not materially affect the usability of the product may be tentatively accepted subject to verification of such deviations by local USDA meat grading personnel. Disposition of products with such verified minor deviations will be at the option of the consignee. All deviations from the specifications noted at the point of delivery must be reported promptly to local USDA meat grading personnel who are instructed to investigate all such reports without delay.

Beef

Grade

Carcasses, Sides, or Quarters. The purchaser must specify a quality grade and a yield grade.

Cuts and Roasts. The purchaser must specify a quality grade, and may also specify a yield grade.

Portion Cuts and Diced Beef. The purchaser must specify a quality grade for Item Nos. 137 and 1137. However, a quality grade shall not be specified for other ground beef items. The upper half or lower half may be specified, otherwise the full range of the grade is acceptable.

When yield grade is specified for forequarters or forequarter cuts, any such item may not be derived from a carcass or side that was yield graded after the removal of more than minor amount of kidney and pelvic fat.

Yield grades 1 through 5 are applicable to all quality grades. However, those yield grades indicated by an "X" are in largest supply.

The yield grades reflect differences in yields of boneless, close trimmed, retail cuts. As such, they also reflect differences in the overall fatness of carcasses and cuts. Yield Grade 1 represents the highest yield of retail cuts and the least amount of fat trim. Yield Grade 5 represents the lowest yield of retail cuts and the highest amount of fat trim.

Weight Range or Size

Carcasses, Sides, Quarters, and Cuts. See weight ranges under beef specifications.

Roasts. If desired, purchasers may specify that roasts be further reduced in size.

Ground Beef Patties. Either the individual patty weight or the number of patties per pound must be specified.

Ground Beef Patty Weight Tolerances

For patties with a specified weight of 3 oz or less, a tolerance of +2 patties from the projected number in a 10 lb unit will be permitted. For patients with diets having a specified weight of more than 3 oz, a tolerance of +1 patty from the projected number in a 10 lb unit will be permitted. (When patties are specified by a number per pound, this shall be converted to patty weight to determine tolerances, e.g., six to the pound = 2.67 oz.)

Example: When 2 oz patties are specified, 10 lb units containing 78 to 82 patties are acceptable.

Portion-Cut Items

For portion-cut items, either the portion weight or thickness desired, not both, must be specified. If weight is specified, see the weight range tables in the section on Institutional Meat Purchase Specifications. If thickness is specified, the actual thickness desired must be indicated (not applicable to cubed steaks). Moreover, in order to control uniformity of portion sizes, the weight range of the IMPS cut from which the portions are to be produced may also be specified. In this case, the fat thickness of the referenced IMPS cut should be the fat thickness specified for the portion cut.

Portion-Cut Weight Tolerances

If portion weight is specified, the following tolerances will be permitted:

WEIGHT SPECIFIED	TOLERANCES (OVER AND UNDER)
Less than 6 oz	1/4 oz
6 oz but less than 12 oz	1/2 oz
12 oz but less than 18 oz	3/4 oz
18 oz or more	1 oz

Example: When 8 oz steaks are specified, individual steaks weighing 7½–8½ ounces are acceptable.

Portion-Cut Tolerances

If thickness is specified, the following tolerances will be permitted.

THICKNESS SPECIFIED	TOLERANCES (OVER AND UNDER)
1 in. or less	3/16 in.
More than 1 in.	1/4 in.

Example: When 1¼ in. steaks are specified, individual steaks measuring 1–1½ in. are acceptable.

Fat Limitations

Cuts and Roasts. Except when yield grade is specified, the purchaser must specify one of the following maximum average thickness of surface fat, unless definite fat limitations are indicated in the detailed specifications.

MAXIMUM AVERAGE THICKNESS	MAXIMUM AT ANY ONE POINT
1 in.	1¼ in.
3/4 in.	1 in.
1/2 in.	3/4 in.
1/4 in.	1/2 in.

When average fat thickness is specified in item descriptions, the appropriate "Maximum at Any One Point" limitations shall apply.

When tying is required, roasts must be made firm and compact and held intact by individual loops of strong twine, uniformly spaced at approximately 2 in. intervals girthwise. In addition, some roasts may require tying lengthwise. In lieu of string tying, it is permissible to enclose roasts in stretchable netting, or by any other equivalent method. Purchasers may specify that roasts be tied, when this requirement is not specified in the detailed item specification.

Steaks. Unless otherwise specified by the purchaser, or unless definite fat limitations are indicated in the detailed item specifications, on surfaces where fat is present, the fat must not exceed 1/2 in. in thickness, and the thickness at any one point must be not more than 3/4 in.

Chops, Cutlets, and Fillets. Unless otherwise specified by the purchaser, surface fat, where present, must not exceed an average of 1/4 in. in thickness, and the thickness at any one point must not be more than 3/8 in. Defatting must be done by smoothly removing the fat by following the contour of the underlying muscle surface. Beveling of the edges only is not acceptable. In determining the average thickness of fat at any one point on steaks and roasts, which have an evident, natural depression of the lean, only the fat above the portion of the depression that is more than 3/4 in. in width will be considered.

State of Refrigeration

1. Chilled
2. Frozen

Aged Beef

The purchaser may specify aged beef. Unless otherwise specified, bone-in cuts may be dry aged or aged in plastic bags. Boneless cuts must be aged in plastic bags. Meat that is dry aged must be trimmed to remove meat that is dry and discolored and/or that has an odor foreign to fresh beef. When examining beef for compliance with these specifications, USDA meat graders will take into consideration the deviation of color from that of fresh-chilled meat that is normal for aged meat.

Material

Beef products described in these specifications must be derived from beef carcasses or wholesale cuts. Cuts that have been excessively trimmed in order to meet specified weights, or that do not meet the specification requirements for any reason, are excluded. The beef shall be of good color normal to the grade and be practically free of bruises, blood clots, bone dust, ragged edges, and discoloration. The spinal cord, thymus glands, and heart fat must be removed. Except as otherwise provided herein, the meat shall show no evidence of freezing or defrosting. In addition, the product shall show no evidence of mishandling and shall be in excellent condition at the time of delivery.

Portion-cut items to be delivered frozen may be produced from frozen meat cuts that have been previously accepted in the fresh-chilled state, provided such cuts are in excellent condition and in their original shape. Products thus produced shall be packaged, packed, and promptly returned to the freezer.

Cutting Steaks

Unless otherwise specified in the individual item specification, steaks must be cut in full slices, in a straight line reasonably perpendicular to the outer surface, and at an approximate right angle to the length of the meat cut from which steaks are produced. Butterfly steaks are not acceptable.

Boning

Boning shall be accomplished with sufficient care to allow each cut to retain its identity, and to avoid objectionable scores in the meat.

INSTITUTIONAL MEAT PURCHASE SPECIFICATIONS GENERAL REQUIREMENTS

IMPS NO. 100 CARCASS

MBG NO. 100
WEIGHT RANGE (in pounds):
RANGE A: 500–600
RANGE B: 600–700
RANGE C: 700–800
RANGE D: 800 and up

A beef carcass is the four quarters from a single carcass. Separating the forequarters from the hindquarters by cutting between the 12th and 13th ribs produces the quarters, the 13th rib remaining with the hindquarters. The diaphragm may be removed, but if not removed, the membranous portion shall be removed close to the lean. The thymus gland and heart fat shall be closely removed.

IMPS NO. 101 SIDE

MBG NO. 101
WEIGHT RANGE (in pounds):
RANGE A: 250–300
RANGE B: 300–350
RANGE C: 350–400
RANGE D: 400 and up

A side of beef consists of one matched forequarter and hindquarter from one-half the carcass prepared as described in Item No. 100.

IMPS NO. 102 FOREQUARTER

MBG NO. 102
WEIGHT RANGE (in pounds):
RANGE A: 131–157
RANGE B: 157–183
RANGE C: 183–210
RANGE D: 210 and up

The forequarter is the front portion of the side, after severance from the hindquarter, as described in Item No. 100. The forequarter shall be trimmed as described in Item No. 100.

IMPS NO. 1100 CUBED STEAKS

Regular MBG NO. 1100

URMIS and COMMON NAMES: Cube Steak, Minute Steak, Swiss Steak

COMMON USES: Braise, cook with moist heat, closed and open-faced sandwiches, Swiss steak

SUGGESTED PORTION SIZES: 3 oz, 4 oz, 6 oz, 8 oz

Cubed steaks may be produced from any boneless meat from the beef carcass that is reasonably free of membranous tissue, tendons, and ligaments. The meat shall be made into cubed steaks through use of machines designed for this purpose. Knitting of two or more pieces, and folding of the meat when cubing, are permissible. Cubed steaks shall be reasonably uniform in shape, that is, practically square, round, or oval. After cubing, surface fat on the edge of the cubed steaks shall not exceed ½ in. in width at any one point, when measured from the edge of the lean. Surface and seam fat shall cover not more than 15 percent of the total area on either side of the steak. The cubed steak shall not break when suspended from any point ½ in. from the outer edge of the steak.

IMPS NO. 1101 CUBED STEAK, SPECIAL

MBG NO. 1101

URMIS and COMMON NAMES: Cube Steak, Minute Steak, Swiss Steak

COMMON USES: Braise, cook with moist heat, open and closed-faced sandwiches, Swiss steak

SUGGESTED PORTION SIZES: 3 oz, 4 oz, 6 oz, 8 oz

Special cubed steaks shall meet all the requirements for item No. 1100, except that they shall be produced only from the muscles contained in the round, loin, rib, or square-cut chuck. Knitting of two or more pieces, or folding of the meat, is not acceptable.

IMPS NO. 102A FOREQUARTER, BONELESS

URMIS and COMMON NAMES: Beef Chuck Shoulder Pot Roast Boneless, Forequarter

COMMON USES: Cook in moist heat, center shoulder roast, chuck roast, pot roast, Swiss steak

WEIGHT RANGE (in pounds):

RANGE A: 104–125

RANGE B: 125–146

RANGE C: 146–168

RANGE D: 168 and up

The boneless forequarter is prepared from Item No. 102. Meat with dark discoloration, all bones, cartilage, backstrap, exposed large blood vessels, and the prescapular lymph gland shall be removed. The thick tendinous ends of the shank shall be removed by cutting back until a cross-sectional cut shows at least 75 percent lean. The clod shall be removed without undue scoring.

IMPS NO. 1102 BRAISING STEAKS, SWISS STEAK

MBG No. 1102

URMIS and COMMON NAMES: Beef Chuck Shoulder Steak Boneless, Beef Chuck Arm Steak Boneless, Cube Steak, Potting Steak, Round Steak

COMMON USES: Braise, panfry, potted Swiss steak, country fried steak, stew meat

SUGGESTED PORTION SIZES: 4 oz, 6 oz, 8 oz, 10 oz, 12 oz, 14 oz, 16 oz

Braising steaks shall be produced from any part of any one, or any combination of, the following cuts of beef: Item No. 112—Rib Eye Roll; Item No. 114—Shoulder Clod; Item No. 167—Knuckle; Item No. 168—Top Round; Item No. 170A—Bottom Round, Heel Out; Item No. 180—Strip Loin, Short Cut, Boneless; Item No. 184—-Top Sirloin Butt; Item No. 186—Bottom Sirloin Butt, Trimmed.

Each braising steak shall be practically free of fat on at least half the circumference, and the surface fat on the remaining half of the circumference shall not exceed ½ in. at any one point. If specified by the purchaser, each steak shall be mechanically tenderized, using machines designed for this purpose. Knitting of two or more pieces of meat, or folding of the meat, is unacceptable.

IMPS NO. 103 RIB, PRIMAL

MBG NO. 103

URMIS and COMMON NAMES: Prime Rib of Beef, Standing Rib Roast, Rib Roast Oven-Ready, Oven-Ready Rib

COMMON USES: Roast, prime rib of beef, standing rib roast

WEIGHT RANGE (in pounds):

RANGE A: 24–28

RANGE B: 28–33

RANGE C: 33–38

RANGE D: 38 and up

The primal rib is that portion of the forequarter remaining after the removal of the crosscut chuck and short plate, the skeletal part of which contains parts of seven ribs (6th to 12th inclusive), the section of the backbone attached to the ribs, and the rear tip of the blade bone (scapula). The crosscut chuck is removed by a straight cut, perpendicular to the split surface of the backbone between the 5th and 6th ribs. The short plate shall be removed by a straight cut across the ribs, from a point on the 12th rib that is not more than 10 in. from the center of the inside protruding edge of the 12th thoracic vertebra, through a point on the 6th rib that is not more than 10 in. from the center of the inside protruding edge of the 6th thoracic vertebra. The portion of the diaphragm, and practically all the fat remaining on the side surface of the vertebrae, shall be removed.

IMPS NO. 1103 RIB STEAKS

MBG NO. 1103

URMIS and COMMON NAMES: Beef Rib Steak, Beef Rib Steak Bone In

COMMON USES: Broil, pan broil, panfry, beef steak

SUGGESTED PORTION SIZES: 8 oz, 10 oz, 12 oz, 14 oz, 16 oz

Rib steaks shall be prepared from a Rib, Primal—Item No. 103. The short ribs on individual steaks shall be removed at a point that is not more than 3 in. from the outer tip of the rib eye muscle. All muscles above the major rib eye muscle, fat overlying these muscles, the blade bone and cartilage, the feather bones, and the backstrap shall be removed.

IMPS NO. 1103A RIB STEAKS BONELESS

MBG NO. 1103A

URMIS and COMMON NAMES: Beef Rib Steak, Rib Eye, Spencer, Delmonico Beauty Steak, Fillet Steak

COMMON USES: Broil, pan broil, panfry, beef rib steak
SUGGESTED PORTION SIZES: 4 oz, 6 oz, 8 oz, 10 oz, 12 oz

Boneless rib steaks shall be prepared as described in Item No. 1103, except that all bones and rib fingers (intercostal meat) shall be removed.

IMPS NO. 107 RIB, OVEN-PREPARED

MBG NO. 107
URMIS and COMMON NAMES: Beef Rib Roast, Prime Rib, Standing Rib Roast, Rib Roast Oven-Ready, Spencer Roll
COMMON USES: Roast, standing rib roast, prime rib, roast beef
WEIGHT RANGE (in pounds):

RANGE A: 17–19

RANGE B: 19–23

RANGE C: 23–26

RANGE D: 26 and up

The oven-prepared rib is prepared from Rib, Primal. A straight cut is made across the ribs from a point on the 12th rib that is not more than 3 in. from the outer tip of the rib eye muscle, through a point on the 6th rib that is not more than 4 in. from the outer tip of the rib eye muscle. The chine bone shall be removed by a straight cut, along a line at which the vertebrae join the feather bones, exposing the lean, but leaving the feather bones attached to the oven-prepared rib. The blade bone and related cartilage shall be removed.

SAUSAGE

The History of Sausage*

It is estimated that more than 200 kinds of sausage are made in the United States by nearly 3,000 meat processors. This sausage-making industry is by no means a twentieth-century innovation, for sausage was probably the world's first "convenience" food. Born of necessity centuries ago as a means of preserving meats, sausage was prechopped or ground, preseasoned, stuffed into a serviceable "package" or casing, and precooked, smoked, or dried. These services, developed because there was no refrigeration, met with approval not only because of their practicality, but also because of the flavors they produced. The ancient art of sausage making predates recorded history. Homer, in his *Odyssey*, mentions sausage as a favorite food of the Greeks, and we learn from other writers as well that the Greeks liked sausage immensely. They served what we would now call "wieners" as an appetizer before meals. By the time of Julius Caesar, the arts of seasoning and preserving meats had advanced to a high level. In his military campaigns, Caesar gained advantages over his barbarian enemies by issuing preserved meats to his legions. His enemies, meanwhile, lost precious hours hunting game in the forests or seizing domestic animals and preparing them to eat. The Romans' liking for sausage was so great that no festive occasion was considered complete without it. When a strong lobby of reformers put through a prohibition law against sausage, the Romans, as impatient of reformers then as today, smuggled sausage past the prohibition agents, and the law finally was stricken from the books.

The book on cooking by Epicius, one of the oldest in existence, listed a number of sausage recipes for various dishes. A small sausage, called Botellum, was made of pork, yolks of hard-boiled eggs, pignolia nuts, onions, leeks, rosemary, and fine pepper and then stuffed into casings. It was usually cooked in wine. Lucanian sausage was prepared in much the same way, except that the seasonings were cumin, rue, and laurel berries.

By the Middle Ages, "wurst" was popular throughout Europe and, served with beer and wine, was a symbol of conviviality and the joy of eating. The mighty medieval butchers' guilds did much to maintain the quality of their wares and the honor of their craft during those long centuries of slow progress.

The word "sausage" is derived from the Latin *salsus*, meaning salted or preserved meat. Although sausage was originally made of pork, during the last 700 years it has been made of mixed meats of all kinds, seasoned with spices gathered from around the world. Modern sausage was developed mainly in the Germanic countries and in Italy. There the people realized how enticing meat could be made by the skillful blending of different kinds of meats with various spices, and by curing and aging.

The warm Italian climate encouraged the development of the so-called dry sausage. It was preserved with an abundance of salt and pungent spices, such as pepper and garlic, then thoroughly dried, generally without smoking. Treated in this way, the sausage could be kept for long periods and stored against months of meat scarcity. Many a delectable sausage today bears the name of an Italian city—Milano, Romano, Genoa, Bologna.

In Germany, the much cooler climate and the cooler storage cellars accounted for the development of fresh and cooked sausage. These are the predecessors of our domestic sausage today; included are frying sausages, "bratenwurst," many styles of liver sausage, head cheese, blood sausage, and various cooked, smoked sausages. Of all these, the wiener or frankfurter, the "hot dog," has become the most popular sausage in the United States.

*(Portions of the material that appears in this section are adapted from *Sausage and Smoked Meats*, a publication of the Oscar Mayer Company. Reprinted with permission.)

The Cure that Flavors

Meats selected for most sausage making acquire their first distinctive characteristics in the curing process. Until a century ago, curing was a principal method of preserving meat, but today it is used primarily to develop flavor and color in meats. Although most curing agents are essentially the same, sausage-making companies use their own well-guarded techniques for applying them. Curing solutions vary, depending on the meat to be cured, the flavor desired, and the length of cure. The solution is basically a brine mixture that has been blended with other preservatives and flavoring agents. The curing solution is applied either by hand-operated equipment or by automatic methods. For example, a side of pork is cured by an "injecto-cure" machine and so becomes bacon. The meat passes under a row of descending needles, and the curing solution is pumped into the meat. Ham curing is a vein-pumping operation in which a solution is injected into the natural circulatory system of the ham. This method ensures that the curing solution penetrates every section of the ham for a thorough, even cure. Meat that has been treated internally with the pickle solution, as it is sometimes called, and salted on the outside surface, then goes to a curing cooler to allow the solution to penetrate.

Temperature controls are used to either speed up or retard the curing process, depending on product requirements for flavor. Time and temperature are most crucial in developing unique flavors in the curing process. After the meat has been cured, it is held at temperatures between 32° and 40° F. Cured meats are not frozen, as the brine solutions invite the development of rancidity, as well as texture changes in the meat, at lower temperatures.

Spices

One of the secrets of sausage flavor is the delicate blending of pure spices with fine meats. The old "sausage maker" measured out spices with a well-trained eye and an experienced hand. Today's exacting formulas call for precise quantities and strict quality to ensure uniformity. This is not easy to acquire. Mere weight or volume measures are not enough. Natural spices vary considerably in flavor effect, because of variance in crop conditions and soil conditions in different locations. The handling of spices from the time of harvest until the time of use is another important factor. When spices are extracted for their active flavor components, the extractive will also differ, depending on the methods used.

There have been times in history when pepper was worth its weight in silver and gold. After the Goths conquered Rome in the fifth century, they demanded a ransom of pepper as well as precious metals. Peppercorns were often employed as money. Adventurers such as Marco Polo, Christopher Columbus, and Vasco da Gama voyaged in search of shorter routes to the spice-laden East. Countries controlling the spice trade became the richest and most powerful nations. Piracy and plundering among these major Western powers marked nearly four centuries of struggle for the control of spice-producing lands in the Middle East and the Orient.

Today spices come from all over the world. Black pepper comes from India, white pepper from Borneo, nutmeg from the East Indies, marjoram and thyme from France, paprika from Spain, cloves from Zanzibar, cardamom from Guatemala, allspice from Jamaica, and cinnamon from Java. Natural spice materials include herbs, seeds, and parts of aromatic plants, in addition to essential oils. Although spices are recognized principally as flavor ingredients, some also have antibiotic properties that aid in preservation. Spices may also affect the color and aroma of sausages. Salt is almost universally used as a preservative for sausage. It also affects texture—a lack of salt results in a rubberlike product. Sugar has some use as a preservative and has a minor effect on color fixation. Spice control is so essential in maintaining a uniform flavor standard that sausage processors often maintain a special spice department. They begin by analyzing the basic materials even before purchase. Samples of natural spices and essential oils, as the liquid spice extracts are called, are tested thoroughly in the flavor laboratory. Spice blends are prepared from basic spice materials in a centralized production unit. Blending spices in large lots, such as 500 lb batches, permits consistent seasoning of sausages even though they are made in different plants.

Sausages and Smoke

The aroma of smoke has tantalized people through the ages, and the more civilized they pretend to be, the greater the appeal of a fireplace, a grill, or an open fire with its pleasant smell of smoke.

The idea of smoking meat developed from the need to preserve meat, just as the curing process did, and smoking for flavor is by no means an innovation. Centuries ago, after curing meat with salt, people smoked it over a smoldering wood fire. Smoking began as a very ordinary procedure, but today it is a technical, processing plant operation. Not all processed meat products are smoked, nor are all of them cured. However, with few exceptions, meats to be smoked must first be cured.

After the curing process is complete, meats and processed products are routed to the smokehouse on a conveyor system or on a "smoking tree" assembly. Production employees transfer the meat to the smokehouse interior, where the penetrating effect of time- and temperature-controlled smoke determines the final flavor and appearance.

In "rotary" smokehouses, six or seven stories high, meat products hung on racks circulate through the smoke chambers on a moving frame resembling a Ferris wheel. The products travel first in an upward direction and then downward for a thorough exposure to the permeating smoke. It requires almost

1½ hours for one rack to complete a rotation. Hams and bacon sides are usually smoked in the large rotary houses.

In "stationary" smokehouses, meats are hung on standing frames or "trees," rather than on rotating racks. These houses are smaller and have less capacity, but require much less floor space and are easier to clean and maintain than the multistory rotary smokehouses. They are used for smoking smaller sausages and specialties. Stokers feed a combustion chamber with hardwood sawdust, selected for its pleasant smoke aroma, and the resulting clouds of smoke are funneled into the smokehouse area. Large fans keep the air circulating so that the smoke reaches every corner of the smokehouse.

Smoking times vary according to product, some requiring 4 hours of smoke, others requiring 24 hours. Temperature control is critical. The internal temperature of a pork product to be labeled "fully cooked" must reach at least 137° F, according to federal inspection regulations. Any trichinae would be destroyed at this temperature. Actually, many smoked meat products are processed to temperatures of 148° F or higher to develop flavor and firmness. Other products that are to be cooked by the consumer may not require as high a temperature, but these products also depend on precise temperature controls to ensure unique flavor qualities. Smoking is said to give processed meat products their "second helping" of flavor, but according to research specialists, you don't taste the smoke, you smell it.

Up to Dry

One of the most interesting and exacting processes in sausage making is drying. The practice probably originated near the Mediterranean Sea centuries before Christ, as one of the earliest methods of preserving foods. Many names of dry sausages are taken from the city of origin, as each community developed its own particular style. There were differences in the kinds of meat and casing used, methods of chopping meat, seasonings, even in the wrapping of twine around the sausage for hanging to dry. Dried sausage may or may not be smoked after curing.

Sausage drying rooms have carefully controlled humidity, temperature, and air circulation. The drying process must be exact to produce even drying throughout the sausage, without producing a crust too hard or internal "soft" spots. The air is continuously washed and redried to eliminate airborne spores and molds that may produce undesirable characteristics in the sausage.

Drying may begin in the smokehouse, and some semidry sausages are processed entirely by smoking. Lebanon bologna, for example, is smoked for as long as two weeks to develop its texture and flavor. The distinctive tangy flavor typical of most dry sausages, as well as the semidry varieties, results from a bacterial fermentation. In addition to traditional methods, pure starter cultures have been developed to control and speed fermentation. Dry and semidry sausages are usually eaten cold, although they have not been fully cooked in processing. There are various precautions taken to ensure wholesomeness in this type of sausage. Either controlled refrigeration or controlled curing are effective treatments in destroying trichinae, as is heating to 137° F. Although controlled freezing is an approved method of destroying trichinae, meats to be used in dry sausage are not generally certified this way. Frozen pork is not considered desirable for dry sausage making. Federal requirements for the refrigeration methods are specific for minimum temperatures and time in relation to the size of the meat pieces, to ensure the necessary penetration of freezing temperatures. For example, meat pieces requiring 20 days at 5° F need only 10 days at –10° F and 6 days at –20° F.

Classification of Sausages

Sausages are usually classified according to processing procedures after stuffing. Although there are differences between sausages within a group, these differences are due primarily to meat combinations and seasonings used, and not to basic processing procedures. They are also basically alike in regard to preparation procedures and typical serving uses. The major classifications are:

Fresh Sausage
Uncooked, Smoked Sausage
Cooked, Smoked Sausage
Cooked Meat Specialties
Cooked Sausage
Dry, Semidry Sausage
Fresh Sausage

As the name suggests, fresh sausage is made of meats that have not been cured, selected cuts of fresh pork and, sometimes, beef. It is the most perishable of all sausage products. Its taste, texture, tenderness, and color are directly related to the ratio of fat to lean. Makers of quality fresh sausage use "trimmings" from primal cuts—that is, the pork loin, ham, or shoulders. As with other sausage products, the spice formulation varies with the meat processor. Expert sausage makers say the product should be seasoned delicately, with a view toward enhancing the natural meat flavors, not masking them.

Fresh sausage must always be kept under refrigeration. It must be cooked thoroughly before serving and is usually fried or grilled. Pork sausage and bratwurst are popular varieties in this classification.

Uncooked, Smoked Sausage

This class of sausage has all the characteristics of fresh sausage, with one major difference: It is smoked, which produces a different flavor and color. Sometimes it contains fresh meat. These sausages, too, must be cooked thoroughly before serving. Smoked pork sausage and kielbasa are two of the few examples of uncooked, smoked sausage.

Cooked, Smoked Sausage

Sausage in this category, usually made from cured meats, is chopped or ground, seasoned, stuffed into casings, smoked, and cooked. The use of cured meats contributes to flavor, color, and preservation of the product. Cooked, smoked sausage comes in all shapes and sizes and is the largest and most popular of all the categories. The "skinless" varieties have been stripped of their casings after cooking. Wieners and smokie links are included in this grouping. Within this category there are two basic classes: fine-cut sausage, such as wieners and bologna, and coarse-cut sausage, such as Berliner or New England.

Cooked Meat Specialties

As the name implies, the primary difference between this classification (cooked meat specialties) and cooked sausage is that specialty items, such as meat loaves, head cheese, souse, and scrapple, are included. These prepared meat products are cooked or baked and are always ready to serve.

Cooked Sausage

Cooked sausage is usually prepared from fresh, uncured meats, although occasionally cured meats are used. Often variety meat or organ meat, such as liver, is included, so sausages in this classification may be especially nutritious. In many instances, the product is smoked; however, the essential difference between sausage in this classification and cooked, smoked sausage is that the smoking is done after cooking has been completed. This product is always ready to serve. Liver sausage is the most popular variety in this classification and has enjoyed increasing consumer acceptance in recent years.

Dry, Semidry Sausage

All dry sausages are characterized by a bacterial fermentation. A lactic acid bacterial growth is intentionally encouraged because it is useful as a meat preservative, as well as in producing the typical tangy flavor. The meat ingredients, after being mixed with spices and curing materials, are generally held for several days in a curing cooler. The meat is then stuffed into casings and started on a carefully controlled air-drying process. Some dry sausage is given a light preliminary smoking, but the key production step is the relatively long, continuous air-drying process. Principal dry sausage products are salamis and cervelats. Salamis are coarsely cut; cervelats, finely cut—with few exceptions. They may be smoked, unsmoked, or cooked. Italian and French dry sausages are rarely smoked; other varieties usually are.

Dry sausage requires more production time than other types of sausage and results in a concentrated form of meat. Medium-dry sausage is about 70 percent its "green" weight when sold. Less dry and fully dried sausage range from 60 to 80 percent of original weight at completion. Logically, many varieties in this group are more expensive per pound than those in other classifications. Semidry sausages are usually heated in a smokehouse to cook the product fully and partially dry it. Semidry sausages are semisoft sausages, with good keeping qualities due to their lactic acid fermentation. Although dry and semidry sausages were originally produced in the winter for use in the summer, and were considered summer sausage, the term "summer sausage" now refers to semidry sausages, especially Thuringer Cervelat.

Smoked Meats

Cuts of fresh meat, usually pork, may be cured with salt or brine and then smoked to give a distinctive flavor and aroma. Current smoking techniques result in a product that is considerably less dry than the smoked products of the last century. The original drying-out in smoking meat contributed greatly to its keeping qualities. Modern refrigeration facilities have reduced the need for this effect of smoking, and methods have been developed that emphasize flavor and aroma rather than drying. Kitchen preparation of today's fully cooked hams is quite different from the cooking methods necessary for the heavily salted, slowly smoked hams of the past. Fully cooked hams are heated in the oven, without simmering, to serving temperature (130° F) in about half the baking time required for the cook-before-eating-type ham. This is a time-saving factor of considerable importance. Curing and smoking meat inhibit the development of bacterial growth, thus increasing the refrigeration storage life of the product. However, the smoke residues have no effect on mold growth. Curing produces chemical changes in meat pigments; these are stabilized during the heat of smoking. This accounts for the characteristic deep pink color of the lean muscle of bacon, ham, and other smoked meats.

Sausage Varieties Examples

Apennio (dry sausage). Coarsely chopped pork and beef are seasoned with garlic, then stuffed in the cap end of a hog middle. This Italian-style salami is air-dried, but not smoked. Serve without heating. Ingredients: beef, pork. Seasonings: garlic, salt, sugar, mustard, pepper. Availability: limited distribution.

Bacon (smoked meat). Pork "sides" are cured, and smoked, usually with hardwood chips. Nearly all bacon is now sold sliced and packaged in 1 or ½ lb units. Thick-sliced bacon is also packaged in 2 lb units. Bacon is packaged as "sliced bacon," "thick-sliced bacon," "thin-sliced bacon," and "ends and pieces." Typically, sliced bacon has 20 to 24 slices per pound; thick sliced, 14 to 18 slices per pound, and thin sliced, 25 to 35 slices per pound. Actually, thin sliced does not mean "thinner" but, rather, smaller slices. It is usually made from small pork sides that are shorter and narrower, thus yielding a higher slice count per pound. The finest-quality bacon is chosen from the center slices of selected bacon sides. These "sides" are cut from lean-type hogs, marketed at weights of 200 to 220 lb. The fat is firm and white, evenly ribboned with lean. Too much lean may result in tough bacon. Second grades of bacon may be sliced from the same sides as the processor's first quality, but the slices will be less uniform in appearance. Other grades usually come from the larger bacon sides. "Ends and pieces" are the irregularly shaped pieces from either end of the bacon side (only the center slices are uniform in length), as well as part slices. This type of bacon may offer good value for flavoring and for use as cooked, crumbled bacon. Bacon flavor and aroma are very elusive, and sliced bacon exposed to air and light loses quality rapidly. The innovation of vacuum packing for sliced bacon was an important development for the consumer. Under refrigeration in an unopened vacuum package, bacon shows little or no loss of flavor and quality even after several days' storage. Bacon should be cooked at a low temperature in a skillet or griddle or baked on a rack in a shallow pan in a preheated 350° F oven for 10–15 minutes. Availability: widespread distribution.

Baked Black Pudding (cooked sausage). See *Blood Sausage*. A type of blood sausage. The mixture is baked in shallow pans, then cooled. To serve, it is sliced, then heated in a skillet or deep fat.

Beerwurst, Beer Salami (cooked, smoked sausage). Meat is chopped and blended with curing ingredients and seasonings. After 48 hours, the mixture is stuffed into beef bungs and cooked at high temperatures in a smokehouse. When a round shape is desired for uniform slicing, large artificial casings are used. Slice for sandwiches, cold meat platters, snacks. Ingredients: beef, pork. Seasonings: garlic, pepper, salt, sugar. Availability: packaged slices in supermarkets or bulk to slice in sausage shops, delicatessens.

Berliner (cooked, smoked sausage). A coarsely cut sausage made of cured lean pork. May also include a small amount of beef. After curing, the meat is stuffed into large artificial casings or beef casings. Berliner is smoked from two to four hours, then cooked in water from four to six hours. Contains no spice. Serve for sandwiches, cold meat platters, salads. Ingredients: lean cured pork, or lean cured pork and small amount of beef. Seasonings: salt, sugar. Availability: generally sold as sliced cold meat. National distribution.

Berliner Blood Sausage (cooked sausage). See *Blood Sausage*. A variety of blood sausage containing diced bacon instead of ham fat, snouts, and lips. After cooking, the sausage is surface dried and smoked.

Black Pudding (cooked sausage). See *Blood Sausage*. English version of blood sausage. Ingredients, which include cereal, are stuffed into large sheep casings and tied into 1 lb rings. Typical seasonings are allspice, black pepper, coriander, ground mustard, celery seed, salt.

Blood Sausage (cooked sausage). This specialty sausage is undoubtedly of high food value, but its acceptance in this country is limited. The ingredients, except blood and ham fat, are cooked before grinding, then combined with seasonings, and blood (and ham fat, if used), and ladled into casings, usually beef bungs or hog middles. After additional cooking in water, the casings are pricked to release any air, then chilled in cold water. Blood sausage is sometimes made into links 4–5 in. in length. Ingredients: pork skins, pork jowls, blood (usually beef). May also include sweet pickled ham fat, snouts, and lips. Seasonings: allspice, black pepper, ground cloves, onion, salt. Availability: chiefly consumed in England and Scotland.

Blood and Tongue Sausage (cooked sausage). Cooked and cured pork or lamb tongues are included with a blood sausage mixture in beef casings 10–12 in. long. The tongues are so inserted that they will be in the center of the sausage when it is sliced. Tongues constitute about one-third of the finished weight of the sausage. Slice for sandwiches, cold meat platters. Ingredients: blood sausage mixture, cured pork or lamb tongues. Seasonings: allspice, black pepper, ginger, marjoram, salt, sugar. Availability: sausage shops, delicatessens.

Bockwurst (fresh sausage). A very perishable, sausage-like delicacy, typically identified with Eastertime. Traditionally, it is made in the spring during the bock beer season, from which it acquires its name. The ingredients are finely chopped and stuffed into wide sheep rounds. The links, 5 in. in length, are light colored. Bockwurst is parboiled or scalded, then chilled. Constant refrigeration is essential. Must be thoroughly cooked before serving. Ingredients: freshly ground pork and veal, milk; may also include raw egg. Seasonings: chopped green onions or chives, cloves, ground mace, parsley, sage, white pepper. Availability: very limited distribution because of perishability.

Bohemian Presky (cooked, smoked sausage). Pork trimmings are cured only with salt, then ground and seasoned. The meat is stuffed into beef weasand and smoked. Baking is done in smokehouses, using high finishing temperatures. Ingredients: cured pork. Seasonings: garlic, pepper, salt, sugar. Availability: specialty item in limited distribution.

Bologna (cooked, smoked sausage). A finely cut sausage, generally stuffed into large casings, but other shapes are also made. This sausage, second only to wieners in popularity, originated in Bologna, Italy, during the Middle Ages. Quality bologna is carefully mixed and stuffed to give it a uniform color and even texture free of air bubbles. "Regular" bologna is always a blend of beef and pork, but an all-beef style is gaining in popularity. Garlic, a typical spice in either version, is a more prominent flavor in all-beef bologna.

SLICED BOLOGNA—Cut from large round "sticks." Usually 4–5 in. in diameter.

RING BOLOGNA—Meat is stuffed into casings about 2 in. in diameter and cut into pieces of about 1 lb. Ends are tied together, forming a horseshoe shape or a "ring."

CHUB BOLOGNA—A very smoothly blended bologna mixture of beef, pork, and smoked bacon is processed in 1 lb units in plastic wrapped "tubes." An especially convenient style to slice for snacks. Serve slices in sandwiches, with salads, in cold meat platters. Heat ring bologna to serve with potato salad. Ingredients: beef and pork, or all beef. Some processors may also use dry milk solids, cereals. Seasonings: black pepper, cloves, coriander, garlic, ginger, nutmeg. Availability: undoubtedly the most popular cold sliced meat with general distribution.

Boneless Pork Loin. See *Dewey Ham.*

Boterham Wurst (cooked, smoked sausage). This Dutch-style sausage is made of veal and pork that is coarsely chopped, cured, and then finely chopped. It is blended with coarsely chopped pork fat and seasonings, then stuffed into ox bungs, smoked, and cooked. Ingredients: pork, veal. Seasonings: white pepper, ginger, nutmeg, mace, salt, sugar. Availability: specialty item in limited distribution.

Bratwurst (fresh sausage). *Bratwurst* is German for "frying sausage." Originally a fresh sausage made of coarsely ground beef and pork, zestfully seasoned. Some processors now make it in a precooked version. Bratwurst is especially popular for grilling, so production is greatest during the cookout season. Links are large, typically about $1^{3}/_{8}$ in. in diameter. Usually six to seven links per pound. Must be thoroughly cooked. This sausage is very popular for charcoal broiling, especially in the Midwest. Grill and serve with buns or with German potato salad. Ingredients: beef and pork, or pork and veal. Seasonings: coriander, ginger, mustard, pepper, sage, salt, sugar, or thyme. Spice formulas vary greatly. Availability: distribution greatest during the outdoor grilling season. Popular in the Midwest; fully cooked versions also available; check label carefully.

Braunschweiger Liver Sausage (cooked sausage). A good grade of smoked liver sausage may be called Braunschweiger; it was in the town of Braunschweiger, Germany, that the art of smoking liver sausage originated. A desirable flavor may be developed either by careful smoking of the cooked liver sausage at a low temperature or by including proper amounts of smoked meat in the formulation. Slice for sandwiches, spread on crackers for appetizers. Ingredients: beef, pork livers, smoked bacon. Seasonings: coriander, ginger, marjoram, onion, black pepper, salt. Availability: national distribution.

INTERNET RESOURCES

www.ams.usda.gov/howtobuy/meat.htm

Topics include wholesomeness, quality, nutritive value, cost, convenience, and informative labeling, and points to consider when making meat purchasing decisions.

www.ams.usda.gov/lsg/ls-mg.htm

Topics include grading services, certification services, student programs, employment opportunities, certified beef and pork programs, and history and grading data, including quality and yield.

www.namp.com

North American Meat Processors Association (NAMP).

CHAPTER

9

Religious Dietary Laws

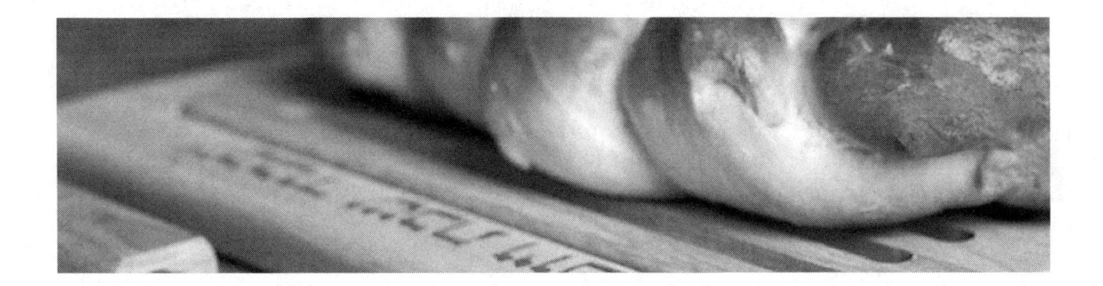

KOSHER

The Hebrew word *kosher* means "fit" or "proper" when referring to Jewish dietary laws or *kashruth*. The basis for these laws is biblical, and their application has been determined by rabbinic interpretation. Kosher food must be made from kosher ingredients, using kosher equipment. The laws of kashruth are extensive and can be complicated, and it is not always easy to determine if food is kosher.

Mammals that have split hooves and chew their cud, such as cows, are kosher. There are no specific characteristics that indicate whether a bird is kosher; rather, the Bible lists birds that are forbidden, and those not mentioned, including most of those eaten in the Western world, are considered kosher. There are many steps in the preparation of kosher poultry and meat. The animals must be slaughtered in a specific manner, called *shechita*, by a *shochet*, a trained Jewish slaughterer. After an animal has been properly slaughtered, the internal organs must be inspected to ensure that the animal had no injuries before slaughtering. The lungs, in particular, are inspected for adhesions. If an animal has no adhesions to the lungs, it is called *glatt kosher*. The word *glatt* means "smooth," referring to a smooth lung. Recently, the term *glatt kosher* has been used more loosely to mean that something is definitely kosher. Meat and poultry must also be *koshered*, or soaked and salted to remove all blood, which is forbidden.

Only fish that have both fins and scales are kosher. This excludes, among other things, all shellfish. Fish require no special preparation.

Kashruth also forbids cooking or eating milk and meat together. Meat and dairy products are not even eaten at the same meal, so buttered bread is not served at a meat meal and milk is not served with coffee after a meat meal. Meat and dairy foods are not prepared using the same equipment, nor eaten on the same dishes. It is not always easy to tell that meat and dairy foods have been mixed. For example, emulsifiers, which are often used in packaged foods, can be made from either animal or vegetable oils. If a dairy product, such as ice cream, contains animal-based emulsifiers, it is not kosher.

Some foods, such as vegetables, fruits, and eggs, are not considered meat or dairy; they are called *pareve*. Pareve foods can be prepared and served with meat or dairy foods using either meat or dairy equipment.

It is not really possible to determine whether packaged food is kosher simply by reading the ingredients. This is because ingredient lists often include general terms, such as "spices," "flavors," or "emulsifiers," and these can include items that are not kosher. The USDA does not require ingredient lists to include all trace ingredients. Furthermore, any equipment used to prepare the food must be kosher. For this reason, there are several certification organizations that authorize food as kosher.

These organizations do not print lists of kosher foods, because conditions change too frequently. For example, a company that makes cookies that are kosher might introduce a new product that contains lard, which comes from pigs. Pigs are not kosher animals. It may be clear from reading the ingredients that the new cookie is not kosher, but it is not clear from reading ingredients that all products made using the same equipment as the new cookies are not kosher. The organizations that certify kosher food stay in close contact with the production companies and are kept informed of changes like this.

In the United States, the four largest kosher certification organizations are the Orthodox Union (OU), the Organized Kashrus Laboratories (OK), Star-K Kosher Certification, and the Kof-K.

- For 80 years the Orthodox Union has maintained the highest standard of kosher supervision and today certifies more than 275,000 products.
- The Organized Kashrus Laboratories (OK Labs) was founded in 1935 by Abraham Goldstein, a kosher food expert and chemist, to meet the religious needs of a rapidly growing American Jewish community.
- Star-K Kosher Certification has provided kosher supervision for more than 50 years. Formerly known as the Orthodox Jewish Council of Baltimore / Vaad Hakashrus, it originally served the Jewish community of Baltimore. As the demand for kosher foods began to explode in the 1970s, the Council expanded both its services and staff. Since then, Star-K has assumed a more prominent role on the international kashruth scene and its roster of local companies has grown to include food manufacturers and distributors worldwide.
- The KOF-K is a worldwide organization with an international network of regional coordinators and rabbinic representatives, all of whom are strictly Orthodox in their personal practices and synagogue affiliations. The KOF-K's rabbinic representatives are based in major Jewish communities throughout North America, Europe, Israel, Mexico, and the United States. They visit the principal sources of kosher food and ingredient suppliers. A panel of kashruth experts has carefully reviewed every aspect of a product or service under KOF-K certification.

There are also many regional and local organizations that authorize foods, such as the Chicago Rabbinical Council and the Rabbinical Council of New England. A symbol is included on the packaging to indicate that the food has been certified kosher. The Orthodox Union symbol is the letter *U* inside the letter *O* (see Figure 9.1). The Star-K has the letter K inside a star. If a product is dairy, it is usually indicated by putting the word *dairy* or the letter *D* next to the symbol. Outside the United States and Canada, supervising organizations may operate a little differently.

HALAL

In Arabic, *halal* means lawful or permitted, and *haram* means unlawful or prohibited. These terms are used to refer Muslim dietary laws, which are described in the Qur'an. Items that are not clearly halal or haram are called *mashbooh*, which means doubtful or questionable. These items need to be investigated before being classified as halal or haram.

Those who observe *halal* do not consume swine/pork and its by-products, carnivorous animals, birds of prey, land animals without external ears, most reptiles and insects, blood and blood by-products, and alcohol and intoxicants.

Those animals that are permitted must be slaughtered in the name of Allah by a Muslim in a specific manner. These animals are called *zabihah*.

Figure 9.1 The Orthodox Union logo, used to indicate that food is authorized as kosher. (By permission of the Orthodox Union.)

As with kosher food, there are organizations that certify food as halal, because ingredient lists do not include enough information to determine whether a product is halal or haram. The Islamic Food and Nutrition Council of America (IFANCA) is the largest certifying organization in the United States. The products it certifies are marked with a 1/8 crescent moon with a capital M symbol.

According to *Kosher World*, the kosher products industry has been on a steady growth curve for years, consistently showing a 10–15 percent annual growth. Now at $6.5 billion annually, the industry is attracting all types of consumers. Reaching a broader audience than just the Jewish community, kosher foods are appropriate for large cross section of consumers, ranging from lactose-intolerant to health-conscious people, as well as those of other philosophical beliefs. Both kosher and halal certifying agencies have websites that include recent changes in products they certify and products that are newly certified. Because of the difficulty in determining whether an item is kosher or halal (or both), certifying agencies can be contacted and will answer questions about an item's status.

INTERNET RESOURCES

www.ou.org
Orthodox Union.

www.okkosher.com
OK Kosher Certification.

www.kof-k.com
Kof-K.

www.star-k.org
STAR-K Kosher Certification.

judaism.about.com/cs/shoponline
Lists kosher food distributors.

www.kashrut.com
Information about kashruth; includes list of certification organizations.

www.IFANCA.org
The Islamic Food and Nutrition Council of America.

www.islamicbulletin.com
Includes links to numerous sites that discuss halal.

www.understanding-islam.com
Includes articles that explain halal.

islam.about.com/od/dietarylaw
Information about halal.

CHAPTER

10

Convenience Foods

Convenience foods is a relatively recent term in foodservice. As with all new broad-based terms, its meaning is not always precise. A commonly accepted definition is that convenience foods are those items to which some or all of the labor of preparation has been added at the time of purchase.

Clearly, although this definition is accurate, it is broad enough to cover almost every food item an operator purchases, whether canned, dried, fresh, or frozen. In addition, it can apply to poultry, fish, fruits, vegetables, baked goods, and other food products. For purposes of this chapter, the term *convenience foods* means prepared entrées, customarily frozen, usually packed in multiple portions. There is a vast range of such convenience food products now being packed for foodservice operations. Moreover, there is no longer a problem anywhere in the United States in acquiring them. There are, at this writing, in excess of 100 companies involved in national distribution of frozen foods.

The problem in setting up for the use of convenience products lies in the short shelf life of refrigerated foods, such as prepared salads and desserts. However, a good convenience food program can utilize prepared foods for all aspects of service. Operators need convenience products for different reasons. Those in large cities have abundant potential labor, but it tends to be expensive. City and rural operators have the opposite problem—a lower wage scale, but a small pool of potential workers. Thus, the metropolitan operator needs convenience foods to reduce its labor costs, whereas the rural operator may need them to fill the gap between its labor supply and production needs.

The key to the most effective day-to-day operation for a foodservice organization is work simplification. The logic is simple: The more details to be handled and the more people involved, the greater the opportunities for things to go wrong and to have a higher labor expense. Every job should be reduced to its basic components. Once a job is reduced, tasks should be eliminated from the job through the use of convenience foods and automated equipment. Convenience foods do not usually replace the regular food served by establishments; they supplement them.

EVALUATING A CONVENIENCE FOOD PROGRAM

In evaluating a convenience food program, there are three options to consider. Option 1, described in the next section, is by far the most preferable, though it is practical only for a really high-volume user.

Option 1: Tailored Convenience Foods

To tailor the use of convenience foods to the needs of your operation, you begin by standardizing all recipe cards for on-premise prepared items, not only as to products but also as to procedures. Where possible, specify quality and type of ingredients per USDA specifications. Then calculate, from production records, the number of portions needed and the maximum number of portions for which storage is available. Take these requirements to the nearest two or three producers of convenience foods and ask for bids for these products. This eliminates many problems that occur when convenience foods are purchased without reference to storage limitations. A common problem is packaging size. Companies in national distribution package their products in different packages. Even foodservice operators disagree on a standard packaging size. A commercial cafeteria or public restaurant may well prefer the few portions that can be packed in quarter- or half-size steam table pans because its individual orders come along slowly. The noncommercial high-volume operator, such as an in-plant caterer, hospital, college, or airline, may prefer to use full-size pans. By specifying its own package, an operator can tailor the best method and time of reheating for a specific product with its facilities.

A hospital with microwave ovens at the floor stations will find convenience products delivered in the frozen state perfectly acceptable (so long as the containers are not metallic, of course). However, where a high volume of foods must be served in a short period, a shallower package would be required for a convection oven. If there is enough refrigeration and conventional ovens are to be used, a bulk pack that can be reheated from the refrigerated state, after a long defrost cycle, may be preferred.

Having a product packaged to your specification guarantees a product consistent with your tastes. Among the minimum standards for user convenience, products should be free of disease-producing organisms, list nutritional content, and be accurate as to the manufacturer's claims for number of portions per container. These standards should also be applied to the new frozen, dehydro-frozen, freeze-dried, chilled, or concentrated items that appear on the market almost daily. Manufacturers and purveyors have special problems. For instance, chicken cordon bleu, veal Marsala, stuffed cabbage, and codfish cakes are prepared differently in various regions of the country. What passes for good clam chowder in San Diego may not be acceptable in Boston.

Option 2: General Marketplace

When an operation is not big enough to have its own convenience food manufactured, the operator may enter the general marketplace. The major shortcoming of convenience food packaging, besides varying sizes, is the failure of the container to communicate information about the contents and their use to the average unskilled kitchen employee. There are usually no receiving specifications on boxes to guide the receiver in examining the product for damage. There is no mention of dangerous temperature zones or shelf life. The advertising logo occupies the primary space, even though this space should be allocated to instructions. Sometimes instructions are on the bottom of the box, which means the worker may have to damage the product to read them. Instruction copy is confusing. Few companies provide line drawings. Those manufacturers who do publish product description sheets or merchandising brochures or, more important, nutritional data, generally make them available "upon request only." Such sheets should be included with each case. Instructions should cover the following subjects: storage, including specific and numerical temperature figures, not simply the words "room temperature" or "freezing." Instructions should also include the method of removal from the master container, the number and size of portions, alternate uses of the product and the handling of leftovers, alternate preparation methods and service equipment that can be used, and any other appropriate directions and warnings. The instruction language should be clearly organized. It should be short and simple, with illustrated directions.

Option 3: Make Your Own

A third choice, often termed a "ready-foods" system, is for an operation to manufacture and freeze its own convenience food products on premise. This system uses off-hour time periods to prepare, package, and freeze bulk quantities of foods.

HANDLING CONVENIENCE FOODS

Regardless of the option chosen, every operator is left with the complicated business of reheating. When food is removed from the freezer, it must be carefully handled. Various types of food must be reheated differently. The only rule that always applies is, "Never defrost at room temperature." Instead, defrosting should be done under controlled conditions, either by heating or by controlled refrigeration. If your volume is large enough to warrant it, consider the use of a tempering refrigerator, now available from several manufacturers. Such refrigerators apply heat or refrigeration, as necessary, to maintain any size load within a limited, safe temperature range. Whatever is done about tempering foods (bringing the product from the frozen to the refrigerated state), it must be done carefully. Foods thawed slowly over two or three days generally retain more stability and "eat" better than those thawed quickly. There is no doubt that, if the product permits, a properly adjusted, forced-air convection oven does an effective job in terms of load, efficiency, and cost in reheating frozen convenience foods. If microwave ovens are used, heating times must be adjusted to the thickness or shape of the products. Improper timing can produce mushiness and a steamed or ozone flavor in meat. Microwave ovens make their greatest contribution in bringing refrigerated foods to serving temperatures. Combination convection-microwave ovens, properly used, should offer a solution to many problems. Ovens (deck, convection, revolving, and microwave) are referred to as dry heat. Steamers (high pressure, low pressure, and atmospheric) and conventional and pressure fryers provide moist heat. These are effective, if slower, methods of reheating and/or cooking. Here, too, serving temperature is crucial. What holds true for freshly prepared foods applies equally to convenience foods. Hot food must be served hot, and cold food must be served cold. Any variance from this rule may prove disastrous for the food's palatability, freedom from pathogens, and nutritional value.

Whereas the facts of heat loss for liquids are generally known, new data are needed for most other foods regarding time requirements related to specific states and mass in order to set standards for reheating and holding during service. The possibilities convenience foods offer for increasing productivity are almost limitless. A kitchen planned for total convenience food use may take perhaps half as much space and capital as a conventional kitchen. Labor requirements may be reduced by as much as 80 percent, although 40 percent is more common. Today almost any conceivable entrée is available in convenience form. These items range in quality from unsatisfactory to the finest of haute cuisine. In addition, sauces, sauced and seasoned vegetables, cakes, and fancy pastries abound in the convenience food marketplace.

Although it is difficult to purchase pre-prepared salad plates, it is possible to buy fresh vegetables and fruits already peeled, chopped, diced, sliced, or in any form needed for salad preparation. Salad mixtures, including potato salad, coleslaw, macaroni salad, lobster salad, and tuna fish salad, are readily available. Puddings and gelatins of high quality are also available.

Most foodservice operations are not in a position to adopt a total convenience system with the requisite new equipment. Thus, it is advisable to gradually modify conventional systems when and where possible. If an operation still butchers meat, labor can be effectively reduced by purchasing cuts prefabricated to the specifications of the Meat Buyer's Guides of the National Association of Meat Purveyors. Instead of operating a bakeshop, labor can be eliminated by purchasing commercial fresh or frozen goods. One means of evaluating convenience foods is a three-part objective scale for rating convenience food products on subjective values. It includes three forms, each on a separate card. On the first form, commercial pre-prepared products are rated against an in-house prepared product on six interrelated, yet individual, quality factors. The closer to 0 a commercial product scores, the greater its potential for replacing the house-prepared product. On the second form, the goal is to equal the house product, not necessarily replace it. The third form measures the cost, in labor dollars, of the on-premise product as compared with the tested convenience product. These cards are kept for future reference. The important data on these cards is transferred to a card that records similar data on various manufacturers' samples of a particular product. When an adequate number of samples have been tested, a decision based on comparative quality versus cost and time savings can be made. In this manner it is possible to test products over an extended period of time without a sophisticated research staff in the kitchen. The labor task analyst should keep a list of man-minutes saved by using a selected product.

Such lists should be arranged by category, such as cold products used by salad preparers and hot items prepared by cooks, and the minutes that can be saved through purchase rather than preparation of each product should be recorded. Then, when a position becomes vacant, and the time saved by using convenience foods in that work area is determined to be roughly equal in man-minutes to the output of the newly vacated position, the operator should leave the job empty and begin using the convenience products on the list. The three simple forms can be filled in and maintained quickly and easily. Their use can prevent costly mistakes in customer satisfaction, as well as money. This system, if used as outlined, can keep budget-oriented operations from cost overruns.

CODE OF RECOMMENDED PRACTICES FOR THE HANDLING OF FROZEN FOODS

To safeguard frozen food quality for the consumer, 11 related trade groups some years ago established a code of recommended handling practices. The subjects covered by these practices relate to merchandising aspects of frozen foods. The groups that joined in subscribing to these practices did so in an organized effort to ensure that new technological developments will continually be made available and to update good practices for the care and handling of frozen foods. These recommended practices are based on extensive research in frozen food time-temperature tolerance by the Western Utilization Research and Development Laboratory of the USDA, with which the Refrigeration Research Foundation concurred. These practices do not replace the more demanding company or industry practices that may be in effect. The industry's goal is to maintain reasonably uniform frozen food product temperatures of 0° F or lower and to ensure the proper care of these products, from packer to consumer. The development of these recommendations was based on the principle that voluntary action by industry members would result in more rapid advancement of, and greater attention to, good care and handling practices than would be produced with compulsion by laws and regulations.

Foods for Freezing

1. Raw products should be harvested at optimum maturity, then delivered promptly to the plant, where they should be prepared for freezing, frozen, and packaged (or placed in proper bulk storage) with all reasonable speed. Similarly, frozen products to be used as ingredients in prepared frozen foods should be of best quality for the intended purpose, handled at temperatures of 0° F or lower at all points. They are permitted to thaw only for the time and to the extent necessary for their incorporation into the end product.
2. Similar care should be used by processors without freezing facilities in moving prepared-for-packaging products to refrigerated warehouses for freezing.
3. Where the processor has its own freezer and warehouse, the product leaving the warehouse should be at 0° F or lower.
4. In movement from a processor that freezes but does not have sufficient warehouse space to complete freezing, the product should leave the plant without delay, at 10° F or lower, in an insulated and refrigerated vehicle. Such movement to the primary warehouse for reduction of a temperature to 0° F or lower should not exceed eight hours.
5. Product temperatures should be reduced to 0° F or lower promptly upon reaching the primary warehouse.

Warehouse Equipment

1. Each warehouse should be of adequate capacity and/or should be equipped with suitable mechanical refrigeration to provide, under extreme conditions of outside temperature and peak load conditions, for maintaining an air temperature of 0° F or lower in all rooms in which frozen foods are stored.
2. Each storage room should be equipped with an accurate temperature-measuring device so installed as to reflect correctly the average air temperature of the room. Each day the warehouse is open, the temperature of each room should be recorded and dated, and a file of such temperatures maintained for a period of at least two years.

Warehouse Handling Practices

1. The warehouse operator should record the product temperature of each lot of frozen foods received and should accept custody in accordance with good commercial practice. The operator should retain lot arrival temperature records for a period of at least one year.
2. Whenever frozen foods are received with product temperatures of 15° F or higher, the warehouse person should immediately notify the owner or consignee and request instructions for special handling. These procedures may consist of any available method for effectively lowering temperatures, such as blast freezing, low-temperature rooms with air circulation, and proper use of dunnage or separators in stacking.
3. Before a lot of frozen food is placed in storage, it should be code marked for effective identification.
4. Frozen foods should be moved over dock areas promptly to minimize exposure to elevated temperatures, rainfall, or other adverse weather conditions.
5. If frozen foods are purchased for resale directly to consumers, such products should be stored in the purchaser's own premises so that the purchaser has complete control of all the conditions for which it is responsible in adhering to the code. The first-in, first-out method of inventory control is desirable.
6. During defrosting of overhead coils in storage rooms, stacks of frozen foods should be effectively protected by tarpaulins or other protective covering, or by removal from beneath the coils.
7. Frozen foods going into a separate breakup room for order assembly must be moved out promptly unless the breakup room is maintained uniformly at 0° F or lower.

Transportation

1. All vehicles should be
   (a) Properly constructed and insulated refrigeration units, capable of maintaining a product temperature of 0° F or lower throughout the load in all movements.

(b) Equipped with an appropriate temperature measurement device to indicate accurately the air temperature inside the vehicle. The dial or reading element of the device should be mounted in a readily accessible position outside the vehicle.
(c) Equipped with air-leak-proof cargo spaces, including tight-fitting doors and suitable closures for drain holes to prevent air leakage.
(d) Racked, stripped, baffled, or otherwise so constructed as to provide clearance for air circulation around the load, unless of cold-wall or envelope-type construction.
(e) Entirely free of any dirt, debris, or offensive odors when placed for loading.

2. Route delivery trucks should comply with all the provisions of section 1 and, in addition, should be equipped with curtains or flaps in the doorway area, or with port doors, to minimize loss of refrigeration during delivery stops.
3. Self-refrigerated containers and other self-contained units utilized in making small shipments of frozen food should be so constructed as to give the product adequate protection against physical damage in transit. A unit should be equipped with a refrigerant or refrigerating system capable of maintaining a product temperature of 0° F or lower during the anticipated movement. All such containers should be free of dirt, debris, and offensive odors when offered for loading.

Handling Practices for Line-Haul or Over-the-Road Transportation

1. All vehicles should be precooled to an inside air temperature of 20° F or lower prior to loading and after completing pretripping procedures.
2. Frozen foods should be securely packaged before they are offered for transportation.
3. Product temperature should be 0° F or lower when tendered to a carrier for loading. The carrier should not accept product tendered at a temperature higher than 0° F. The shipper, consignor, or warehouse person should not tender to a carrier any container that has been damaged or defaced to the extent that it cannot be sold.
4. Carriers should provide their personnel with appropriate testing thermometers and instructions in proper procedures to determine that the products they receive are at 0° F or lower. Arrival temperatures of products should be taken inside the vehicle within a reasonable time after arrival and prior to any unloading. However, the carrier must continue to protect the product until such time as the consignee is ready to accept that which the carrier is ready to tender.
5. No product should be loaded in such manner in any vehicle that it will interfere with the free flow of air into or out of the refrigeration unit, or with the free flow around the load in vehicles of other than envelope or cold-wall type construction, or those using freon or liquid refrigerant.
6. Vehicles should be loaded and unloaded within allowable free time as provided for in the governing tariffs to prevent accrual of detention charges.
7. The vehicle's refrigeration unit should be turned on and the doors kept closed during any period when unloading operations cease.
8. The thermostat on the vehicle's refrigeration unit should be set at 0° F or lower.
9. All frozen foods shall be held, tendered, and transported at an air temperature of 0° F or lower, except for defrost cycles, loading and unloading, or other temporary conditions beyond the immediate control of the person under whose care and supervision the frozen food is held. The internal product temperature of frozen food shall be maintained at 0° F as quickly as possible.
10. After loading has been completed and the vehicle doors closed, the carrier's equipment should be checked prior to departure to ensure that the refrigeration system is in proper working order.

Handling Practices for Route Delivery

1. All applicable instructions in the preceding section on line-haul or over-the-road haul transportation should be followed in the case of route delivery.
2. In addition, each lot for individual consignment should be refrigerated by means of mechanical refrigeration, or by any other method of maintaining product temperature.
3. Vehicles or containers should be precooled to a temperature of 20° F or lower before being loaded with frozen foods.
4. Doors of route delivery trucks should be kept closed during any period of loading or unloading operations. In addition, door curtains or flaps should be used during actual unloading, if the vehicle is not equipped with port doors, to minimize loss of refrigeration.

Storage Facilities for Retail Stores

1. Frozen food storage facilities should be capable of maintaining a product temperature of 0° F or lower.
2. Cabinet-type frozen food storage facilities should be defrosted as frequently as necessary to maintain refrigeration efficiency, and should be equipped with an accurate thermometer indicating a representative air temperature.
3. Frozen food storage facilities should have provision for circulation of refrigerated air and should be defrosted as frequently as necessary to maintain refrigeration efficiency. Such facilities should be equipped with an accurate thermometer, the sensing element of which should be located within the upper third of the distance between the floor and ceiling. It should not be placed in a direct blast of air from the cooling unit, cooling coils, heat exchange units, or near the entrance door.

Retail Display Cases

1. Display cases should be capable of maintaining an air temperature of 0° F or lower.
2. Frost on the refrigerated coils and in air passages of display cases should be removed as frequently as necessary to maintain refrigeration efficiency.
3. Each display case should be equipped with an accurate thermometer, the sensing element of which is located within the path of refrigerated air being returned to the evaporator coils.
4. The manufacturer of display refrigerators shall clearly mark on the walls of each refrigerator the proper food-product load levels.
5. Each display case should be equipped with separators to provide false walls to ensure free circulation of refrigerated air around the display product.

Retailer Handling Practices

1. Frozen foods should not be accepted by a retail outlet when the product temperature exceeds 0° F. The department or store manager should approve any product rejection.
2. All frozen foods at a retail outlet should not be held except in frozen-food storage or display cases having the characteristics described in the preceding section.
3. Each retail outlet should be equipped with frozen food storage facilities of sufficient cubic capacity to accommodate those frozen foods (except those to be sold in thawed or semithawed condition) that are not placed directly in a display case at time of delivery.
4. Retailers should not place frozen food products outside the designated load-limit lines.
5. Retail outlets should employ a method of inventory control.

Storage Facilities for Foodservice Installations

1. Frozen food storage facilities should be capable of maintaining an air temperature of 0° F or lower.
2. Total-storage facilities should be of sufficient cubic capacity to easily accommodate frozen foods in quantities anticipated for the operation of the installation, taking into account the frequency of deliveries, probable peak requirements, ordering practices, and related factors.
3. Cabinet-type frozen food storage facilities should be equipped with an accurate thermometer indicating a representative air temperature and should be defrosted as frequently as necessary to maintain refrigeration efficiency.
4. Walk-in-type storage facilities should have provisions for circulation of refrigerated air and should be defrosted as frequently as necessary to maintain refrigeration efficiency. Such facilities should be equipped with an accurate thermometer, the sensing element of which should be located in the upper third of the distance between the floor and ceiling, and away from any entrance door or cooling unit or evaporator coil.

Foodservice Installation and Handling Practices

1. Frozen foods should not be accepted by a foodservice installation when the product delivery temperature exceeds 0° F. The installation manager or his or her assistant should approve any rejection.
2. All frozen foods received at a foodservice installation should be placed promptly in storage facilities having the characteristics described in the preceding section. A product should be removed from storage in quantities sufficient only for immediate use.
3. Foodservice installations should rotate frozen food inventories on a first-in, first-out basis.

Product Temperature

1. Product temperature is that steady temperature that may be determined in two ways. The first is by:
   (a) Opening the top of the case,
   (b) Removing two corner packages,
   (c) Punching a hole through the case wall,
   (d) Proceeding from the inside at a point coincident with the center of the first stack of packages and the first and second layer of packages,
   (e) Inserting the sensing element of an accurate dial thermometer, or other appropriate means of temperature measurement, about 3 in. from the outside so that it will fit snugly between the packages,
   (f) Replacing the two corner packages,
   (g) Closing the case, and
   (h) Placing a couple of cases on top to ensure good contact with the sensing element of the thermometer.

   The second method calls for using a sharp blade and partially cutting out a small section of the case wall in the approximate area of the first stack of packages and the first and second layer of packages, slitting the cut section to allow for insertion of the sensing element, and then proceeding as in the first method.
2. Only when an accurate determination of temperature is impossible without sacrificing packages of frozen foods should representative packages or units be opened to allow for insertion of the sensing element for temperature measurement at the approximate center of the packages in question.
3. All temperature-measuring equipment should be of high quality and subject to periodic checking for accuracy, with the use of methods recommended by the manufacturer.

Note: It is recommended that frozen food handlers concerned with taking temperatures of frozen foods refer to Technical Service Bulletin No. 7, *Frozen Food Temperatures—Their Meaning and Measurement*, by the American Frozen Food Institute. This bulletin outlines in detail the correct methods for taking product temperatures, describes appropriate equipment for the purpose, discusses certain consideration for proper care and handling of frozen food, and cites certain pertinent provisions of the Association of Food & Drug Officials of the U.S. (AFDOUS) Model Frozen Food Code. This code is endorsed and subscribed to by the American Frozen Food Institute, Frozen Potato Products Institute, International Foodservice Manufacturers Association, National Fisheries Institute, National Association of Refrigerated Warehouses, National Food Brokers Association, National Frozen Food Association, National Institute of Locker and Freezer Provisioners, National Prepared Frozen Food Processors Association, and National Restaurant Association.

SAMPLE MEAT PRODUCTS SPECIFICATION

(All percentages of meat are given on the basis of fresh, uncooked weight unless otherwise indicated.)

Baby Food. High Meat Dinner: At least 30 percent meat. Meat and Broth: At least 65 percent meat. Vegetable and Meat: At least 8 percent meat.

Bacon (cooked). Weight of cooked bacon cannot exceed 40 percent of cured, smoked bacon.

Bacon and Tomato Spread. At least 20 percent cooked bacon.

Bacon Dressing. At least 8 percent cured, smoked bacon.

Barbecued Meats. Weight of meat when barbecued cannot exceed 70 percent of that of fresh uncooked meat. Must have barbecued (crusted) appearance and be prepared over burning or smoldering hardwood or its sawdust. If cooked by other dry-heat means, product name must mention the type of cookery.

Barbecue Sauce with Meat. At least 35 percent meat (cooked basis).

Beans and Meat in Sauce. At least 20 percent meat.

Beans in Sauce with Meat. At least 20 percent cooked, or cooked and smoked meat.

Beans with Bacon in Sauce. At least 12 percent bacon.

Beans with Frankfurters in Sauce. At least 20 percent frankfurters.

Beans with Meatballs in Sauce. At least 20 percent meatballs.

Beef and Dumplings with Gravy (or Beef and Gravy Dumplings). At least 25 percent beef.

Beef and Mushrooms. At least 25 percent beef, 7 percent mushrooms.

Beef and Pasta in Tomato Sauce. At least $17\frac{1}{2}$ percent beef.

Beef Burger Sandwich. At least 35 percent hamburger (cooked basis).

Beef Burgundy. At least 50 percent beef; enough wine to characterize the sauce.

Beef Carbonade. At least 50 percent beef tenderloin.

Beef Sausage (raw). No more than 30 percent fat. No by-products, no extenders.

Beef Stroganoff. At least 45 percent fresh uncooked beef or 30 percent cooked beef, and at least 10 percent sour cream or a "gourmet" combination of at least $7\frac{1}{2}$ percent sour cream and 5 percent wine.

Beef with Barbecue Sauce. At least 50 percent beef (cooked basis).

Beef with Gravy. At least 50 percent beef (cooked basis). [*Gravy with Beef.* At least 35 percent beef (cooked basis).]

Breaded Steaks, Chops, and the like. Breading cannot exceed 30 percent of finished product weight.

Breakfast (frozen product containing meat). At least 15 percent meat (cooked basis).

Breakfast Sausage. No more than 50 percent fat.

Brown and Serve Sausage. No more than 35 percent fat, and no more than 10 percent added water.

Brunswick Stew. At least 25 percent of at least two kinds of meat and/or poultry. Must contain corn as one of the vegetables.

Burgundy Sauce with Beef and Noodles. At least 25 percent beef (cooked basis); enough wine to characterize the sauce.

Burritos. At least 15 percent meat.

Cabbage Rolls with Meat. At least 12 percent meat.

Cannelloni with Meat and Sauce. At least 10 percent meat.

Cappelletti with Meat in Sauce. At least 12 percent meat.

Chili con Carne. At least 40 percent meat.

Chili con Carne with Beans. At least 40 percent meat in chili.

Chili Macaroni. At least 16 percent meat.

Chili Pie. At least 20 percent meat; filling must be at least 50 percent of the product.

Chili Sauce with Meat or *Chili Hot Dog Sauce with Meat.* At least 6 percent meat.

Chopped Ham. Must be prepared from fresh, cured, or smoked ham, plus certain kinds of curing agents, and seasonings. May contain dehydrated onions, dehydrated garlic, corn syrup, and not more than 3 percent water to dissolve the curing agents.

Chop Suey Vegetables with Meat. At least 12 percent meat.

Chorizos Empanadillos. At least 25 percent fresh chorizos, or 17 percent dry chorizos.

Chow Mein Vegetables with Meat. At least 12 percent meat.

Chow Mein Vegetables with Meat and Noodles. At least 8 percent meat, and the chow mein must equal two-thirds of the product.

Condensed, Creamed Dried Beef, or *Chipped Beef.* At least 18 percent dried or chipped beef (figured on total content).

Corned Beef and Cabbage. At least 25 percent corned beef (cooked basis).

Corned Beef Hash. At least 35 percent beef (cooked basis). Must contain potatoes, curing agents, and seasonings. May contain onions, garlic, beef broth, beef fat, or other fat. No more than 15 percent fat; no more than 72 percent moisture.

Corn Dog. Must meet standards for frankfurters, and batter cannot exceed the weight of the frankfurter.

Country Ham. A dry-cured product frequently coated with spices.

Cracklin' Corn Bread. At least 10 percent cracklings (cooked basis).

Cream Cheese with Chipped Beef (sandwich spread). 12 percent chipped beef.

Crepes. At least 20 percent meat (cooked basis), or 10 percent meat (cooked basis) if the filling has other major ingredient, such as cheese.

Croquettes. At least 35 percent meat.

Curried Sauce with Meat and Rice (casserole). At least 35 percent meat (cooked basis) in the sauce and meat part; no more than 50 percent cooked rice.

Deviled Ham. No more than 35 percent fat.

Dinners (frozen product containing meat). At least 25 percent meat or meat food product (cooked basis) figured on total meal minus appetizer, bread, and dessert. Minimum weight of a consumer package is 10 oz.

Dumplings and Meat in Sauce. At least 18 percent meat.

Egg Foo Yong with Meat. At least 12 percent meat.

Egg Rolls with Meat. At least 10 percent meat.

Enchilada with Meat. At least 15 percent meat.

Entrées. Meat or Meat Food Product and One Vegetable must have at least 50 percent meat or meat food product (cooked basis). Meat or Meat Food Product, Gravy or Sauce, and one Vegetable must have at least 30 percent meat or meat food product (cooked basis).

Frankfurter, Bologna, and similar Cooked Sausage. May contain only skeletal meat. No more than 30 percent fat, 10 percent added water, and 2 percent corn syrup. No more than 15 percent poultry meat (exclusive of water in formula).

Frankfurter, Bologna, and similar Cooked Sausage with By-products or Variety Meats. Same limitations as in preceding entry for fat, added water, and corn syrup. Must contain at least 15 percent skeletal meat. Each by-product or variety meat must be specifically named in the list of ingredients. These include heart, tongue, spleen, tripe, stomach, and the like.

Frankfurter, Bologna, and similar Cooked Sausage with By-products or Variety Meats and which also contain nonmeat binders. Product made with the formula in the preceding entry and also containing up to $3\frac{1}{2}$ percent nonmeat binder (or 2 percent isolated soy protein). These products must be distinctively labeled, such as "Frankfurters with by-products, nonfat dry milk added." The binders must be named in their proper order in the list of ingredients.

Fried Rice with Meat. At least 10 percent meat.

Fritters. A breaded product. At least 35 percent meat.

German Style Potato Salad with Bacon. At least 14 percent bacon (cooked basis).
Goulash. At least 25 percent meat.
Gravies. At least 25 percent meat stock or broth or at least 6 percent meat.
Ham, Canned. Limited to 8 percent total weight gain after processing.
Ham, Cooked or Cooked and Smoked (not canned). Must not weigh more after processing than the fresh ham weighs before curing and smoking; if it contains up to 10 percent added weight, must be labeled "Ham, Water Added"; if more than 10 percent, must be labeled "Imitation Ham."
Ham à la King. At least 20 percent ham (cooked basis).
Ham and Cheese Spread. At least 25 percent ham (cooked basis).
Hamburger, Hamburg, Burger, Ground Beef, or Beef. No more than 30 percent fat; no extenders.
Ham Chowder. At least 10 percent ham (cooked basis).
Ham Croquettes. At least 35 percent ham (cooked basis).
Ham Salad. At least 35 percent ham (cooked basis).
Ham Spread. At least 50 percent ham.
Hash. At least 35 percent meat (cooked basis).
Hors d'Oeuvre. At least 15 percent meat or 10 percent bacon (cooked basis).
Jambalaya with Meat. At least 25 percent meat (cooked basis).
Knishes. At least 15 percent meat (cooked basis), or 10 percent bacon (cooked basis).
Kreplach. At least 20 percent meat.
Lasagna with Meat and Sauce. At least 12 percent meat.
Lasagna with Sauce, Cheese, and Dry Sausage. At least 8 percent dry sausage.
Liver Products (such as Liver Loaf, Liver Paste, Liver Pate, Liver Cheese, Liver Spread, and Liver Sausage). At least 30 percent liver.
Macaroni and Beef in Tomato Sauce. At least 12 percent beef.
Macaroni and Meat. At least 25 percent meat.
Macaroni Salad with Ham or Beef. At least 12 percent meat (cooked basis).
Manicotti (containing meat filling). At least 10 percent meat.
Meat and Dumplings in Sauce. At least 25 percent meat.
Meat and Seafood Egg Roll. At least 5 percent meat.
Meat and Vegetables. At least 50 percent meat.
Meatballs. No more than 12 percent extenders (cereal, including textured vegetable protein). At least 65 percent meat.
Meatballs in Sauce. At least 50 percent meatballs (cooked basis).
Meat Casseroles. At least 25 percent fresh uncooked meat, or 18 percent cooked meat.
Meat Curry. At least 50 percent meat.
Meat Loaf (baked or oven-ready). At least 65 percent meat, and no more than 12 percent extenders including textured vegetable protein.
Meat Pasty. At least 25 percent meat.
Meat Pies. At least 25 percent meat.
Meat Ravioli. At least 10 percent meat.
Meat Ravioli in Sauce. At least 10 percent meat.
Meat Salads. At least 35 percent meat (cooked basis).
Meat Shortcake. At least 25 percent meat (cooked basis).
Meat Soups. Ready-to-Eat: At least 5 percent meat. Condensed: At least 10 percent meat.
Meat Spreads. At least 50 percent meat.
Meat Taco Filling. At least 40 percent meat.
Meat Tacos. At least 15 percent meat.
Meat Turnovers. At least 25 percent meat.
Meat Wellington. At least 50 percent cooked tenderloin spread with a liver pâté or similar coating and covered with not more than 30 percent pastry.
Mincemeat. At least 12 percent meat.
Oleomargarine or Margarine. If product is entirely of fat, or contains some animal fat, it is processed under federal inspection. Must contain, individually or in combination, pasteurized cream, cow's milk, skim milk, a combination or nonfat dry milk and water, or finely ground soybeans and water. May contain butter, salt, artificial coloring, vitamins A and D, and permitted functional substances. Finished product must contain at least 80 percent fat. Labels must clearly state which types of fat are used.
Omelet with Bacon. At least 9 percent bacon (cooked basis).
Omelet with Dry Sausage or with Liver. At least 12 percent sausage or liver (cooked basis).
Omelet with Ham. At least 18 percent ham (cooked basis).
Pan Haus. At least 10 percent meat.
Pâté de Foie. At least 30 percent liver.
Peppers and Italian (type) Sausage in Sauce. At least 20 percent sausage (cooked basis).
Pepper Steak. At least 30 percent beef (cooked basis).
Petcha. At least 50 percent calves feet.
Pizza Sauce with Sausage. At least 6 percent sausage.
Pizza with Meat. At least 15 percent meat.
Pizza with Sausage. At least 12 percent sausage (cooked basis), or 10 percent dry sausage, such as pepperoni.
Park Sausage. No more than 50 percent fat; may contain no by-products or extenders.
Pork with Barbecue Sauce. At least 50 percent pork (cooked basis).

Pork with Dressing and Gravy. At least 30 percent pork (cooked basis).
Prosciutto. A flat, dry-cured ham coated with spices.
Salisbury Steak. At least 65 percent meat, and no more than 12 percent extenders including textured vegetable protein.
Sandwiches (containing meat). At least 35 percent meat in total sandwich; filling must be at least 50 percent of the sandwich.
Sauce with Chipped Beef. At least 18 percent chipped beef.
Sauce with Meat or Meat Sauce. At least 6 percent meat.
Sauerbraten. At least 50 percent meat (cooked basis).
Sauerkraut with Wieners and Juice. At least 20 percent wieners.
Scalloped Potatoes and Ham. At least 20 percent ham (cooked basis).
Scaloppine. At least 35 percent (cooked basis).
Scrambled Eggs with Ham in a Pancake. At least 9 percent cooked ham.
Scrapple. At least 40 percent meat and/or meat byproducts.
Shepherd's Pie. At least 25 percent meat; no more than 50 percent mashed potatoes.
Sloppy Joe (sauce with meat). At least 35 percent meat (cooked basis).
Snacks. At least 15 percent meat (cooked basis), or 10 percent bacon (cooked basis).
Spaghetti with Sliced Frankfurters and Sauce. At least 12 percent frankfurters.
Spanish Rice with Beef or Ham. At least 20 percent beef or ham (cooked basis).
Stews. At least 25 percent meat.
Stuffed Cabbage with Meat in Sauce. At least 12 percent meat.
Stuffed Peppers with Meat in Sauce. At least 12 percent meat.
Sukiyaki. At least 30 percent meat.
Sweet and Sour Pork or Beef. At least 25 percent meat and 16 percent fruit.
Sweet and Sour Spareribs. At least 50 percent bone-in spareribs.
Swiss Steak with Gravy. At least 50 percent meat (cooked basis).
[*Gravy and Swiss Steak.* At least 35 percent meat (cooked basis).]
Tamale Pie. At least 20 percent meat.
Tamales. At least 25 percent meat.
Tamales with Sauce (or with Gravy). At least 20 percent meat.
Taquitos. At least 15 percent meat.
Tongue Spread. At least 50 percent tongue.
Tortellini with Meat. At least 10 percent meat.
Veal Birds. At least 60 percent meat, and no more than 40 percent stuffing.
Veal Cordon Bleu. At least 60 percent veal, 5 percent ham, and containing Swiss, Gruyère, or Mozzarella cheese.
Veal Fricassee. At least 40 percent meat.
Veal Parmagiana. At least 40 percent breaded meat product in sauce.
Veal Steaks. Can be chopped, shaped, cubed, frozen. Beef can be added with product name shown as "Veal Steaks, Beef Added, Chopped, Shaped, and Cubed" if no more than 20 percent beef, or must be labeled.
Veal and Beef Steak, Chopped and Cubed. No more than 30 percent fat.
Vegetable and Meat Casserole. At least 25 percent meat.
Vegetable and Meat Pie. At least 25 percent meat.
Vegetable Stew and Meatballs. At least 12 percent meat in total product.
Won Ton Soup. At least 5 percent meat.

INTERNET RESOURCES

www.affi.com
American Frozen Food Institute (AFFI).

www.ifmaworld.com
International Foodservice Manufacturers Association.

www.nfi.org
National Fisheries Institute. Represents companies in the fish and seafood industry. Includes Glossary, Species Guide, Top Ten Seafoods, Recipes and Nutrition, and Industry Links.

www.hvacmall.com
International Association of Refrigerated Warehouses (IARW). Directory of heating, ventilating, air-conditioning, and refrigeration companies.

www.restaurant.org
National Restaurant Association.

CHAPTER

11

Miscellaneous Groceries

Miscellaneous groceries take up a large percentage of the total time required for food purchasing. Although a few of the products in this classification have federal specifications, for example, olives and pickles, some have only trade and industry specifications. We have tried to include in this chapter some miscellaneous grocery items that are available on the American market and all the information available about these products, yet the number of items grows daily.

In selecting products from this group, the same basic purchasing principles prevail. The buyer needs accurate knowledge as to the final use that will be made of the product to be purchased and information as to the products that are available, their descriptions, and the range of quality offered for selection.

SALAD DRESSINGS

A salad dressing is an emulsified, semisolid food prepared from oils, acidifying ingredients, one or more egg-containing ingredients, and a cooked starchy paste prepared with food-grade starches. Salad dressings may be seasoned or flavored with one or more of the following ingredients: salt, sugar, mustard, paprika, other spice, or spice oil. An exception is that no turmeric or saffron is used and no spice oil or spice extract is used that imparts to the salad dressing a color simulating the color imparted by egg yolk. They contain not less than 30 percent by weight of vegetable salad oil and not less than 4 percent by weight of liquid egg yolk.

Mayonnaise—An emulsified, semisolid food prepared from edible vegetable oil, acidifying ingredients, and one or more of egg yolk–containing ingredients, and flavored with salt, sugar, spices, and any suitable imitations. Contains no vegetable oil.

Sandwich Spread—Mayonnaise or salad dressing base, usually with the addition of pickle relish and may or may not contain cooked starch paste.

Refrigerated Dressing—An emulsified, semisolid food of a perishable nature that requires refrigerated temperatures at all times. May contain ingredients such as blue cheese, Roquefort cheese, sour cream, or pickle relish, in combination with mayonnaise and salad dressing.

Spoon-Type—All other emulsified, semisolid types prepared with either a mayonnaise or salad dressing base in combination with other suitable food-grade ingredients.

French Dressing—A separable, liquid food or emulsified, viscous, fluid food prepared from a mixture of vinegars, and optional acidifying acid. May be seasoned or flavored with one or more of the following ingredients: salt, sugar, mustard, spice oils, monosodium glutamate, and any oil, vinegar, paprika, spice or suitable harmless food seasoning other than imitations, and tomato paste or tomato puree, and emulsifying ingredients. Contains not less than 35 percent by weight of vegetable oil.

Oil and Vinegar—A separable, clear liquid meeting the requirements of a French-type dressing, not emulsified or containing paprika, tomato paste, or tomato puree. Includes Italian garlic, herb, and spice liquid dressing.

Cheese—An emulsified, pourable French dressing with cheese and egg yolk added.

Low Calorie and Dietetic—A separable, liquid food that may or may not be emulsified. Contains selected ingredients and is normally a product having low caloric content.

Pourable-Type—All other separable, liquid dressings containing selected ingredients that may or may not be emulsified, with a French-type dressing base.

FATS AND OILS

Crude fats and oils, as they are obtained from vegetable or animal sources, contain varying but relatively small amounts of naturally occurring, nonglyceride materials. Many of these substances, such as free fatty acids, phosphatides, mucilaginous substances, and resins, are undesirable elements in that they contribute color, flavor, odor, instability, foaming, and other unwanted characteristics to the fat. These materials are removed through a series of processing steps referred to as "refining, bleaching, and deodorization." It should be pointed out, however, that not all the nonglyceride materials are undesirable elements. Tocopherols, for example, perform the important function of protecting the oils from oxidation and provide vitamin E. Processing is usually carried out in such a way as to encourage retention of these substances. Hydrogenation is frequently employed to improve the keeping qualities of fats and oils and to provide increased usefulness by imparting a semisolid consistency to the fat for many food applications. Most nutritionists and food technologists generally agree that the modern processing of edible fats and oils is the single factor most responsible for upgrading the quality of the fat consumed in the U.S. diet today.

The term *refining* generally refers to any purifying treatment that is intended to remove free fatty acids, phosphatides, proteinaceous and mucilaginous substances, or other gross impurities from the fat or oil. By far the most important and widespread method of refining is by treatment of the fat or oil with an alkali—usually caustic soda or soda ash. This results in almost complete removal of free fatty acids through their conversation into oil-insoluble soaps.

The term *bleaching* refers to the treatment that is given to remove color-producing substances and to further purify the fat or oil. It is normally accomplished after the oil has been refined. The usual method of bleaching is by adsorption of the color-producing substances on an adsorbent material. Bleaching earth or clay, sometimes referred to as fuller's or diatomaceous earth, is the adsorbent material that has been used most extensively. These substances consist primarily of hydrated aluminum silicate. Activated carbon is also used as a bleaching adsorbent to a limited extent.

Deodorization is the treatment of fats and oils for the purpose of removing trace constituents that give rise to undesirable flavors and odors. This is normally accomplished after refining and bleaching. The deodorization of fats and oils is simply a removal of the relatively volatile components from the fat or oil with the use of live steam. This is feasible because of the great differences in volatility between the substances that give flavors and odors to the fats and oils, and the triglycerides. Normally, deodorization is carried out under reduced pressures to facilitate the removal of the volatile substances, to avoid undue hydrolysis of the fat, and to make the most efficient use of the steam. Deodorization, as it is usually accomplished by the industry, does not have any significant effect on the fatty acid composition of the fat or oil. In the case of vegetable oils, most of the tocopherols remain in the finished oils after deodorization.

Hydrogenation is the process by which hydrogen is added directly to points of unsaturation in the fatty acids. Hydrogenation of fats has developed as a result of a need:

1. To convert liquid oils to the semisolid form for greater utility in certain food uses, and
2. To increase the stability of the fat or oil to prevent rancidity.

In the process of hydrogenation, a fat or oil is treated with hydrogen gas at a suitable temperature and pressure in the presence of a catalyst. The catalyst is frequently finely divided nickel which is, of course, removed from the food after the hydrogenation processing is completed. The hydrogenation process is easily controlled and can be stopped at any desired point. As hydrogenation progresses, there is a gradual increase in the melting point of the fat or oil. If the hydrogenation of cottonseed or soybean oil, for example, is stopped after only a small amount of hydrogenation has taken place, the oil remains liquid. More hydrogenation produces soft, but solid-appearing fats that still contain appreciable saturated fatty acids. This degree of hydrogenation is frequently employed in the preparation of shortening and margarine. However, if hydrogenation is continued considerably beyond this point, a fat results that is too hard to produce an acceptable shortening. If an oil were completely hydrogenated, the unsaturated double bonds would be entirely eliminated, and the resulting product would be a hard, brittle solid at room temperature, composed entirely of saturated triglycerides. The hydrogenation conditions can be varied by the manufacturer to meet the characteristics desired in the finished selection, including the proper temperature, time, catalyst, and starting oils. Both positional and geometric isomers are formed to some extent during hydrogenation, the amount depending on the conditions chosen.

The Process of Frying

When fresh oil is heated in a deep-fat fryer and the frying operation begins, three general chemical reactions occur simultaneously:

1. *Hydrolysis*—This is the reaction of the oil with moisture wherein the fatty acids are separated from the glycerol-forming free fatty acids. This change, when analyzed, is expressed as the percentage of free fatty acids. A fresh oil will have a free fatty acid level of about 0.05 percent or less, but during the course of frying this value may go as high as 5 percent, if the oil is badly abused and broken down. Usually a value of 1 percent free fatty acids is considered maximum for most frying operations. The smoke point of a fresh oil is usually in excess of 425° F. As the percentage of the free fatty acids rises, the smoke point correspondingly goes down, so that at 0.8 percent free fatty acids, the smoke point is about 325° F. Metals containing copper will cause an appreciable increase in the percentage of free fatty acids. Drip-back from the exhaust system will also cause increases in fatty acids.
2. Oxidation—Frying oils have a bland flavor, but upon heating and frying become less bland and develop burned, scorched, or off-flavors through abuse. During the frying operation, a blanket of steam arising from the foods being fried protects the surface of the oil. During slack periods, air comes in contact with the surface of the oil, causing deterioration of the oil by oxidation. Reducing the temperature of the oil to 150°–200° F when it is not being used will reduce the rate of oxidation. Aeration of the oil by allowing a pump to pull air through or allowing the oil to cascade from one area to another, results in rapid oxidation. It is best to have the oil enter the fryer or holding tank near the bottom to minimize the aeration. Holding tanks should be designed, if possible, to be vertical with a small surface area, rather than horizontal or flat with a large surface area. Light catalyzes the development of off-flavors in oil and should be excluded wherever practical.
3. Polymerization—The union of two or more molecules of oil to form a larger molecule is known as polymerization. Thermal polymers cannot be detected by odor or flavor as can oxidative polymers. The decrease in iodine number, a measure of unsaturation or double bond, is an indication of polymer formation. Polymer formations are visible on frying equipment that is not properly cleaned. This polymerized material should be removed at frequent intervals, and the proper technique used in filtering the oil to remove as much of these polymers as possible.

Products Prepared from Fats and Oils

A wide variety of products based on edible fats and oils are offered to the consuming public. Shortening, margarine, butter, salad and cooking oils, and other specialty salad dressings are some of the widely available products that are either based entirely on fats and oils or contain fat or oil as a principal ingredient. Many of these products are also sold in commercial quantities to food processors, bakeries, restaurants, and institutions.

Salad and Cooking Oils

Salad and cooking oils are prepared from vegetable oils that are usually refined, bleached, deodorized, and sometimes lightly hydrogenated. Cottonseed, corn, and soybean are the principal oils sold in this form, although peanut, safflower, and olive oils are also used. Because of the presence of substantial quantities of linoleic acid naturally occurring in soybean oil, it is frequently desirable to lightly hydrogenate this oil when it is to be used as a salad and cooking oil.

Shortenings

Shortenings are fats used in the preparation of many foods. Because they impart a "short" or tender quality to baked goods, they are called shortenings. For many years, lard and other animal fats were the principal edible fats used in this country, but during the last third of the nineteenth century large quantities of cottonseed oil became available as a by-product from the growing of cotton. The first semisolid shortenings using vegetable oil were prepared by blending liquid cottonseed oil with certain animal fats. These were referred to as "compound" shortenings. The introduction of the hydrogenation process in this country, in the early 1900s, initiated a new era in the manufacture of shortening. For the first time semisolid shortenings could be prepared entirely from vegetable oils, and shortening manufacturers were no longer restricted by the availability of animal fats. Many types of vegetable oils, including soybean, cottonseed, corn, and others, are now used in shortening products.

Hydrogenated shortenings may be made from a single hydrogenated fat, but they are usually made from blends of two or more hydrogenated fats. The conditions and extent of hydrogenation may be varied for each to achieve the characteristics desired. Thus, in the manufacture of hydrogenated shortenings, considerable flexibility is possible, providing a wide choice of finished product characteristics. Until 1961 most hydrogenated vegetable shortenings were processed under conditions that substantially reduced the polyunsaturate content of the fats to levels ranging from 5 to 12 percent. These products had excellent consumer acceptance and were noted for their high degree of stability. Most manufacturers of vegetable shortenings are now producing these products with substantially higher levels of polyunsaturated fatty acids, in response to research findings that suggest the advisability of a greater intake of these fatty acids. These shortenings contain 22 to 33 percent polyunsaturated fatty acids. Some dieticians have referred to these products as "special" shortenings because, in addition to their use in normal diets, they are also suitable for use in "special" diets where an increased level of linoleic acid is desired. Lard and other animal fats and mixtures of animal and vegetable fats are also used in shortening.

Mixtures of animal and vegetable fats are frequently hydrogenated to some extent to obtain the physical characteristics desired. Lard is used extensively in some commercial applications such as pastry and bread. Some shortening manufacturers have developed and marketed liquid shortening products that are based on liquid or lightly hydrogenated liquid vegetable oils. These products have polyunsaturated fatty acid contents ranging from 30 to 50 percent and iodine values ranging from 95 to 125. These products have also had usage in commercial bread and other baking applications.

Margarine

Margarines are fatty foods blending fats and oils with other ingredients such as milk solids, salt, flavoring materials, and vitamins A and D. By federal regulation, margarine must contain at least 80 percent fat. The fats used in margarine may be of either animal or vegetable origin, although vegetable oils are by far the most widely used. The fats may be prepared from single hydrogenated fats, from two or more hydrogenated fats, or from a blend of hydrogenated fat and an unhydrogenated oil. Many manufacturers of margarine are now producing these products with substantially higher levels of polyunsaturated fatty acids, in response to medical research findings that suggest the advisability of higher levels of these fatty acids in the diet. These margarines contain fats having 22–60 percent polyunsaturated fatty acids and iodine values ranging from 92 to 130. In contrast, regular margarines contain 10–20 percent polyunsaturated fatty acids and have iodine values from 78 to 90. Some dieticians have referred to these products high in polyunsaturated fatty acids as "special" margarines because, in addition to their use in normal diets, they are also suitable for use in special diets where an increased level of linoleic acid is desired.

Cooking Oils

Although fried foods are generally cooked at 375° F or less, it is important to specify frying oil with a smoking point of no less than 426° F. Hydrogenated vegetable shortenings, peanut oil, and soybean oils meet this standard. Corn oil and animal fats smoke at lower temperatures. The following fats and oils are commonly used in foodservice operations:

Corn Oil. Refined oil from corn germ. Used in salad oils and for deep frying. High smoking point.

Cottonseed Oil. Refined oil from cottonseed. Used in making margarine, salad oils, and vegetable cooking fats. High smoking point.

Hydrogenated Fats. Vegetable oils, usually cottonseed oil, solidified by hydrogen. Used as shortening. Smoking point is 440°–460° F.

Lard. Melted pig fat. Best is from the abdomen or around the kidneys. The highest grade is kettle-rendered leaf lard. Used firm when cold. Kettle-rendered lard is derived from back and leaf fat. Prime stream lard is derived from fats taken when preparing cuts for market; most lard is of this type.

Olive Oil

Virgin olive oil—Taken from ripe olives under low pressure.

Second grade olive oil—Most olive oil. Taken under pressure from the remaining pulp after the first press.

Third grade olive oil—The press cake remaining from the second pressing is reheated in water and then pressed.

Bleached olive oil—Dark oil bleached to resemble better-grade oils.

Cloudy olive oil—May become thick and white when stored in a refrigerator. Will melt if kept at room temperature.

French olive oil—Fruity taste and golden color. Used in mayonnaise, salad dressings, batters, marinades, and for frying foods.

Grade A—Color: Typical greenish to light yellow color. Free fatty acid content: Not more than 1.4 percent calculated as oleic acid. Defects: Entirely free of cloudiness at 60° F, due to stearin, and free of sediment. Odor: Has a typical olive oil odor and is practically free of off-odors of any kind. Flavor: Has a typical olive oil flavor and is practically free of off-flavors of any kind.

Grade B—Color: Typical greenish to light yellow color. Free fatty acid content: Between 1.4 percent and 2.5 percent, calculated as oleic acid. Defects: Reasonably free of cloudiness at 60° F, due to stearin, and from sediment; and no water or other liquid immiscible with the olive oil is present. Odor: Has a typical olive oil odor and is reasonably free of off-odors of any kind. Flavor: Has a typical olive oil flavor and is reasonably free of off-flavors of any kind.

Grade C—Color: Typical greenish to light yellow color. Free fatty acid content: Between 2.5 percent and 3 percent, calculated as oleic acid. Defects: May have cloudiness at 60° F, due to stearin, and may have sediment, but these may not impair the quality of the product; and no water or other liquid immiscible with the olive oil may be present. Odor: Has a typical olive-oil odor and is fairly free of off-odors of any kind. Flavor: Has a typical olive oil flavor and is fairly free of off-flavors of any kind.

Peanut Oil—Oil taken from peanuts and used in cooking and salad dressings.

Suet—Fat taken from around the kidneys and loins of sheep and bullocks. Bought in lumps or shreds. Used in mincemeat and other products.

NUTS

Almond

There are two kinds of almonds (*Prunus amygdalus*): hard shell and soft shell. They are in season in September. They can be eaten plain and used in baking, cooking, and confectionery. Sizes: No. 10 cans (4 lb), 5 lb boxes, 25 lb boxes, 100 lb drum, 160 lb bag, 200 lb drum.

Almonds in the Shell

U.S. No. 1 Almonds are similar in shape and degree of hardness of the shells to each other (bitter almonds are not mixed with sweet almonds); are free of loose foreign material; have shells that are clean, fairly bright, fairly uniform in color, and free of damage due to discoloration, adhering hulls, broken shells, or other means; and have kernels that are well dried, free of decay, rancidity, and damage caused by insects, mold, gum, shriveling, discoloration, or other means; and are not less than 28/64 inch in thickness. Tolerances: Shell defects—10 percent; dissimilar varieties—5 percent, including not more than 1 percent for bitter almonds mixed with sweet almonds; size—5 percent; foreign material—2 percent, including not more than 1 percent that can pass through a round opening 24/64 in. in diameter; kernel defects—10 percent.

U.S. No. 1 Mixed: Meet the requirements of U.S. No. 1 grade, except that two or more varieties of sweet almonds are mixed.

U.S. No. 2: Same as U.S. No. 1, except that an additional tolerance of 20 percent is allowed for almonds with shells damaged by discoloration.

U.S. No. 2 Mixed: Meet the requirements of U.S. No. 2 grade, except that two or more varieties of sweet almonds are mixed.

Unclassified: No definite grade is applied.

Almonds, Shelled

U.S. Fancy: Similar varietal characteristics that are whole, clean, and well dried, and that are free of decay, rancidity, insect injury, foreign material, doubles (shells containing two kernels), split or broken kernels, injury caused by chipped and scratched kernels, and damage caused by any means.

U.S. Extra No. 1: Same as U.S. Fancy, except for increased tolerances.

U.S. Select Sheller Run: Same as U.S. Fancy, except for increased tolerances.

U.S. Standard Sheller Run: Same as U.S. Fancy, except for increased tolerances.

U.S. No. 1 Whole and Broken: Have similar varietal characteristics; are clean and well dried and are free of decay, rancidity, insect injury, foreign material, doubles, particles, dust, and damage caused by any means. Not less than 30 percent of the kernels are whole, and the minimum diameter shall not be less than 20/64 in.

U.S. No. 1 Pieces: Not bitter; clean and dried and free of decay, rancidity, insect injury, foreign material, particles, dust, and damage caused by any means. The minimum diameter shall not be less than 8/64 in.

Mixed Varieties

A mixture of two or more dissimilar varieties that meet the other requirements of the grades shall be designated as U.S. No. 1 Mixed, U.S. Select Sheller Run Mixed, U.S. Standard Sheller Run Mixed, or U.S. No. 1 Whole and Broken Mixed. No lot may contain more than 1 percent of bitter almonds mixed with sweet almonds.

Brazil Nut

The Brazil nut (*Bertholletia incisa* or *Treculia africana*) is also known as butternut and cream nut. Brazil nuts become rancid quickly.

Brazil Nuts in the Shell, Fresh

Grade: U.S. No. 1: Well-cured whole Brazil nuts that are free of loose, foreign material and meet one of the size classifications given later in this section. The shells are clean and free of damage caused by splits, breaks, punctures, oil stain, mold, or other means and contain kernels that are reasonably well developed, free of rancidity, mold, decay, and damage caused by insects, discoloration, or other means.

PERMITTED TOLERANCES

For Defects of Shell—10 percent may fail to meet the requirements of the grade, including not more than 6 percent for serious damage.

For Defects of the Kernel—10 percent may fail to meet the requirements of the grade, including not more than 7 percent for serious damage, provided that not more than 5 percent is allowed for damage by insects, including not more than 0.5 percent with live insects in the shell.

For Foreign Material—1 percent.

SIZE CLASSIFICATIONS

Extra Large—Not more than 15 percent pass through a round opening 78/64 in. in diameter, or count does not exceed 45 nuts per pound.

Large—Not more than 15 percent pass through a round opening 73/64 in. in diameter, or count does not exceed 50 nuts per pound.

Medium—Not more than 15 percent pass through a round opening 59/64 in. in diameter, or count is 51 to 65 nuts per lb.

Carob

Carob (*Ceratonia siliqua*) is also known as St. John's bread. Can be eaten plain soon after harvesting but becomes brittle and acidic after storage. Milled carob is used in baking as a filler.

Cashew

Cashews (*Anacardium occidentale*) can be eaten plain and used in baking, confectionery, and cooking; sometimes available in butter form or paste. Size: 25 lb tins.

Chestnut

Chestnuts (*Castanea sativa*) are usually available dried. They are in season in November and are used in cooking, baking, and confectionery.

Filberts in the Shell, Fresh

Grade: U.S. No. 1 Filbert (*Corylus*). Round or oblong. Filberts are in season in September; they can be eaten plain. Sizes: 5 lb cartons, 100 lb bags. Filberts are similar to each other in type and dry. The shells are well formed; clean and bright; free of broken or split shells and kernels filling less than one-quarter the capacity of the shell, and free of damage caused by stains, adhering husks, or other means. The kernels are reasonably well developed; not badly misshapen; free of rancidity, decay, mold, and insect injury; and free of damage caused by shriveling, discoloration, or other means. The size is specified in connection with the grade.

PERMITTED TOLERANCES

For Mixed Types—20 percent of which are of a different type. Defects: 10 percent for filberts below the requirements of this grade, provided that not more than 5 percent are kernels filling less than one-quarter of the shell, and not more than 5 percent have rancid, decayed, moldy, or insect-injured kernels, including not more than 3 percent for insect injury.

For Off-Size—15 percent, but not more than 10 percent are undersized filberts.

Hickory Nut

The hickory nut (*Juglandacea carya*) is also known as kingnut, mocker nut, pig nut, shagbark, and water hickory. Round to oblong in shape.

Lotus Seed

The lotus seed (*Nelumba nucifera*) is also known as rattle nut and water nut. It has a less delicate flavor, but is firmer, than macadamia nuts. Available canned or dried.

Macadamia Nut

The macadamia nut (*Macadamia turnifolia*), also known as Australian hazelnut and Queensland nut, is a creamy white, round, tender kernel with a flavor similar to a filbert's.

Mixed Nuts in the Shell

Mixed nuts in the shell that meet the requirements of a U.S. grade are required to conform to the applicable mixture, sizes, and grades set forth in one of the following grades:

Grades: U.S. Extra Fancy, U.S. Fancy, U.S. Commercial, and U.S. Select.

Peanut

(*Arachis hypogaea*, Spanish peanut)—also known as goober, groundnut, and monkey nut. Peanuts are in season in November; they can be eaten plain and used in cooking. Size: 85 lb bag, roasted.

Peanut, Cleaned Virginia Type in the Shell

U.S. Jumbo Hand Picked: Peanuts that are mature, dry, and free of loose peanut kernels, dirt, and other foreign material, fully developed shells that contain almost no kernels (pops), peanuts that have very soft and/or very thin ends (paper) ends, and free of damage caused by cracked or broken shells, discoloration, or other means. The kernels are free of damage. The peanuts may not pass through a screen having 37/64 in. by 3 in. perforations and shall not average more than 176 count per pound.

Permitted Tolerances: 10 percent for pops, paper ends, damaged shells, loose kernels, or other foreign material; 5 percent for peanuts that pass through the prescribed screen, but do not have pops or paper ends; and 3.5 percent for damaged kernels.

U.S. Fancy Hand Picked: Mature, dry, and free of loose kernels, dirt, and other foreign material, pops, paper ends, and damage. The peanuts may not pass through a screen having 32/64 in. by 3 in. perforations and shall not average more than 225 count per pound.

Permitted Tolerances: 11 percent for pops, paper ends, or damaged shells and kernels, or other foreign material; 5 percent for peanuts that pass through the prescribed screen, but do not have pops, paper ends, or damaged shells; and 4.5 percent for damaged kernels.

Unclassified: No definite grade.

Peanuts, Shelled Runner Type

U.S. No. 1 Runner: Have similar varietal characteristics; are whole; are free of foreign material, damage, and minor defects due to discoloration, sprouts, and dirt; and will not pass through a screen having 16/64 in. by 3/4 in. openings.

U.S. Runner Splits: Have similar varietal characteristics; may be split or broken; but are free of foreign material, damage due to rancidity, decay, mold, or insects, and minor defects due to discoloration, sprouts, and dirt; and will not pass through a screen having 17/64 in. round openings.

U.S. No. 2 Runner: Have similar varietal characteristics; may be split or broken; but are free of foreign material, damage, and minor defects; and will not pass through a screen having 17/64 in. round openings.

Peanuts, Shelled Spanish Type

U.S. No. 1 Spanish: Whole and free of foreign material, damage, and minor defects, and will not pass through a screen having 15/64 in. by 3/4 in. openings.

U.S. Spanish Splits: Kernels may be split or broken, but are free of foreign material, damage due to decay, rancidity, mold, or insects, and minor defects due to discoloration, sprouting, and dirt. They will not pass through a screen having 16/64 in. round openings.

U.S. No. 2 Spanish: Same as U.S. Spanish Splits, except for tolerances.

U.S. Extra Large Virginia: Kernels of similar varietal characteristics that are whole; are free from foreign material, damage due to rancidity, decay, mold, or insects, and minor defects due to discoloration, sprouts, and dirt; and will not pass through a screen having 20/64 in. by 1 in. openings. There may not be more than 612 peanuts per pound.

U.S. Medium Virginia: Kernels of similar varietal characteristics that are whole; are free of foreign material, damage, and minor defects; and will not pass through a screen having 18/64 in. by 1 in. openings. There may not be more than 640 peanuts per pound.

U.S. No. 1 Virginia: Same as U.S. Extra Large Virginia, except the kernels will not pass through a screen having 15/64 in. by 1 in. openings, and there are not more than 864 peanuts per pound.

U.S. Virginia Splits: Kernels of similar varietal characteristics that are free of foreign material, damage, and minor defects; and will not pass through a screen having 20/64 in. round openings. Not less than 90 percent, by weight, shall be splits. (*Split* means the separated half of a peanut kernel.)

U.S. No. 2 Virginia: Kernels of similar varietal characteristics that may be split or broken, but that are free of foreign material, damage, and minor defects, and that will not pass through a screen having 17/64 in. round openings.

Pecan

Carya pecan (*Carya olivaeformis*) and Hicoria pecan. The pecan is the most popular native nut; hard-shell or soft-shell; small, medium, and large. It is in season in November and December.

SIZES

Shelled—3 oz can, 8 oz can, 5 lb box, 25 lb box, 50 lb box, 180 lb bushel.

Unshelled—25 lb box, 50 lb box, 140 lb bag, 175 lb bushel.

Pecans in the Shell—Size Classification: Specified in connection with the grade.

COLOR CLASSIFICATION

1. Light—Outer surface is mostly golden brown or with not more than 25 percent darker than golden, none of which is darker than light brown.
2. Light Amber—More than 25 percent of outer surface is light brown, with not more than 25 percent darker than light brown, none of which is darker than medium brown.
3. Amber—More than 25 percent of outer surface is medium brown, with not more than 25 percent darker than medium brown, none of which is darker than dark brown.
4. Dark Amber—More than 25 percent of outer surface is dark brown, with not more than 25 percent darker than dark brown.

Grades

Grade No. 1: Free of loose hulls, empty broken shells, and other foreign material. Shells are fairly uniform in color and free of damage; but 5 percent damaged shells, including not more than 2 percent seriously damaged shells, are allowed. Kernels are free of damage; but 12 percent damaged kernels, including not more than 5 percent seriously damaged, and 8 percent discolored kernels are allowed.

Grade U.S. Commercial: Same as Grade No. 1, except that there is no requirement for uniformity of color; and 10 percent damaged shells, including 3 percent seriously damaged shells; and 30 percent damaged kernels, including 10 percent seriously damaged kernels, are allowed; 0.5 percent foreign material may be present in both grades.

Pecans, Shelled

U.S. No. 1 Halves: Quality: Well dried; fairly well developed; fairly uniform in color; not darker than "amber"; free of damage or serious damage by any cause; free of pieces of shells, center wall, and foreign material; and comply with tolerances for defects. Size: Halves are fairly uniform in size; conform to size classification or count specified; and comply with tolerances for pieces, particles, and dust.

U.S. No. 1 Halves and Pieces: Quality: Same as U.S. No. 1 Halves. Size: Same as U.S. No. 1 Halves, except that at least 50 percent are half-kernels; both halves and pieces will not pass through a 5/16 in. round opening, and comply with tolerances for undersize.

U.S. No. 1 Pieces: Quality: Same as U.S. No. 1 Halves, except that there is no requirement for uniformity of color. Size: Same as U.S. No. 1 Halves, except that there is no requirement for percentage of half-kernels; they conform to any classification and comply with applicable tolerances for size.

U.S. Commercial Halves: Quality: Same as U.S. No. 1 Halves, except no requirement for uniformity of color and increased tolerances for defects. Size: Same as U.S. No. 1 Halves, except no requirement for uniformity of size.

U.S. Commercial Halves and Pieces: Quality: Same as U.S. No. 1 Halves and Pieces, except no requirement for uniformity of color and increased tolerances for defects. Size: Same as U.S. No. 1 Halves and Pieces.

U.S. Commercial Pieces: Same as U.S. No. 1 Pieces, except for increased tolerances for defects.

SIZE CLASSIFICATIONS FOR HALVES	NUMBER OF HALVES PER POUND:
Mammoth	250 or less
Junior Mammoth	251–300
Jumbo	301–350
Extra Large	351–450
Large	451–550
Medium	551–650
Small (Topper)	651–750
Midget	751 or more

Pine Nut

Also known as Indian nut, pignolia, and pinon. Used in Levantine and barbecue dishes.

Pistachio

The pistachio (*Pistacia vera*) has a green color. Sizes: 25 lb tin, 27 lb tin.

Walnuts

Walnut (English) (*Juglans regia*). English walnuts have a round to slightly elongated shape, medium light to hard shell, and range in size from small to large. Fresh nuts deteriorate a few months after harvesting but are good before this time. Meats are high in oil but quickly oxidize when exposed to air. In season in December. Sizes: 3 oz can, 8 oz can, 5 lb carton, 25 lb carton.

Walnut (Black) (*Juglans nigra*). Also known as American walnuts, black walnuts have a hard shell that is difficult to crack. Sizes: 50 lb bushel, 100 lb bag.

Walnuts in the Shell: Size Specifications: Specified in connection with the grade.

Mammoth Size—Not more than 12 percent pass through a round opening 96/64 in. in diameter.

Jumbo Size—Not more than 12 percent pass through a round opening 80/64 in. in diameter.

Large Size—Not more than 12 percent pass through a round opening 77/64 in. in diameter, except for the Eureka type, which limits the opening to 76/64 in. in diameter.

Medium Size—At least 88 percent pass through a round opening 77/64 in. in diameter, of which not more than 12 percent pass through a round opening 73/64 in. in diameter.

Baby Size—At least 88 percent pass through a round opening 74/64 in. in diameter, of which not more than 10 percent pass through a round opening 60/64 in. in diameter.

Grades

U.S. No. 1: Dry, practically clean, bright, and free of splits (walnuts with the seam opened completely around the nut so that the halves of the shell are held together only by the kernel), injury by discoloration, and free of damage due to broken shells, perforated shells, adhering hulls, or other means. Kernels are well dried, free of decay, discoloration, rancidity, and damage caused by mold, shriveling, insects, or other means. At least 70 percent have kernels that are no darker than "light amber," provided that at least 40 percent have kernels that are not darker than "light" on the USDA Walnut Color Chart.

U.S. No. 2: Same as U.S. No. 1, except that at least 60 percent have kernels that are not darker than "light amber" on the USDA Walnut Color Chart.

U.S. No. 3: Dry, fairly clean, and free of splits, damage caused by broken shells, and serious damage caused by any means. The kernels are well dried and free of decay, dark discoloration, rancidity, and damage caused by any means.

Unclassified: No grade has been applied.

Walnuts, Shelled (Juglans regia)

U.S. No. 1: Portions of kernels that are well dried, clean, and free of shell, foreign material, insect injury, decay, rancidity, and damage due to shriveling, mold, and discoloration.

U.S. Commercial: Same as U.S. No. 1, except for increased tolerances. Color may not be darker than "amber."

INTERNET RESOURCES

www.ams.usda.gov/standards/nutpdct.htm

U.S. Department of Agriculture Agricultural Marketing Service (USDA AMS). Quality Standards—Nuts and Specialty Products.

CHAPTER

12

Storage and Handling

The flow of materials through a foodservice operation begins in the receiving and storage areas. Careful consideration should be given to receiving and storage procedures, as well as to the construction and physical needs of both areas. In planning, there should be a straight line from the receiving dock to the storeroom and/or refrigerators, with everything preferably on the same level as the kitchen. A short distance between receiving and storage will reduce handling labor, lessen pilferage hazards, and allow the least amount of deterioration in food products. Moreover, it is also advisable from a cost standpoint.

This chapter is concerned with the following topics:

QUALITY LOSS AND SPOILAGE

Satisfactory methods of handling foods are designed to overcome the two major causes of food deterioration:

1. Chemical changes within the food that result in loss of quality and reduction in nutrient value
2. Spoilage organisms (mold, yeast, bacteria) that get into the food and produce undesirable or even dangerous effects

Changes in the surface of foods, such as softening, darkening, or discoloration, as well as losses of flavor and vitamins, are caused by chemical changes. As food travels from producer to consumer, deterioration begins. Chemical changes do not, however, affect the quality of the food to the point where it must be discarded, nor do they cause illness when the food is consumed. Chemical changes result primarily in a loss of vitamins and bring about obvious signs of quality loss, such as the wilting of lettuce or the graying of meat. Spoilage begins when food that is held too long before use begins to develop off-odors, a condition that leads to the foods being unfit for human consumption. This spoilage is caused by molds, yeast, and bacteria, all organisms that have a definite role in nature—that of decomposing plants and animals to the basic elements of soil. These microorganisms not only result in a waste of food, but can also cause illness. It is impossible to completely protect against them, as they are everywhere and are not visible to the naked eye. It is only when they develop in large numbers that signs of their presence become apparent.

Molds

Molds require air and thus appear only on the surface of tart and bland foods. When massed, they grow in the form of odd-colored patches—white, yellow, green, or black—and they also develop a "whiskery" appearance when allowed to grow. Molds are usually undesirable, because of a characteristic musty odor and flavor, except in foods such as Roquefort cheese, in which the flavor and odor are enhanced by particular molds.

Yeast

Yeast grows best in tart foods that are liquid or semisolid in consistency. Within this medium, yeasts form carbon dioxide gas that appears on the surface and throughout the food material. "Bulged" cans are examples, indicating improper canning techniques in which yeast spores have not been killed. Yeast gives off a typical odor and changes sugars to alcohol, which adds a "fermented" flavor to the food. Although the fermentation process is useful in the making of wines, it is undesirable in most foods.

Bacteria

Bacteria grow and develop in all foods that are low-acid or bland, such as vegetables, meats, poultry, and dairy products, but not in tart foods, such as fruits, pickled products, and so on. Unlike molds and yeasts, bacteria can cause varying degrees of illness. Some bacteria are not harmful, and it is difficult to distinguish between them and those that are harmful, particularly because they grow in the same conditions. Bacteria are most obvious when massed on the surface of a food, as a glossy sheen is evident. The food may also have a "sulfur" odor, especially in meat or poultry. Cans of low-acid foods may also be "bulged," thereby indicating bacteria spoilage. As bacteria grow, various acids are formed that cause an abnormally sour taste in usually bland food. This taste is a caution signal. The operator should be aware of changes caused by both chemical and microorganism activity. Signs that indicate quality loss, but not necessarily harmful effects, include:

1. Changes in texture, such as softening or shriveling
2. Discoloration
3. Loss of flavor
4. Yeast odor or flavor
5. Moldy odor or flavor
6. Mold formation

Signs that indicate spoilage and changes that may cause varying degrees of illness include:

1. Putrid, sulfur, or tainted color
2. Unnatural sourness
3. Gas bubbles
4. Bulged can or loose cover

Whether the food is merely unappealing because of deterioration or whether it is actually unsafe for human consumption, the operator must exercise caution to protect all consumers from hazardous foods.

RECEIVING CONTROL

In both small and large foodservice operations, it is vitally important that meats, vegetables, and all other foods be carefully checked, weighed, and compared with the invoice accompanying the merchandise. Receiving personnel should be trained to weigh meats, fresh produce, eggs, and other items, to verify quantity as well as quality. Scales capable of weighing large meat items should be available to assure the operation that it is paying only for the merchandise it has received. Products that are spoiled or damaged should be reported to the operator so that the defective merchandise may be returned and credit may be received from the vendor. When vendors are aware that all merchandise is being carefully inspected, there will be less chance of receiving inferior products. The operator should send a printed copy of quality specifications used in the operation to its vendors to inform them of the operation's quality standards. If a receiving sheet is kept, the receiving clerk should also make certain that the unit price listed on the receiving sheet is the same as that on the invoice. Occasionally, vendors will quote one price and charge another, a discrepancy that can easily be overlooked within the foodservice operation.

DRY STORAGE CONTROL

Food is normally stored in dry or refrigerated areas. Both need to be carefully planned so that buying can be controlled and food can be stored efficiently, which ultimately will save the operation many dollars. The storeroom should be well ventilated in order to remove odors; it should also be well lighted, dry, and clean. Good lighting is achieved by the use of 2 or 3 watts for each square foot of floor area. Control of temperature in storage areas is vitally important in planning storage spaces. Too often, dry storage space is given little attention, and whatever space remains after all other functions have been assigned is given over to it; frequently a dry storage space is fitted with water heaters, hot water pipes, and other heat-producing devices. Ideally, no equipment requiring ongoing maintenance should be located in a dry storage area. Temperatures should be controlled within a range of 40° F to 70° F. In no case should they exceed 70° F, as this may damage food products and later cause spoilage. The floors of the storeroom should be of heavy concrete. Walls should be constructed of a material that can be washed easily. The door should be secured with a lock and opened only when receiving and storing or issuing items for use. The storeroom should be kept neat and orderly at all times to ensure that items are not "lost" and that control is maintained. The area should look very much like a small supermarket. Stored items represent an investment and, in addition to the initial outlay of cash for these products, there are other cost factors that add to the original cost. These include interest on the money invested, insurance, cost of storage, and possible losses due to shrinkage and deterioration.

These factors influence the purchase quantities. Some operators purchase according to a "par" stock level, whereby each item in the storeroom has a set minimum and maximum level based on the usage of that item. As the quantity of an item reaches its minimum level, it is reordered up to its maximum level. Using this method, the operator seldom runs out of necessary items. Yet this system of purchasing to par stock on every item used may have its limitations by resulting in excessive inventories and overbuying. Some operators employ a combination of two methods of purchasing:

1. Acquiring par stock levels for those items used daily that are not greatly affected by the menu, and
2. Purchasing meats, vegetables, and other more expensive and perishable items on an as-needed basis.

This system tends to control inventory costs more satisfactorily while providing the security of not running out of needed items.

In determining the amount of space required for dry storage, the operator needs to give thorough consideration to the menu, the type of operation, the volume of business, the purchasing policies, and the frequency of deliveries. There are varying procedures for calculating storage space requirements. A simple and effective method is to determine the total meal load for the heaviest day expected and divide this number by 2, which results in the number of square feet needed for a storeroom in an average operation for a 30-day supply of dry goods and supplies. For bimonthly deliveries of supplies, the operator divides again by 2, or for weekly deliveries, by 4.

Storage shelves should be labeled with the names of the items routinely found there, as well as their container sizes. Items should be placed in the same order as they appear on the inventory form in order to facilitate taking inventory. Some operators choose to arrange storeroom items alphabetically, and others arrange them according to product groups, such as fruits, vegetables, fats and oils, and so on. Whatever the arrangement, storage on shelving should always be neat and orderly.

Within the storeroom, items should be rotated and issued so that items received first are used before new items. Commonly referred to as "first in, first out," this procedure generally ensures that products will be of the highest quality when served to the consumer. Tables 12.1 and 12.2 offer practical suggestions for storage of dry goods. Food technologists, food processors, and storage experts have been consulted in the preparation of these tables. An operator should consider the limits listed as outside limits. By staying well within those limits, the operator should be able to get the most value from the purchasing dollar. In so doing, the operation will be able to serve its clientele nutritious and safe food.

INTERNET RESOURCES

www.fsis.usda.gov/
USDA Food Safety and Inspection Service

www.restaurant.org/foodsafety/
National Restaurant Association—resources for food safety and nutrition

vm.cfsan.fda.gov/~lrd/advice.html
U.S. Food and Drug Administration—produce and food storage and food safety recommendations

Table 12.1 Dry Stores Chart

Products	Cool Storeroom	Refrig.	Humidity	Signs of Deterioration	Notes
			Baking Stores		
Baking Powder	8 to 12 mo.		Max. 60%	Caking	If stored too long, will lose leavening power.
Chocolate, baking	6 to 12 mo.	2 yr.	Max. 60%	Mold, mustiness, loss of flavor	Keep away from strong-odored products.
Chocolate, sweetened	2 yr. (not over 65° F)		Max. 60% (prefer. not over 50%)	Mold; sugar bloom on surface; stale flavor; webbing	Store off floor, away from walls; refrigerate in hot weather; more expensive grades keep longer due to better quality raw materials, higher fat coatings.
Chocolate Milk	6 to 12 mo.	1 yr.	Same as above	Same as above	
Cornstarch	2 to 3 yr.	3 to 5 yr.	Max. 60%	Caking, foreign, musty odors	Same as above. Begins losing some flavor after one month.
Tapioca	1 yr.	2 yr.	Max. 70%		
Yeast, dry	18 mo.	3 yr.			Keep away from strong-odored products.
Baking Soda	8 to 12 mos.		Max. 60%	Caking	
			Beverages		
Cocoa	1 yr.		Max. 85%	Caking, infestation (will occur if left uncovered for long)	Keep away from strong-odored products. Will become rancid in high temperature.
Coffee, ground (vacuum packed)	7 to 12 mo.	12 to 18 mo.	Max. 60%	Loss of flavor	Store off floor and away from walls and strong-odored products.
Coffee, ground (not vacuum packed)	2 wk.	1 mo.	Max. 60%	Loss of flavor, foreign odors	
Coffee, instant	8 to 12 mo.	3 yr.	Max. 85%	Caking, loss of flavor	Caking tends to occur at high temperatures as well as in dampness if container leaks. Cakes are usually reclaimable.
Tea Leaves	12 to 18 mo.	3 to 4 yr.	Between 50% and 80%; 75% optimum	Loss of sweet aroma, absorption of foreign odors, mold, mustiness, dried out leaves, infestation	Store off floor away from direct sunlight and strong-odored products. High temperatures and less than 50% humidity both tend to dry out the product.
Tea, instant	8 to 12 mo.	3 yr.	Max. 85%	Loss of flavor, caking	Avoid high temperatures and excessive dampness, as either condition will cause caking, particularly if container is not completely sealed. Gradual flavor loss over course of a year at room temperature.
Carbonated Beverages	Indefinitely				
			Canned Goods		
Note: Following information is based on general tolerances.					
Fruits (in gen.)	1 yr.	4 yr.	Max. 60%	Softening, fading color, loss of flavor, can bulge, leakage, pinholing.	Good idea to specify "current pack only" or "last year's pack not acceptable" when buying canned fruits and vegetables. Keep away from direct sunlight, especially in warm climates, as solar radiation can easily push temperature inside cans above 100° F and considerably shorten storage life. Use well-ventilated area. (See other suggestions in text.)

Table 12.1 Dry Stores Chart (*continued*)

Products	Cool Storeroom	Refrig.	Humidity	Signs of Deterioration	Notes
			Canned Goods (cont.)		
Fruits, acid-types, i.e., citrus, berries, sour cherries	6 to 12 mo.	3 yr.	Max. 60%	Same as above, but particularly pinholing due to acid action.	Same as above. Acid factor reduces storage life.
Fruit Juices	6 to 9 mo.	3 yr.	Max. 60%		
Meat, Poultry, Seafood (in general)	1 yr.	2 to 4 yr.	Max. 60%	Same as above. Crystals may form in canned crabmeat. They are not harmful, but annoying, and thus quality is reduced.	Same as above. Temperatures above 90° F hasten deterioration appreciably for these products. Freezing affects texture, particularly of those containing tomato, or other high-acid sauces and have slightly reduced storage life.
Pickled Fish, Fish in Brine	4 mo.	12 mo.	Max. 60%	Mold and off-odor.	If mold and off-odor develop, discard product.
Smoked Fish, light smoking sauce					This is a highly perishable product.
Smoked Fish, heavy smoking	1 mo.	2 to 3 mo.		Mold and off-odor	If mold and off-odor develop, discard product.
Dried Fish	1 mo.	2 to 3 mo.		Same as above	Same as above.
Soups (including bouillon cubes and dehydrated soup mixes)	1 yr.	3 to 4 yr.			Avoid freezing (most canned vegetables freeze at 25° F and all sudden changes in temperature (to minimize condensation). Keep away from direct sunlight (explained under fruits).
Vegetables (except high-acid ones)	1 yr.	4 yr.		Softening, fading color, loss of flavor, can bulge, leakage, pinholing	
Vegetables, high acid, i.e., tomatoes and sauerkraut	7 to 12 mo.	2 to 3 yr.		Same as above, but particularly pinholing due to acid action	Same as above.
			Condiments, Flavors, & Seasonings		
Mustard, prepared	2 to 6 mo.	1 yr.		Gas bubbles, brown surface color, sulfur-like odor	Affected by direct light as well as prolonged storage at high temperatures. Continuous flavor loss to some degree from time of manufacture, so should be used as soon as possible.
Flavoring Extracts	Indefinite				Never refrigerate.
Monosodium Glutamate	Indefinite				
Salt	Indefinite		Max. 60%		Caking will occur in humid conditions, but has no lasting effect on quality.
Sauces (hot pepper, soy, steak, Worcestershire)	2 yr.	2 to 3 yr.		Separation of ingredients, off-odors, color change	Affected by long exposure to light (color change), and should never be frozen.
Spices, Herbs, Seeds, whole	2 yr. to indefinite			Fading color in herbs, loss of aroma	In whole form, the flavor and aroma of spices are protected by cell structure. Bay leaves and other herbs lose bright color within a year.
Spices, Herbs, Seeds, ground (exc. paprika)	1 yr.	2 yr.	Max. 60%	Loss of aroma, faded colors, caking	Keep off floor, away from outside walls. Reseal containers quickly after each use. Purchase in moderate quantities, replace frequently.

(*continues*)

Table 12.1 Dry Stores Chart (*continued*)

Products	Cool Storeroom	Refrig.	Humidity	Signs of Deterioration	Notes
			Condiments, Flavors, & Seasonings (cont.)		
Paprika, Chili Powder, Cayenne, Red Pepper	1 yr.	1 yr.	Max. 60%	Fading color, infestation	Keep away from direct light. During summer in hot climates, may be wise to refrigerate to guard against infestation.
Seasoning Salts	1 yr.	2 yr.	Max. 60%	Caking	Very hygroscopic and once solidly caked, they are usually unredeemable.
Vinegar	2 yr.	3 yr.		Presence of "mother" or vinegar "eel," evaporation or infestation	Both "mother" and "eel" are signs of improper manufacturing; however, "mother" may also indicate storage troubles (usually loose caps), as may evaporation and infestation (fruit flies).
			Dairy Foods & Cheese		
Cheese, hard types (cheddar or American), particularly natural		1 mo. (32° F) 2 wk. (45° F)		Drying out, "oils off," moldiness	Needs tight wrapping. Store away from strong-odored products.
Cheese, soft types, (cream, cottage, limburger, etc.)		1 to 2 wk.		Drying out, "oils off," moldiness	Needs tight wrapping. Store away from strong-odored products. Soft cheeses tend to change body characteristics undesirably when stored at freezing temperatures. Storage life under refrigeration is shorter than for hard cheeses because texture allows spoilage agents to enter easily. Mild types in particular are potential sources of food poisoning if storage is prolonged or improperly handled.
Cheese Spreads	1 yr.			Surface darkening, off-flavor, odor, mold	Those with wine have slightly longer storage life. All spreads become perishable once opened.
Cream, powdered	4 mo.	6 mo.		Stale or tallowy odor, separating of butterfat	Best refrigerated.
Milk, condensed	1 yr. (if av. temp. no more than 60° F; 2 mo. at 70° F)	1 yr.	Max. 60%	Darkening, thickening, change in flavor, "sandiness," swelled cans (from fermentation)	Swelled cans should be discarded.
Milk, dry, nonfat (regular & instant)	1 yr.	2 yr.	Max. 35% containers not hermetically sealed	Flavor changes, lumping and darkening	Extra grade (moisture 4% in regular, 5% in instant) in hermetically sealed tins or glass keeps longer, but foil laminated or polyethylene bags are also good.
Milk, dry, whole	10 to 12 mo.	2 yr.		Darkening, stale flavor, rancidity	
Milk, evaporated	1 yr. (if aver. temp. not more than 60° F; 2 mo. at 70° F)	1 yr.	Max. 60%	Darkening, separation of cream layer, gelling (due to changes in the protein)	Milk that has gelled should not be used because this condition may be due to a defective can as well as to protein changes.
			Fats & Oils		
Butter		1 mo. (35° to 45° F); 6 to 9 mo. (0° F)	Max. 55%	Absorption of foreign odors, rancidity	The sweeter the cream used in making the butter, the longer the product will keep. Use double the normal wrapping (or airtight moisture-proof containers) for freezer storage.

Table 12.1 Dry Stores Chart (*continued*)

Products	Cool Storeroom	Refrig.	Humidity	Signs of Deterioration	Notes
			Fats and Oils (cont.)		
Margarine		1 mo. (35° to 45° F); 3 mo. (0° F)	Max. 55%	Stale flavors, rancidity, surface darkening, mold, foreign odors, oil separation	Keep away from strong-odored products. Aluminum wrapping offers additional storage protection. Higher quality products keep longer.
Salad Dressings (inc. mayonnaise)	2 mo. (max. at 60° F; opt. is 50° F)	2 mo.	Not a factor	Rancidity, off-color, separation of oil from water	Avoid direct light or sunlight and sudden temperature changes.
Salad Oil	6 to 9 mo.		Not a factor	Rancidity (odor similar to that of old lard or grease)	A light high-grade oil will keep better.
Shortenings, vegetable	2 to 4 mo.	1 yr.	Not a factor	Rancidity, absorption of foreign odors	
Shortenings, animal fat (lard, etc.)		2 mo.	Not a factor	Rancidity, absorption of foreign odors	Should be refrigerated because they become very perishable at room temperature. More perishable than vegetable shortenings at room temperature because they are softer (the harder the fat, the longer the storage life).
			Grains & Grain Products		
Cereal Grains (for cooked cereal—any type)	8 mo.		Below 40%	Mold, webbing, infestation; musty, stale, or free-fat odor	Store off floor away from walls. Be especially careful to keep them away from all strong-odored products, as they act like a sponge for odors around them; the same is true of dampness. As one expert put it, "Cereals are best when first made and thereafter deteriorate a little bit each day." In excessive heat or humidity they may last not more than a month.
Cereals (ready-to-eat)	6 mo.		Max. 60%	Loss of crispness and development of a toughness; mold; webbing, infestation, rancidity	Essentially the same as for cereal grains, above.
Cornmeal	8 to 12 mo.		Max. 60%	Rancidity, mold, infestation, caking	Essentially same as above.
Flour, bleached	9 to 12 mo.	2 yr.	Max. 60%	Infestation, caking, mustiness	Essentially same as above. Low temps. (32° to 43° F) best protection against infestation. Refrigerating may be necessary in very hot climates.
Flour, whole grain	2 to 4 mo.	4 to 6 mo.	Max. 60%	Rancidity, infestation, caking, mustiness	Essentially same as above. Because whole-grain flours retain the oil-bearing wheat "germ," they are more susceptible to rancidity than bleached flour.
Macaroni, Spaghetti, and all Pastas	3 mo.		Max. 60%	Mold, infestation, checking, mustiness	Essentially the same precautions as for other cereal grains, above. If pastas are subjected to sudden changes in temperature, they may check, that is, develop fine cracks that will cause disintegration in cooking.
Prepared Mixes	6 mo.	12 mo.	Max. 50%	Infestation, stale odor, discoloration, loss of baking performance	Temperature particularly important; level above 70° F may reduce life to 2 to 3 mo.

(*continues*)

Table 12.1 Dry Stores Chart (*continued*)

Products	Cool Storeroom	Refrig.	Humidity	Signs of Deterioration	Notes
			Grains & Grain Products (cont.)		
Rice, parboiled	9 to 12 mo.			Infestation, color change (to yellowish), rancid odor	Same general precautions as for other grains.
Rice, brown or wild		6 to 9 mo.	Max. 50%	Rancid odor, infestation	Refrigeration is required because these products still have the dark outer covering that is more prone to rancidity and absorption of foreign odors than the rest of the kernel.
			Sweeteners		
Sugar, granulated	Indefinite		Max. 60%		Caking will occur in humid conditions, but has no lasting effect on quality.
Sugar, confectioners'	Indefinite		Max. 60%		Same as above, but slightly less tendency to cake.
Sugar, brown		1 yr.	Max. 70% Min. 60%		Brown sugar should be refrigerated to give it the ratio of humidity it needs to keep soft.
Syrups, corn, honey, molasses, sugar	1 yr.			Mold	Susceptible to mold after container is opened. Refrigerate opened containers.
			Miscellaneous		
Beans, dried (also all lentils)	1 to 2 yr.	2 to 3 yr.	Max. 60% Min. 30%	Mold, musty appearance	More than 60% humidity will cause mold; less than 30% will dry out the product.
Candied Peel—Citron	18 mo.	3 yr.			
Cookies—Crackers	1 to 6 mo.	4 to 12 mo.	Max. 60%	Staling, infestation, rancidity, softening	Softening can be remedied by placing crackers in the oven to restore crispness.
Cracker Meal	1 yr.	3 yr.	Max. 60%	Same as above.	Same as above.
Dried Fruits	6 to 8 mo.		About 75%	Sogginess or hardness, crystallization, infestation	Higher than 75% humidity will result in sogginess; less than that will dry out the fruit to the point of hardness.
Gelatin	2 to 3 yr.	3 to 4 yr.	Max. 70%	Caking	
Dried Prunes		15 mo. (see notes)	About 30%	Infestation, sweating, fermenting, molding, excessive dry out and sugaring	Best temperature is between 40° and 50° F. Store on skids away from walls and away from strong-odored products.
Jams, Jellies	1 yr.	2 to 3 yr.	Max. 60%	Color change, crystallization, caramelization	Red-colored jams and jellies lose color if stored at excessively high temperatures. All these products will caramelize at high temperatures.
Nuts	1 yr.			Mold, infestation, rancidity	If vacuum packed, they will keep better.
Pickles, Relishes	1 yr.	3 yr.	Max. 60%	Pickles soften, develop hollow centers, cloudy brine	Important to keep away from strong light, particularly dill pickles.
Potato Chips	1 mo.		Max. 50%	Staleness, rancidity	May be re-crisped by heating in oven.
Coconut, sweetened	4 to 6 mo.			Browning, mold, off-flavor, sogginess	If only soggy, may be reclaimed by heating in oven.
Dates, pasteurized	8 to 12 mo.	8 to 12 mo.	Max. 65%		
Dates, raw	1 to 2 mo.	8 to 12 mo.	Max. 65%		
Dehydrated Vegetables	7 to 12 mo.	4 yr.	Max. 60%		
Eggs, powdered	4 mo.	1 yr.		Caking, off-odors, mold	If moldy, discard.

Table 12.1 ***(continued)***

Frozen Food Storage Control

The use of frozen foods within a foodservice operation has increased steadily in recent years. Frozen foods offer definite advantages to the operator:

1. Limited waste
2. Year-round availability
3. Less preparation time
4. Long storage life
5. Many sources of supply
6. More menu variety

This increasing use of frozen foods requires the operator to have additional storage space. Frozen food storage requires lower temperature conditions than do other types of storage. Freezers provide storage space at 0° F or lower at all times. Freezers should also be capable of freezing foods quickly, because the faster the process, the smaller the ice crystals that form within the food, thereby maintaining it at a higher quality. Food should be frozen in a three-step process:

1. Refrigeration to cool
2. Sharp freezing
3. Holding at low temperature

Frozen foods are highly perishable and deteriorate quickly if not handled correctly. For this reason, the operator should check frozen items as they are received and store them as quickly as possible in a 0° F freezer. Frozen foods never return to original quality once they begin to break down. Quality in frozen foods is judged by the color, flavor, texture, appearance, and nutritive value of the product. Quantities ordered also affect the quality of the product, as smaller inventories with more turnover generally result in higher-quality products, as compared with large inventories maintained over a long period of time.

It is advisable to store food items in the original shipping cartons. Foods should be packed in vaporproof containers to prevent dehydration, oxidation, discoloration, odor absorption, and loss of volatile flavors. Operators should check the condition of the containers to verify that there are no breaks that will cause freezer burn on food items.

The following points should help the operator to improve the overall efficiency in the use of freezers:

1. When unloading food items in the freezer, move loaded cart directly into the freezer, if possible. If the cart is too large for the freezer, move it as close to the freezer as possible. This will reduce handling time and protect the frozen food against exposure to temperatures above 0° F.
2. Rotate stock by marking new frozen foods to show the date received so that the oldest-dated food will be used first.
3. Arrange frozen foods in the freezer by product groups. Labels should be available so that foods can easily be identified. This reduces handling time and helps to keep the freezer arranged in an orderly manner.
4. Maintain optimum air circulation in the walk-in-type freezer by keeping foods off the floors, away from ceilings and walls. To ensure proper air circulation, platform racks are often used for stacking boxes on floors.
5. Train employees to open freezers only when necessary. Remove as many items at one time as possible, in order to minimize door opening.
6. Install a thermometer to measure the average temperature of the freezer so that it will not be affected by the opening of the door, by the cooling coils, or by direct air from the cooling unit. The thermometer should be placed where it is easy to see. For walk-in freezers, the thermometer should be attached to the outside. An alarm system should also be installed to indicate mechanical failure. Occasionally, the temperature of the products should be checked with a calibrated, dial-type, hand thermometer.
7. Defrost freezers regularly to prevent excessive formation of ice. This will increase efficiency and also reduce operating costs, labor, and damage to the product.
8. Maintain a clean, orderly, and well-organized freezer. This requires daily cleanup but is well worth the time and effort because it will result in better utilization of products and reduction in damage to frozen foods.
9. Establish a regular service schedule to be followed, regardless of the type of freezer being used.

The following tables indicate the length of time that frozen foods may be stored at 0° F without noticeable loss of quality. If stored at 10° F, the foods have approximately ¼ to ½ of the storage life listed. Those stored at –10° F will retain their quality for periods longer than those shown in the tables. The times listed in these tables assume the food products are of high quality when placed into storage.

Approximate Storage Life of Various Foods at Zero Degrees Fahrenheit (0° F)[a]

	No of Months at 0° F
Fruit	
Apricots	12
Peaches	12
Raspberries	12
Strawberries	12
Vegetable	
Asparagus	8 to 12
Beans, Snap	8 to 12
Beans, Lima	12
Broccoli	12
Brussels Sprouts	8 to 12
Cauliflower	12
Corn, on the cob	8 to 12
Corn, cut	12
Carrots	12
Mushrooms	8 to 12
Peas	12
Spinach	12
Squash	12

(continues)

Table 12.1 ***(continued)***

	No of Months at 0° F
Meat	
BEEF	
Roasts, Steaks	12
Ground	8
Cubed Pieces	10 to 12
VEAL	
Roasts, Chops	10 to 12
Cutlets, Cubes	8 to 10
LAMB	
Roasts, Chops	12
PORK	
Roasts, Chops	6 to 8
Ground, Sausage	4
Pork or Ham, Smoked	5 to 7
Bacon	3
Variety Meats	up to 4
Poultry	6 to 12
Fish	
Fatty Fish	
(Mackerel, Salmon, Swordfish, etc.)	3
Lean Fish	
(Haddock, Cod, Ocean Perch, etc.)	6
Shellfish	
SHELLFISH	
Lobsters and Crabs	2
Shrimp	6
Oysters	3 to 4
Scallops	3 to 4
Clams	3 to 4
Precooked Foods	
BREAD	
Quick	2 to 4
Yeast	6 to 12
ROLLS	2 to 4
CAKE	
Angel	4 to 6
Gingerbread	4 to 6
Sponge	4 to 6
Chiffon	4 to 6
Cheese	4 to 6
Fruit	12
COOKIES	4 to 6
PIES	
Fruit	12
Mince	4 to 8
Chiffon	1
Pumpkin	1
POTATOES	
French Fries	4 to 8
Scalloped	1
SOUPS	4 to 6
SANDWICHES	2

[a]Charles E. Eshbach, *Foodservice Management* (Boston: Cahners Books International, Inc., 1976), 97–99.

Equipment

There is a great deal the operator should know when planning the freezer equipment needed for an operation. First, the operator should be familiar with the correct terminology used when referring to various pieces of equipment.

Frozen food storage, also called "low temperature reach-in" or "walk-in space," is equipment designed to store frozen food, usually at 10° F to –10° F.

Processing freezer, the equipment designed to actually freeze food. It operates at –20° F and is sometimes referred to as a "blast freezer," "plate freezer," or "tunnel freezer." After the freezing of food is completed, the food is transferred to frozen food storage. Based on current menu needs and projections of future changes, operators can determine their frozen food processing and storage requirements. The following facts should, however, also be determined:

1. Quality of off-premise prepared foods or ingredients specified during peak periods
2. Frequency of frozen food deliveries
3. Possibility of buying larger frozen quantities to reduce per pound or per serving costs
4. Maximum length of time on-premise that prepared frozen items are held
5. Amount of on-premise frozen foods to be stored
6. Unusual consumption peaks
7. Short-term space needed for pre-dished foods
8. Anticipated use of government commodities
9. Time and equipment factors, if any, involved in proper defrosting

The two major types of storage facilities for frozen foods are reach-in and walk-in freezers. Reach-in units usually have less storage space. They may use either forced air or freezer plates to maintain temperatures. Those with plates cannot be defrosted automatically. Consider the following checkpoints when purchasing reach-in equipment:

Construction—properly sealed, easily cleaned
Exterior—rust resistant; in keeping with decor if location makes this important
Adequate insulation
Doors—well constructed, easily opened, hinges easy to operate
Lights—adequate; consider automatic light switch on door
Shelves—easy to clean and adjust; maximum load provisions
Condensing Unit—sufficient horsepower; type; location; necessary ventilation
Temperature control—visible thermometer or temperature record; automatic alarm system
Defrosting system
Amount of floor space required

Reach-ins can be adapted to a wide variety of institutional requirements because of the optional accessories available, including tray and pan slides, roll-out shelves and drawers. One disadvantage of this type of storage is that it is more difficult to control the merchandise within the reach-in unit.

Walk-in freezer units provide more efficient use of storage space and make handling easier. Unlike reach-in units, walk-ins

Table 12.1 (*continued*)

can function either as holding units or as processing units. They come in a wide range of standard ready-to-install models. Some of the popular walk-in models provide the following cubic content: 4 by 6 ft—117 cu ft; 6 by 8 ft—184 cu ft; 8 by 10 ft—435 cu ft; 10 by 18 ft—1022 cu ft. Walk-ins provide a bulk storage capacity that enables operators to buy ahead. In many operations, frozen food walk-ins are located inside refrigeration walk-ins. These are called dual temperature walk-ins (part cooler, +35° F, and part freezer, 0° F to –10° F). As operators continue to use more frozen food products, prefabricated walk-ins are expected to best serve their space needs, as they may be added either inside or outside the building.

Insulation is important in food holding and processing equipment. The thickness of the insulation varies with different manufacturers. One manufacturer specifies 6 in. of insulation for –5° F temperatures and 8 in. insulation for –20° F. A more recent development by manufacturers is 4 in. of polyurethane for either cooler (+35° F) or freezer (0° to –20° F) units. This material can be used for temperatures between –20° F and 40° F and gives the operator the flexibility of easily converting coolers to freezers by merely changing the refrigeration system.

Consider the following factors when planning walk-in space:

Exterior—material offering long life, easy cleaning
Interior—tight, easily maintained
Insulation—adequate for temperatures required
Assembly—easy, tight, minimum cost
Refrigeration system—sufficient capacity to maintain required temperature; conditions surrounding location where it is to be installed
Amount of cubic space needed for storage
Defrost System—how is it controlled, automatic
Door Construction
Floor Level—should provide easy movement of food
Transportation Cost
Alarm System
Accessories—shelving, racks, display doors, etc.

The importance of the preceding factors cannot be overemphasized, nor can the choice of the manufacturer from whom you purchase the equipment. In the long run, it will pay the operator to purchase from a reputable manufacturer.

When installing reach-in or walk-in frozen holding or processing equipment, locate the cabinets away from any heat-producing equipment. If the unit is self-contained, there should be necessary circulation of air around the equipment. Careful attention should be given to the amperage and voltage supply at the compressor locations to ensure proper electrical service. Floors should be checked for stability to ensure that they will be able to support the weight of freezers and their contents. Furthermore, floors should be level to enable the cabinets to rest evenly on all four corners. Before the door is installed, the operator must determine satisfactory height and width dimensions based on the institution's food-handling procedures. Some questions worth pondering when planning the efficient use of a walk-in unit are whether the door should swing right or left, whether it should open and close automatically, and whether glass panels should be installed in the door in order to check contents.

Because of the increasing quest for space, many operators are solving their frozen storage problem by installing a prefabricated walk-in unit outside. Such units are designed for blast freezing or for holding frozen food, thereby permitting the purchase of larger quantities for maximum discounts while releasing expensive inside space for a more profitable use. An added advantage is that the equipment may be placed on a base, is delivered completely equipped and ready to use, and is designed to be easily expanded in size as the needs for frozen foods change.

Refrigeration Storage

In general, fresh and prepared foods are refrigerated at 30° to 45° F to keep them safe from harmful bacterial. If separate refrigerators are used for each food, they should be kept at the following temperatures:

45° to 50° F	Fruits
40° to 45° F	Vegetables, Eggs, Processed Foods, Pastry
38° to 40° F	Dairy Products
34° to 38° F	Fresh Meats
32° to 36° F	Fresh Poultry, Fish, and Seafood
–10° to 0° F	Frozen Foods

For accurately checking these temperatures, a thermometer should be located in the warmest place in each refrigerator.

The foodservice operator must carefully determine the best equipment and products to use in the establishment. In the case of refrigerated storage, the models to choose from are many:

1. Walk-in
2. Reach-in
3. Roll-in
4. Compartmentalized
5. Pass-thru
6. Counter Refrigeration
7. Display Refrigeration
8. Portable Refrigeration
9. Refrigerated Dispensers

All of the types listed here can be used to help "extend" the amount of refrigerated space in an operation. Operators are purchasing more and more foods that are partially or completely prepared and therefore require immediate refrigeration. How does the operator decide which units can meet his or her needs most economically? Experts in this specialized area of refrigerated storage have suggested that the operator estimate what percentage of a typical meal will consist of fresh food. In addition, the operator should determine how many days' supply of each kind of food must be refrigerated according to delivery schedules. From these calculations, it is recommended that:

20 to 25% of the space be allocated for meat (if portion cuts are used, the space needed would be 10 to 15%)
30 to 35% for fruits and vegetables
20 to 25% for dairy products
5 to 10% for salads, sandwich material, bakery products, and leftovers

(*continues*)

Table 12.1 ***(continued)***

The type of menu used and the type of establishment both influence the amount of refrigeration space required. For example, a school lunch kitchen with limited menu will require less refrigerated space than an expensive gourmet restaurant that serves many more fresh, refrigerated items.

The walk-in refrigerator is used extensively in the foodservice industry. One authority suggests that walk-ins are feasible in operations serving 300 to 400 meals per day. Some of the advantages of this type of storage include:

- The operator's ability to carry a greater variety of food to purchase in greater quantities with discount savings
- Decreases in the number of expensive deliveries
- The availability of more space for leftovers, which reduces food loss

Walk-ins today are usually assembled from prefabricated sections. For this reason, they can be enlarged easily and/or relocated, if required. The metal walls are easily cleaned, protect against rodents, and other pests, and require little maintenance. As with frozen walk-in space, discussed earlier, the operator should check the following points, as summarized by Kotschevar and Terrell in *Food Service Planning*[b], in choosing a particular unit.

1. Proper insulation—at least 3 in. thick
2. Floor on level plane for walk-in
3. Vaporproof walls, ceilings, and floors
4. Sturdy, well-insulated door with heavy-duty lock
5. Opening device on the inside
6. Sturdy, durable, and adjustable shelving
7. Outside thermometer
8. Adequate storage space—1½ to 2 ft on either side of aisle, 42 in. aisles preferable, wide enough to accommodate whatever mobile equipment is used in the walk-in
9. Audiovisual alarm system to alert personnel when a change in holding temperature occurs

The operator has the option of leasing these prefab units from the manufacturer, thereby releasing available funds for other equipment, building projects, and related needs.

To be sure, there are many alternatives when choosing refrigerated storage. Innovations such as refrigerated drawers, refrigerated sandwich units, undercounter units, dispensers, display cases, and many more items give the operator a certain flexibility in planning. Manufacturers of modern refrigeration equipment can provide the operator with whatever units are needed to solve an organization's refrigerated storage problems.

The following guidelines, from *The Complete Book of Cooking Equipment*, may help the operator to attain optimum utilization, whatever type of refrigeration is used in an organization.

WHAT YOU DO	WHY YOU DO IT
1. Pack food loosely.	1. To get circulation.
2. Hang prime cuts of meat away from walls.	2. Cold air needs to circulate to keep food from spoiling.
3. Cover food (below) with wax, or other covering paper.	3. Prevent dripping.
4. Discard things not needed.	4. To prevent crowding and increase circulation.
5. Place new purchases at back.	5. Use older things first.
6. Wash refrigerator frequently.	6. It must be kept clean.
7. Defrost before ¼ in. frost gathers.	7. Frost slow cooling process.
8. Open door only when necessary.	8. Open door raises temperature.

The following reference chart gives the operator information on how to store various foods and familiar food combinations. The all-important time limitations affecting the refrigeration of food should serve as a guide from which to start. Many factors actually affect the storage life of food; however, it is vitally important that the operator be able to distinguish between quality loss in food that would not ordinarily cause illness, and danger signals in those foods that warn of potential health hazards.

[b]L. Kotschevar and M. Terrell, *Food Service Planning* (New York: John Wiley & Sons, Inc., 1961).

Table 12.2 Storage Chart

	How to Store	Approx. Time Limit for Storage: Refrig.	Freezer	Evidence of Quality Loss
Sandwiches				
Meat, poultry, fish, and egg salad	Moisture-proof wrapper in refrig. or other cold place. Refrigerate 2 to 3 hr. before prolonged exposure to summer heat, e.g., if to be used for a picnic.	2 days	6 months	If merely dried out, reclaim by grilling, if practicable. Discard if off-flavor.
Other	Moisture-proof wrapper.	5 to 6 days	6 months	If dried out or soggy, reclaim by grilling. If moldy, discard.
Soups, Gravy, and Stuffing				
Gravy	Allow to cool until just warm to the touch. Refrigerate in covered dish.	3 to 4 days	1 year	
Soups				
Left-over	Allow to cool until just warm to the touch. Refrigerate in original container or covered dish.	3 to 4 days	1 year	
Canned				
Unopened	Original container.			
Opened	See left-over soups.			
Frozen	Original container.	7 days	1 year	
Stuffing	Remove from cavity. Refrigerate in covered container.	3 to 4 days	1 year	
Fish and Shellfish				
Fresh	Moisture-proof wrapper or container. Refrigerate. Leave shellfish in shell. Plan to hold only briefly if to be eaten raw.	1 to 2 days	1 year	
Cooked Fish and Fish Salads	Covered container in refrigerator.	3 to 4 days	1 year	
Bisques, Broths, Chowders, Stews, and Soups	Original or other covered container in refrigerator.	1 to 2 days	1 year	
Canned				
Unopened	Original container. Kitchen shelf.			
Opened	Original or other covered container.	3 to 4 days	1 year	
Frozen	Original container. Freezer.	3 days	1 year	
Smoked				
Light	Orig. container. Refrigerator.	1 to 2 days	1 year	If moldy and/or off-odors develop, not reclaimable. Discard.
Heavy	Original container. Cool place.	2 to 3 months	1 year	Same as above.
Dried	Orig. container. Cool place.	2 to 3 months	1 year	Same as above.
Pickled, with vinegar, wine, and/or sour cream	Orig. container. Refrigerator.	2 to 3 months	1 year	Same as above.
Eggs and Egg Dishes				
Fresh				
In Shell	Clean if necessary. Covered container. Large end of egg up. Refrigerator.	10 days		Slight off-odors can be masked by seasonings in cooking.
Separated Yolks	Covered container or cover with water. Refrigerator.	1 to 2 days	1 year	If runny, or is shape has been lost, discard.
Separated Whites	Covered container. Refrigerator.	1 to 2 days	1 year	If whites become watery, discard.
Cooked				
In Shell	Uncovered. Refrigerator.	10 days	Usable if merely dried out, otherwise discard.	
Shelled	Covered container. Refrigerator.	10 days	Usable if merely dried out, otherwise discard.	

(continues)

Table 12.2 Storage Chart (*continued*)

		Approx. Time Limit for Storage		
	How to Store	**Refrig.**	**Freezer**	**Evidence of Quality Loss**
		Eggs and Egg Dishes (cont.)		
Prepared Dishes, such as egg salads, stuffed, fondue, etc.	Covered dish. Refrigerator.	3 to 4 days	3 to 4 months	
Frozen Yolks or Whites	Original container. Freezer.	3 to 4 days	1 year	If not moldy, slight off-color can be masked by seasonings in cooking. If mold is present, discard.
Pickled	Original container. Refrigerator.	3 months		Usable if not moldy and if texture is acceptable. Otherwise discard.
Dried (powdered)	Original container. Kitchen shelf.	1 year	1 year	If merely caked, slight off-odors and flavors can be masked by seasonings in cooking. If moldy, discard.
		Cheese and Cheese Dishes		
Hard Cheese, such as cheddar, Swiss, grated, etc.	Original or other covered container in refrigerator.	3 to 9 months	1 to 2 years	If dried, use as grated. If moldy, remove mold, and if flavor of remainder is acceptable, it is usable. Otherwise discard.
Soft Cheese, such as cream, cottage, limburger, and cheese spreads	Original or other covered container in refrigerator.	1 to 2 weeks	1 year	
Cheese and Cheese Spreads in cans or jars or in wine				
Unopened	Original container. Kitchen shelf.			Remove surface discoloration. Remove mold. If the flavor of the remainder is acceptable, it is usable. Otherwise discard.
Opened	Original or other covered container in refrigerator.	2 to 3 weeks		If surface darkens, remove discoloration and/or mold. Remains are usable if the flavor is acceptable. Otherwise discard.
Prepared Dishes, such as fondue, rarebit, etc.	Covered dish in refrigerator.	2 to 3 days	1 year	
		Poultry		
Fresh				
Whole	Remove entrails of undressed poultry. Covered dish or original wrapping. If bloody, wipe with damp cloth and store in fresh wrapping material or covered dish. Refrigerator.	6 to 8 days	6 months	If dried out, grease skin; if slightly moldy or of slightly tainted odor, wash with lightly salted water and scald body cavity.
Uncooked, stuffed	Cool dressing before stuffing the fowl. Moisture-proof wrapper. Refrigerator.	1 day	6 months	
Cut-up	If bloody, wipe with damp cloth and store in fresh wrapping material or covered dish. Refrigerator.	4 to 5 days	6 months	If slightly moldy or of slightly tainted odor, wash with lightly salted water.
		Poultry and Stuffing		
Left-over Cooked Poultry	Covered dish or moisture-proof wrapping. Refrigerator. Remove stuffing, refrigerate separately.	4 to 5 days	6 months	
Poultry Pies, Stews, à la King, etc.	Cool until just hot to the touch, then refrigerate	1 to 2 days	6 months	

Table 12.2 Storage Chart (*continued*)

	How to Store	Approx. Time Limit for Storage: Refrig.	Freezer	Evidence of Quality Loss
Poultry and Stuffing (cont.)				
Canned				
Unopened	Kitchen shelf.			
Opened	Original or other covered container. Immerse in own liquid. Refrigerator.	4 to 5 days	6 months	
Frozen	Original container or moisture-proof wrapper. Refrigerator.	10 days	6 months	If dried out, grease skin. If slightly moldy or of slightly tainted odor, wash with lightly salted water and scald body cavity.
Smoked	Original or other moisture-proof container. Cool place or refrigerator.	10 to 20 days	6 months	If dried out, grease skin. If slightly moldy or of slightly tainted odor, wash with lightly salted water and scald body cavity.
Meat (Beef, Lamb, Pork, and Veal)				
Raw Roasts, Steaks, Chops, and Stew Meat	Covered dish or original wrapper. If bloody, wipe with damp cloth and store in fresh wrapping material or covered dish. Repeat if necessary during long storage. Refrigerator.	6 to 8 days	6 months to 1 year	Darkening and drying out.
Raw Boned and Rolled Roasts	Covered dish or original wrapper. If bloody, wipe with damp cloth and store in fresh wrapping material or covered dish. Repeat if necessary during long storage. Refrigerator.	4 to 5 days	6 months to 1 year	Darkening and drying out.
Raw Livers, Hearts, Kidneys, etc.	Covered dish or original wrapper. If bloody, wipe with damp cloth and store in fresh wrapping material or covered dish. Repeat if necessary during long storage. Refrigerator.	3 to 4 days	6 months to 1 year	Darkening and drying out.
Raw Ground Meats	Covered dish or wrapping material that will not stick to the meat. Refrigerator.	1 to 2 days	6 months to 1 year	Graying and drying out.
Smoked, Corned, Salt-Cured, Pickled, and Dried	Covered dish or original wrapper. If bloody, wipe with damp cloth and store in fresh wrapping material or covered dish. Repeat if necessary during long storage. Refrigerator.	10 to 20 days	1 year	Excessive drying.
Cold Cuts	Covered dish or original wrapper. Separate slices with waxed paper for longer storage life. Refrigerator.	7 days	6 months to 1 year	Darkening and drying out.
Left-over Broiled, Fried, or Roasted Meat	Cool to room temperature. Covered container. Refrigerator.	5 to 6 days	6 months to 1 year	Darkening and drying out. Irridescence on cooked ham or lamb harmless.
Left-over Casseroles, Meat Pies, and Stews	Cool until just hot to the touch. Covered container. Refrigerator.	3 to 4 days	6 months to 1 year	
Canned				
Unopened	Original container. Kitchen shelf.			
Opened	Original or other covered container. Immerse in own liquid if any is available. Refrigerator.	4 to 5 days	6 months to 1 year	
Frozen	Original container or moisture-proof wrapper. Freezer.	10 days	6 months to 1 year	Thawing, darkening and drying out.
Fruits and Vegetables				
Fresh				
Soft-textured fruits and vegetables such as berries and tomatoes	Uncovered. Avoid bruising or crushing. Refrigerator.	3 to 14 days	1 year[a]	If softening, discoloration, mold, and/or general decay is present, separate damaged produce for immediate use. Remove soft, moldy, and/or decayed parts. Partially cook for further storage.

(*continues*)

Table 12.2 Storage Chart (*continued*)

	How to Store	*Approx. Time Limit for Storage* Refrig.	Freezer	Evidence of Quality Loss
		Fruits and Vegetables (cont.)		
Leafy Vegetables	Wash and remove undesirable parts. Cover or wrap to retain moisture. Refrigerator.	3 to 8 days	1 year[a]	Discard decayed parts. Recrisp remainder by placing in cold water.
Firm or Hard Fruits and Vegetables, such as apples, oranges, carrots, and potatoes.	Uncovered. Cool place such as cold cellar or refrigerator. Do not refrigerate potatoes, because they become sweet.	1 to 4 months	1 year[a]	Separate damaged produce and use as soon as possible. Remove soft or bad parts. Some may be recrisped in cold water. Partially cook for further storage.
Cooked Fruits and Vegetables, including creamed vegetables	Cover with accompanying sauce or immerse in cooking liquid. Refrigerator.	4 to 5 days	1 year	If curdled, or if mold and/or off-odor and flavor develop, product is not reclaimable. Discard.
Canned Fruites and Vegetables				
Unopened	Original container. Kitchen shelf.			If can is bulged or cover is loose, if bubbles appear on surface, and/or mold and off-odors develop, fruits may be used if flavor is acceptable. Vegetables, particularly home canned, should be brought to a boil and boiled for 15 minutes before tasting when spoilage is suspected.
Opened	Original or other covered container. Immerse in own liquid. Refrigerator.	4 to 5 days	1 year	If moldy, curdled, and/or off-odor and flavor, not reclaimable. Discard.
Frozen Fruits and Vegetables	Original container. Freezer.	7 days	1 year	If merely thawed, refreeze but expect some softening when thawed again for use. Usable if not moldy and if flavor is acceptable. Otherwise discard.
Dried Fruits and Vegetables	Original or other covered container. Kitchen shelf.	Not advisable because of moisture absorption		Remove mold if practicable; if the flavor of the remainder is acceptable, it may be used.
Pickled				
Fruits	Original containers. Refrigerator.	2 to 3 weeks	1 year[b]	If food darkens or softens, or if mold develops, remove mold if practicable, and if flavor is satisfactory, may be used.
Vegetables, including pickles and relishes	Original container. Immerse in pickling liquid. Cool place or refrigerator.	5 to 6 months		Remove mold if practicable, and if the flavor of the remainder is acceptable, use.
		Spaghetti, Macaroni, Noodles and Rice		
Dry	Original or other covered container. Kitchen shelf.			If odor and/or flavor are stale or rancid, slight off-odor and/or flavor can be masked by seasonings.
Prepared Dishes, plain or with meat, fish, chicken, etc., added	Covered container. Refrigerator.	4 to 5 days	1 year	
		Salads, Dressings, and Sauces		
Salads				
Vegetable and fruit (raw, canned, cooked, or frozen)	Plain to hold only briefly. Refrigerator.	4 to 6 hours		If there is an excess of liquid and the garnish is limp, drain and replace garnish.

[a]Special preparation prior to freezing is required for most fruits and vegetables.

Table 12.2 Storage Chart (*continued*)

	How to Store	Approx. Time Limit for Storage: Refrig.	Freezer	Evidence of Quality Loss
		Salads, Dressings, and Sauces (cont.)		
Cooked meat, poultry, fish, and eggs	Plan to hold only briefly. Refrigerator.	3 to 4 hours		
Left-over Salads	Discard garnish. Refrigerator.	(See chart for particular ingredient in question)		
Thickened Salad Dressing	Original or other covered container. Cool place or refrigerator.	1 week to 1 year		If liquid separates, stir or beat to reblend. If off-flavor, discard.
Oil and Vinegar Dressings	Original or other covered container. Cool place.	1 week to 1 year		If off-flavor is evident, discard.
White Sauce and those sauces having a white sauce base	Covered container. Refrigerator.	2 to 3 days	1 year. Frequently separates at these temperatures	If liquid separates, stir. If mold and/or off-flavors develop, discard.
Sweet or Sour Cream Sauces and their variations	Covered container. Refrigerator.	2 to 3 days	1 year	If liquid separates, stir. If mold and/or off-flavors develop, discard.
Hard Sauce or Special Flavor Sauces, such as chocolate and butterscotch	Covered container. Refrigerator.	1 to 6 months		If merely separated, stir or beat to reblend. If moldy and/or off-flavor, discard.
		Fats and Oils		
Butter and Margarine	Covered or wrapped. Refrigerator.	7 to 30 days	6 to 8 months	If off-odors and/or flavors are noticed, they can be masked by seasonings in cooking.
Lard	Covered or wrapped. Refrigerator.	15 to 60 days	10 to 12 months	Slight off-odor and/or flavor can be masked by seasonings in cooking. If used for deep fat frying, some off-odors can be removed by frying a few pieces of potatoes first.
Firm Vegetable Shortenings	Original container, covered. Kitchen shelf.	No definite limit		Mask slight off-odor and/or flavor by seasonings in cooking. If used for deep fat frying, fry few potatoes first to remove some off-odors.
Fat Drippings	Any covered container. Refrigerator.	7 to 30 days	6 to 8 months	If off-odors and/or flavors develop, not reclaimable. Discard.
Cooking and Salad Oils	Original container, covered. Cool place or refrigerator, depending on the directions supplied with the material.	No definite limits		If merely cloudy, warm to restore clarity. If off-odors and/or flavors occur, mask by seasonings in cooking.
		Breads, Biscuits, Muffins, and Rolls		
Breads, Biscuits, Muffins and Rolls, with or without fruit and/or nuts	Moisture-proof wrapper or container. Cool place.	2 weeks	5 to 6 months	If moldy or dried out, or if yeasty odor or stale flavor develops, dry completely and use for bread crumbs, or discard if moldy or yeasty.
Rolls with Special Fillings	(See chart information for particular type of filling.)			If mold, drying out, yeasty odor, or stale flavor develops, use for bread crumbs if reclaimable, but discard filling if moldy, yeasty, or sour.
Prepared Mixes				
Dry	Original or other covered container. Fold bag or liner to surface of contents. Kitchen shelf.	5 to 6 months		If odor and/or flavor are stale or rancid, slight off-odor and/or flavor can be masked by adding flavoring or seasoning.
With added liquid	Covered dish. Refrigerator.	3 to 4 days		If moldy, curled, or of sour or yeasty odor, not reclaimable. Discard.

(*continues*)

Table 12.2 Storage Chart (*continued*)

	How to Store	Approx. Time Limit for Storage: Refrig.	Freezer	Evidence of Quality Loss
		Breads, Biscuits, Muffins, and Rolls (cont.)		
"Ready-for-the-oven" Rolls (baked but not browned)	Original container. Refrigerator.	2 weeks	5 to 6 months	If mold, drying out, yeasty odor or stale flavors develop, dry completely and use for bread crumbs, or discard if moldy or yeasty.
Refrigerator Doughs or Unbaked Doughs	Moisture-proof wrapper or container. Refrigerator.	5 days	5 to s 6 month	If moldy and/or soured, not reclaimable. Discard.
Dry Bread Crumbs	Original or other covered container. Kitchen shelf.			Slight off-odor and/or flavor can be masked by using in highly seasoned food.
		Cereals, Flour, Pancake and Waffle Mixes		
Dry	Original or other covered container. bag or liner to surface of contents. Kitchen shelf.	5 to 6 months		If odor is stale or rancid, can be masked by adding flavoring or seasoning.
Uncooked Batter or Cooked Left-overs	Covered dish. Refrigerator.	3 to 4 days		If batter thickens or there is a loss of leavening power in mixes, add liquid or leavening agent as needed. If moldy, curdled or soured, it is not reclaimable. Discard.
		Baking Chocolate and Cocoa		
Bar Chocolate, Dry Cocoa, and Prepared Dry Cocoa Mix	Wrapped or in original covered container. Kitchen shelf.	1 year		If an off-flavor is evident, discard.
Chocolate Beverage Prepared with Milk or Water	Covered container. Refrigerator.	7 days		If curdled and/or off-flavor, discard.
		Sugar and Other Sweetening Agents		
Sugar and Maple Sugar	Original or other covered container. Fold bag or liner to surface of contents. Kitchen shelf. Refrigerate brown sugar.	(Brown Sugar) 1 year		If caked or hardened, dry out by brief heating in a very low oven.
Honey, Maple Syrup, Sugar Syrups and Corn Syrup				
Unopened	Original container. Cool place.			If mold and/or off-flavor develop, remove mold. If flavor of remainder is satisfactory, it is usable. Otherwise discard.
Opened	Recover tightly. Original container. Refrigerator.			If sugar crystals are present, or mold and/or off-flavors are evident, heat to dissolve crystals. Remove mold. If the flavor of the remains is satisfactory, it is usable. Otherwise discard.
		Cakes, Cookies, and Crackers		
Cakes and Cookies Containing No Fruit, with or without frosting	Covered container. Store with piece of apple or fresh bread. Kitchen shelf.		6 months	If only stale and dried out, use for pudding. If moldy, discard.
Cakes and Cookies with Fruit and/or Nuts	Covered container. Store with piece of apple or fresh bread. Kitchen shelf.	3 weeks to 6 months	1 year	If merely dried out, use for pudding. Otherwise, discard.
Cakes and Cookies with Custard Topping or Filling	Moisture-proof wrapper or container. Refrigerator.	2 days		

Table 12.2 **Storage Chart (*continued*)**

		Approx. Time Limit for Storage		
	How to Store	**Refrig.**	**Freezer**	**Evidence of Quality Loss**
		Cakes, Cookies, and Crackers (cont.)		
Refrigerator Desserts Containing Cake or Cookies and Whipped Cream	Covered container for prolonged storage. Refrigerator.	5 to 6 days	6 months	Usable if merely dried out. Otherwise discard.
Crackers, Snacks, such as potato chips, cheese crackers, etc.	Moisture-proof wrapper or container. Kitchen shelf.			If soggy or tough, recrisp in warm oven. Otherwise discard.
Canned Cakes, such as fruit cake				
Unopened	Original container. Kitchen shelf.			If there is mold on the inside of container, or if the food has a sour odor, discard.
Opened	Original or other covered container. Kitchen shelf.	3 weeks to 6 months	1 year	If moldy and/or sour in odor, or if cobwebs are noticed, discard.
Prepared Mixes				
Dry	Original or other covered container. Kitchen shelf.	5 to 6 months		Slight off-odor and/or flavor may be masked by adding flavoring.
Batter	Covered container. Refrigerator.	3 to 4 days		Usable if flavor is acceptable. More leavening may be added.
Raw Cookie Doughs	Covered container. Refrigerator.	3 weeks	6 months	If moldy and/or sour in odor, discard.
		Pies and Puddings		
Soft Custard and Cornstarch Pies and Puddings, with or without fruit	Refrigerator.	2 to 3 days	6 to 8 months	
Baked Custard Pies and Puddings, with or without fruit. Squash and Pumpkin Pies	Refrigerator.	2 to 7 days	6 to 8 months	If mold and/or sour odor develop, not reclaimable. Discard
Gelatin and Tapioca Pies and Puddings, with or without fruit	Refrigerator.	3 to 7 days	6 to 8 months	If moldy and/or soured, it is not reclaimable. Discard.
Fruit Pies and Puddings, such as apple pie, Brown Betty and mincemeat pie	Cool place or refrigerator.	3 to 7 days	6 to 8 months	If mold and/or sour odor are evident, not reclaimable. Discard.
Other Baked Puddings, such as bread and rice puddings	Refrigerator.	4 to 7 days	6 to 8 months	If mold and/or sour odor develop, not reclaimable. Discard.
Meringue and Whipped Cream Toppings	Refrigerator.			If odor and/or flavor are sour, not reclaimable. Discard.
Unbaked Pie Doughs	Covered container. Refrigerator.	7 to 10 days	6 to 8 months	If moldy and/or rancid, discard.
Baked Pie Shells	Covered container. Kitchen shelf.	3 to 7 days		If merely soggy, reclaim by heating in the oven. If moldy or rancid, discard.

(*continues*)

Table 12.2 Storage Chart (*continued*)

	How to Store	Approx. Time Limit for Storage: Refrig.	Freezer	Evidence of Quality Loss
Ice Cream and Other Frozen Desserts				
Ice Cream				
Insulated bag	Unopened. Refrigerator.	5 to 6 hours		If merely soft, refreeze. Usable if texture and flavor are acceptable.
Not in insulated bag	Original container. Surface should be covered to prevent ice crystals from forming. Refrigerator.	3 to 4 hours	8 months	Refreeze to harden. Usable if texture and flavor are satisfactory.
Ices, Mousses, and Sherbets				
Insulated bag	Unopened. Refrigerator.	3 to 4 hours		If soft, refreeze.
Not in insulated bag	Original container. Surface should be covered to prevent ice crystals from forming. Refrigerator.	1 hour	8 months	If softening has occurred, refreeze.
Frozen Combinations with cake, cookies, or pastry				
Insulated bag	Unopened. Refrigerator.	3 to 4 hours		If frozen part has softened, but texture and flavor are acceptable, refreeze. If stale flavor and/or sogginess have developed in baked material, or if whipped cream garnish is sour, discard. Remove garnish if soured.
Not in insulated bag	Original container. Surface should be covered to prevent ice crystals from forming Refrigerator.	1 to 2 hours	4 months	If frozen part has softened, but texture and flavor are acceptable, refreeze. If stale flavor and/or sogginess have developed in baked material, of if whipped cream garnish is sour, discard. Remove garnish if soured.
Ice Cream Powder	Original container. Kitchen shelf.			If stale in flavor, discard.
Canned Ice Cream Mixes	Original container. Kitchen shelf.			If can is bulged, discard.
Jams, Jellies, Marmalades, and Preserves				
Unopened Container	Original container. Kitchen shelf.			If cover is loose or can is bulged, or if mold and/or fermented odor and flavor are evident, remove mold. If the flavor of the remainder is acceptable, it is usable, otherwise discard.
Opened Container	Original container. Keep covered. Refrigerate for long holding.	3 to 5 weeks	1 year	If mold and/or fermented odor and flavor are evident, remove mold. If the flavor of the remainder is acceptable, it is usable. Otherwise discard.
Candy, Gum, and Nuts				
Plain Chocolate Bars and Chocolate without fruit or nuts	Original container. Kitchen shelf. Refrigerate in hot weather.	1 year		If white spots appear, or softening and/or stale flavor are evident, usable if flavor is acceptable.
Hard Candy, such as lollipops and Christmas candy	Moisture-proof wrapper or container. Kitchen shelf.			If softened, usable in sauces or frostings.
Candies and Chocolates containing fruit and/or nuts	Original container. Kitchen shelf. Refrigerate in hot weather.			Unless cobwebby, usable if flavor is acceptable. Otherwise discard.
All Other Candies, including those with caramel and nougat centers	Original container. Kitchen shelf.	1 year		If dried out and/or stale in flavor, usable if the flavor is acceptable.

Table 12.2 **Storage Chart (*continued*)**

	How to Store	*Approx. Time Limit for Storage* Refrig.	Freezer	Evidence of Quality Loss
		Candy, Gum, and Nuts (cont.)		
Gum	Original container. Kitchen shelf.			If hardened, usable if desired.
Nuts				
In shell	Kitchen shelf.			If moldy, or if worms, moths, or cobwebs are evident, discard.
Shelled	Covered container. Kitchen shelf.		6 months	If merely soggy, warm briefly in the oven or use in cooking.
Vacuum Packed, unopened	Original container. Kitchen shelf.			If stale or rancid in flavor, usable if flavor is acceptable.
Vacuum Packed, opened	Covered container. Kitchen shelf.	6 months		If moldy, stale, rancid in flavor, soggy, and/or cobwebby, warm briefly in oven or use in cooking. Otherwise discard.
		Fruit and Vegetable Juices		
Freshly Prepared	Nonmetallic container. Refrigerator.	5 to 6 days	1 year	If mold and off-odor and/or flavor develop, fruit juices can be tasted, and if the flavor is acceptable, are usable.
Canned				
Unopened	Original container. Cool place.			Fruit juices: If can is bulged or cover is loose, or if mold, off-odor, and/or bubbles on the surface are evident, they can be used. Vegetable juices, particularly home canned, should be brought to a boil and boiled for 15 minutes before tasting, when spoilage is suspected.
Opened	Nonmetallic container. Refrigerator.	5 to 6 days	1 year	
Frozen				
Unopened	Original container. Freezer.	7 days	1 year	
Opened	Nonmetallic container. Refrigerator.	5 to 6 days	1 year	
		Milk, Milk Drinks, and Cream		
Cream, Flavored Milk Drinks and Plain Milk (whole or skimmed)	Original container. Refrigerator.	3 to 4 days		If curdling, mold, and/or sour odor are evident: If of acceptable flavor and not moldy, use for cooking.
Sour, Cultured, or Buttermilk	Original container. Refrigerator.	10 to 14 days		If curdled, moldy, and/or sour in odor, use for cooking, if free of mold and acceptable in flavor.
Condensed and Evaporated				
Unopened	Original container. Kitchen shelf. Invert can after 6 months storage.			If can is bulged, discard. If contents merely thickened, reclaim by adding water and use for cooking.
Opened	Original container, covered. Refrigerator.	10 days		If moldy, discard. If curdled and/or sour in odor, but of acceptable flavor, use for cooking. If thickened, add water.

APPENDIX

A

Quality Controls and Federal Regulations

WHAT IS QUALITY?

Quality is measurable on objective grounds, even though there are many varied subjective attributes. Quality is measured in terms of a product's chemical and physical attributes: flavor, texture, color, appearance, consistency, palatability, nutritional values, safety, ease of handling, convenience, storage stability, and packaging.

The five senses are invaluable. The average layperson can identify about 2,000 odors and tastes; the trained technician can identify some 5,000.

Smell is not really a function of the nose. It is a function of the olfactory bulb, which is located in the back of the nasal cavity. The rest of the nose picks up only the sensations of touch: heat and pain.

Taste is the most complex of the senses, because flavor is a combination of taste, smell, and feelings. The tongue is the receptor of flavor. It registers only four responses: sweet, salty, sour, and bitter. It is the taste buds that pick up three sensations, and the intensity of these sensations is directly affected by the temperature of the food. Sweet tastes are picked up only at the front of the tongue, saltiness is determined at the tip and extends up the sides slightly, sour is on the sides of the back of the tongue, and bitterness is registered in the main body of the tongue.

The temperature of the food tasted greatly affects a person's ability to judge flavor. Food should not be sampled for taste when its temperature is below 40° F or above 100° F. The tongue will not register accurate sensations of sourness or detect subtle flavors in any temperature range except 40° F to 100° F. The sensations of food being tough, mushy, gristly, or hard are really the qualities of firmness, softness, juiciness, texture, and grittiness.

Determining Quality

How does one instruct a taste panel in order to get a valid judgment of a product? According to Marvin Thorner, chemical engineer, and premier food technologist: "Prior to the actual testing, an inexperienced taster should perform a series of examinations using solutions of foods having the four basic tastes. Dilute solutions of pure substances exhibiting the effects of sweetness, sourness, bitterness, and saltiness should be prepared. As each solution is tested, a mental note should be made of the reaction and the location of the stimulus on the tongue."

Before actual testing, the mouth should be rinsed with warm water. This preliminary step will freshen the mouth cavity and the taste buds for sharper perception. The sample should then be drawn into the mouth with a "slurp" or whirling action, so that all areas of the cavity are moistened. Immediately after the mouth is fully moistened and the sense registered, the liquid must be expelled. If the impression was not clear, a second or third test should be made. Between tests of the same or different substances, the mouth should be flushed with warm water.

The sense of touch plays an important role in taste evaluation. The temperature of the material must be noted, and if not in the proper range, it must be adjusted to meet its physical character. The tasting should be done early in the morning because people's tasting senses become dulled later in the day. Limit the number of samples to no more than six, as a taster's perception is rarely accurate after that point. Use two identical samples and one different sample, all blind coded to weed out faulty perception. Use definitive words for judging that will relate to the product differentiation at a later time.

Equipment Used to Determine Quality

There are several pieces of equipment used in the determination of quality.

Scales are needed to check the portion weights, not only as a receiving control but also as a cooking control. (Microwave ovens will cook or reheat properly only when the timings are set in accordance with the product weights.)

Thermometers are essential in the receiving function of convenience foods as a minimal check against product damage. For testing the temperature of plated or plate-ready foods, one should use only the cup-type or digital thermometer because it brings up some of the food with it, which prevents the temperature from changing before the reading is made.

Another valuable tool is the Abbe refractometer. This tool measures the density of liquids through the refraction or bending of light. In most operations, the operator will want a hand refractometer. Tables supplied with USDA Standards enable the operator to measure such things as the amount of sugar in fountain syrups; the amount of total solids in sauces, gravies, and tomato products; and the purity of the fats and oils used in the operation.

The Brix hydrometer is also a helpful tool. This instrument measures the sugar concentration in liquids and is somewhat like the hydrometer used by a gas station attendant to measure the amount of antifreeze in automobile radiators. All the syrups of canned goods have USDA specified Brix points. Only a Brix hydrometer or a hand refractometer will enable the buyer to know if he or she got what was ordered.

A fat analyzer is basic to a quality assurance program. This is a little machine about the size of a briefcase and just as portable. The fat analyzer will, in a 15-minute period, give you an accurate reading as to the percentage of fat in ground beef. Not only is this essential to measuring the quality of fresh meat, but it is invaluable in the determination of the quality of frozen foods such as stuffed cabbage and stuffed peppers.

DETERMINING STANDARDS AND SPECIFICATIONS

The United States Department of Agriculture has determined standards and specifications for virtually every food product on the market. It is important that buyers use these specifications in the purchase of all foodstuffs. Buyers should purchase on quotations inasmuch as the reading of every detailed specifications is very involved. An alternative is to prepare a typewritten list of all your specifications for your operation, and bind them into a booklet. Make divisions for the obvious purchase categories such as milk, butter, eggs and cheese, fresh and frozen vegetables, meats, and so forth. Send each purveyor a copy of the appropriate section with instructions that these are the detailed specifications on which future quotations will be taken. These specifications will mean nothing if the buyer does not have a receiving program that ensures the quality of the goods received. It is difficult to know just what quality level is needed in each product.

Wherever there are USDA A, B, and C grades for canned fruits and vegetables, most operations should find USDA Grade B quite satisfactory. Grade A costs too much extra money for the difference in quality. Grade C generally allows an excess of defects that will affect the plate-worthiness of the foods. If the choice is between Grade A and Grade C—for example, with jams and jellies, for which there is no USDA Grade B—then take Grade A. For canned goods the buyer must insist, at a minimum, that the cans or cases bear the mark of the federal inspector. Because most operations are not large enough to take advantage of having the local USDA office run a grading inspection on a sample of the lot, the buyer can cut its own cans and determine the appropriateness of the drained weight, sieve or count, defects, and Brix of the syrup. The higher the Brix point, the more sugar in the syrup. For canned fruits that will be served in their juices, the buyer will want a heavy syrup of 21 Brix or maybe extra heavy at 25. Fruits for salads need only have a 16 Brix, which is called "slightly sweetened water."

Frozen vegetables that are to be cooked need only be Grade B, but the buyer must make sure that each bears the USDA shield and the grade. Fresh vegetables and fruits, generally served in the raw state, should be Grade No. 1. For some products, certification on the case is available. For meat, buyers should use the Institutional Meat Purchase Specifications (IMPS) or the Meat Buyers Guides (MBG) published by the National Association of Meat Purveyors. The MBG specs are the same as the IMPS, but slightly shortened and accompanied by photographs that are a great aid in making receiving checks. Meat purchasing can be very costly, and the only way the buyer can be sure that he or she is getting a fair deal is to insist on the specs being followed. For all meats that are cooked by dry-heat methods—roasting, broiling, pan-frying—you should use USDA Choice, as the fat in the muscle walls is needed for tenderness. For moist-heat cooking—boiling, stewing, and so forth—you can use USDA Select or Good, as the slow cooking process will break down the long muscle tissues and create a tender product. Note, however, that not all food products are graded.

Standards for meat and poultry products are pertinent guidelines to assist in the task of quality interpretation and cost evaluation. To qualify for such classification, a food must contain a minimum amount of meat or poultry as prescribed by the USDA. For example, ready-to-serve chicken soup must contain at least 2 percent chicken. Condensed chicken soup must contain 4 percent chicken or more, because it would then contain at least 2 percent when diluted with water. But chicken-flavored soup, which is not considered a poultry product, may contain less chicken. The standards for meat ingredients are usually based on the fresh weight of the product, whereas those for poultry are measured on the weight of the cooked, deboned product. Because meat and poultry shrink during cooking, standards take this fact into account. For instance, turkey pot pie must contain 14 percent or more cooked turkey. Chicken burgers must be 100 percent chicken; a product containing fillers must be called chicken patties.

At present there are more than 150 standards for prepared convenience entrées. The following are examples: Beef Stroganoff, at least 45 percent fresh uncooked beef or 30 percent cooked beef, and at least 10 percent sour cream, or a "gourmet" combination of at least 7.5 percent sour cream and 5 percent wine; Chop Suey, at least 25 percent meat; Fritters, at least 50 percent cooked meat; Chicken or Turkey à la King, at least 20 percent poultry meat. We can look forward to such standards being developed for all convenience entrée products.

Improper and careless storage techniques will downgrade quality. Poor stock rotation and prolonged storing in any of the storage areas (dry, cooler, or freezer) may lead to changes in flavor and texture. Positive storage-control procedures will tend to eliminate these problems. Haphazard food assembly methods, such as piling or stacking of food products, spillage, and mixing of liquids, such as sauces and gravies, so that they become contaminated with each other, will help to destroy the inherent character of the food. Spices, condiments, and herbs should be assembled in a neat, orderly fashion until needed for preparation. Sandwich spreads should be kept refrigerated and stored in tight containers during slack periods. Sandwich meats, fish, and other products that have a tendency to form a crust or hard surface when exposed to air should be mixed in small lots or covered. Improper thawing procedures for certain foods will affect quality. Items intended for microwave oven heating may require special thawing techniques to prevent uneven doneness. Foods that are to be cooked in the frozen state, like French fries, should not be allowed to thaw, but should be held in a freezer until used.

Although *cooking* refers to all modes of heating food (fully or partially), factors affecting quality, such as temperature, timing, formulations, and equipment maintenance, are all involved in the final outcome. For example, a faulty timer on a microwave oven will yield inconsistent quality; unfiltered fat for deep frying will produce low-quality fried products; a loose door or a worn gasket on a steam pressure cooker may give unsatisfactory results.

Employment of proper utensils is essential for consistent quality. The use of a 10 gal container to heat 1 gal of a product will cause excessive shrinkage, texture changes, and flavor losses. Salad dressings should be added just prior to serving tossed or mixed salads containing soft-textured components (tomato slices), because they should not be allowed to stand for long periods of time. Lettuce and other salad vegetables should be kept dry to prevent browning and loss of crispness. Tossed or mixed salad ingredients should be stored in coolers until used. Before storage they should be wrapped in plastic film or placed in a covered vessel. Forcing or inducing heating by the use of high temperatures or exposed flames will reduce flavor, decrease tenderness, increase shrinkage, and may produce a burnt taste. Cooking should always be performed at the right temperature and within the prescribed time cycle.

Formulations or standardized recipes should be written clearly and precisely. All quantity measurements should be listed, together with the size of the measuring device to be used, such as ladle or scoop size and number. Heating cycles and equipment should also be posted. Holding food in a steam table, cabinet, or under infrared lamps for excessive periods of time will reduce quality, as it may affect food texture and flavor or increase shrinkage and loss of nutrients. Foods will become mealy, mushy, soggy, or dried out. Overproduction should be held to a minimum. Coffee should not be held more than one hour and only at temperatures of 185°–190° F. Desserts like cream or custard pies require refrigerated storage to prevent spoilage. Leftover foods should be properly refrigerated; packed loosely in shallow, covered pans; labeled; and dated. Foods that will not be used should be thrown away. Fresh produce should always be washed before serving or use in salads or soup. Washing will remove excess soil, dust, insecticides, and surface bacteria.

There is no substitute for experience when it comes to determining the essential variations within each factor. Nevertheless, an effective job can be done if the buyer will keep in mind this general scoring and tolerance outline:

USDA Grade A: Must be practically perfect in every respect, allowing a tolerance of 15–18 percent within the grade that is scored between 85 and 100 points, inclusive.

USDA Grade B: Must be reasonably perfect in every respect, allowing a tolerance of 16⅔–20 percent within the grade that is scored between 75 and 89 points, inclusive.

USDA Grade C: Must be fairly perfect in every respect, allowing a tolerance of 20–25 percent within the grade that is scored between 60 and 74 points, inclusive.

The exceptions are those products for which only two grades are commercially packed. In such instances the general scoring and tolerance outline is:

USDA Grade A: Must be practically perfect in every respect, allowing a tolerance of 15–18 percent within the grade that is scored between 85 and 100 points, inclusive.

USDA Grade B: Must be fairly perfect in every respect, allowing a tolerance of 18–22 percent within the grade that is scored between 70 and 84 points, inclusive.

The relative importance of each factor can be expressed numerically on a scale of 1 to 100, and the USDA has worked out a scoring system, which is used in government grading. In commercial practice, although no particular attempt has been made to use a numerical scoring system, the factors emphasized include the following:

Flavor	Symmetry	Wholeness
General Appearance	Absence of Defects	Cut
Consistency	Color	General Character
Finish	Type	Maturity
Clearness of Liquor	Style	Texture
Count	Firmness	Clearness of Syrup
Uniformity of Size	Tenderness	Syrup Density
	Drained Weight	

Flavor. Because there is such a great variation in individual likes and dislikes, it is exceedingly difficult to score so elusive a factor as "Flavor." In most of the U.S. standards for grades of foods, the term "Normal Flavor" is used. In commercial grading, however, "Flavor" is the prime factor. If what we eat tastes good, we somehow overlook minor deficiencies.

General Appearance. People "eat with their eyes" even before they taste what they see. If a food does not look good, the prejudice may very well affect a person's opinion of its flavor. For this reason, some food experts rate "General Appearance" every bit as important as "Flavor." The relative importance of all other factors depends entirely on the product being judged. Although no attempt is made here to score "Quality" or "Standard" factors for the available grades, the important factors by which to judge the quality of some of the more popular, everyday products are emphasized to help food buyers make certain that deliveries conform to purchase specifications.

Color. "Color" is the chief subdivision of "General Appearance," and to receive proper rating, the color should be typical of the product. Many food experts score color on a par with flavor for some items.

Type. By "Type" is meant distinctive classifications of a specific product. For example, Culturally Bleached Asparagus is one distinct type, and all Green Asparagus is a separate, distinct type.

Style. When we refer to "Style," we think of prevalent approved ideas of form adaptable to popular food items that canners and processors make available to buyers. A good example is:

PEACHES—Sliced or Halves.
COUNT—Actual number of pieces found upon opening and examining the container's contents.

Uniformity of Size. The degree of consistency relative to freedom from variation or difference. Sameness or alikeness.

Symmetry. The degree of consistently harmonious proportions of units in a container.

Absence of Defects. By "Absence of Defects" we mean the degree of freedom from grit, from harmless foreign or other extraneous material, and from damage resulting from poor or careless handling, from insects, or from mechanical, pathological, or other similar injury.

General Character. In regard to the factor "General Character," consideration is given to degree of ripeness or maturity, the texture and condition of flesh, the firmness and tenderness of the product, its tendency to retain its apparent original conformation and size without material disintegration, the wholeness or cut, consistency or finish, and clearness of liquor or syrup.

Maturity. This factor refers to the degree of development or ripeness of the product.

Texture. By "Texture" we mean structural composition or character of the product's tissues.

Firmness. The degree of soundness of the product's structure.

Tenderness. The degree of freedom from tough or hard fibers.

Wholeness. The state of completeness or entirety.

Cut. This refers to the character of the cut; that is, the effect of the cut on the appearance of the product.

Consistency. In some products, such as Fruit Butters, this factor refers to viscosity, that is, stickiness or gumminess. In other products, such as Tomato Catsup and Tomato Puree, the term is applied to density or specific gravity.

Finish. This factor especially refers to the size and texture of particles, the smoothness, evenness, and uniformity of the grain.

Clearness of Liquor. This factor requires no elaboration. The degree of sediment and cloudiness materially affects the score for quality.

Clearness of Syrup. Any degree of sediment or cloudiness materially affects the grading score.

Syrup Density. The degree or percentage, by weight, of sugar going into solution as measured by either the Brix or Balling Scale on hydrometers or a saccharometer.

Drained Weight. The weight of the product after draining the liquor or syrup according to the method prescribed by the National Canners Association or the USDA.

IMPORTANT QUALITY "INSPECTION CHECK" FACTORS

CANNED FRUITS

APPLES

Flavor
General Appearance
Color
Style
Drained Weight
Absence of Defects
Cut
Texture

APPLESAUCE

Flavor
General Appearance
Color
Uniformity of Size
Absence of Defects
General Character
Consistency
Finish
Texture

APRICOTS

Flavor
Symmetry
General Appearance
Absence of Defects
Color
General Character
Style
Maturity
Uniformity of Size
Texture
Count
Degree of Ripeness

BLACKBERRIES

Flavor
Absence of Defects
General Appearance
General Character
Color
Maturity
Uniformity of Size
Firmness

BLUEBERRIES

Flavor
Absence of Defects
General Appearance
General Character
Color
Firmness

BOYSENBERRIES

Flavor
Absence of Defects
General Character
Appearance
Color
Maturity
Uniformity of Size
Firmness

CHERRIES (BLACK SWEET)

Flavor
Symmetry
General Appearance
Color
Absence of Defects
Style
General Character
Uniformity of Size
Texture
Count

CHERRIES (RED SOUR PITTED)

Flavor
Absence of Defects
Texture
General Appearance
General Character
Syrup Density
Color
Firmness

CHERRIES (ROYAL ANNE)

Flavor
Symmetry
General Appearance
Color
Absence of Defects
Style
General Character
Uniformity of Size
Texture

CRANBERRY SAUCE

Flavor
Style
Texture
General Appearance
Absence of Defects
Finish
Color
General Character
Wholeness
Consistency

FIGS (KADOTA)

Flavor
Symmetry
Texture
General Appearance
Count
Firmness
Color
Absence of Defects
Tenderness
Style
General Character
Syrup Density
Uniformity of Size
Maturity
Drained Weight

FIGS (TEXAS SKINLESS)

Flavor
Symmetry
Texture
General Appearance
Count
Firmness
Color
Drained Weight
Absence of Defects
Tenderness
Style
General Character
Syrup Density
Uniformity of Size
Maturity

FRUIT COCKTAIL

Flavor
Absence of Defects
Texture
General Appearance
General Character
Syrup Density
Appearance
Color
Firmness
Clearness of Syrup
Uniformity of Size
Tenderness
Drained Weight
Percentage by Weight of Fruits in Combination

FRUITS FOR SALAD

- Flavor
- Number of Portions
- General Appearance
- Absence of Defects
- Color
- General Character
- Uniformity by Weight of Fruits in Combination
- Tenderness
- Texture
- Syrup Density
- Clearness of Syrup
- Drained Weight

GOOSEBERRIES

- Flavor
- Absence of Defects
- General Appearance
- General Character
- Color
- Texture
- Uniformity of Size
- Tenderness
- Firmness
- Maturity
- Syrup Density
- Drained Weight

GRAPEFRUIT

- Flavor
- Absence of Defects
- General Appearance
- General Character
- Color
- Maturity
- Wholeness
- Firmness
- Tenderness
- Texture
- Syrup Density
- Drained Weight

GRAPES

- Flavor
- Absence of Defects
- General Appearance
- General Character
- Color
- Maturity
- Style
- Texture
- Uniformity of Size
- Firmness
- Tenderness
- Syrup Density
- Drained Weight

HUCKLEBERRIES

- Flavor
- Absence of Defects
- Texture
- General Appearance
- General Character
- Syrup Density
- Color
- Firmness
- Drained Weight

LOGANBERRIES

- Flavor
- Absence of Defects
- Texture
- General Appearance
- General Character
- Tenderness
- Color
- Maturity
- Syrup Density
- Uniformity of Size
- Firmness
- Drained Weight

MIXED ORANGE AND GRAPEFRUIT SEGMENTS

- Flavor
- Absence of Defects
- Tenderness
- General Appearance
- General Character
- Texture
- Color
- Maturity
- Syrup Density
- Wholeness
- Firmness
- Drained Weight
- Percentage by Weight of Each Fruit in Combination

NECTARINES

- Flavor
- Symmetry
- Firmness
- General Appearance
- Count
- Tenderness
- Color
- Absence of Defects
- Clearness of Syrup
- Type
- General Character
- Syrup Density
- Style
- Maturity
- Drained Weight
- Uniformity of Size
- Texture

ORANGE SEGMENTS

- Flavor
- Absence of Defects
- General Appearance
- General Character
- Color
- Maturity
- Wholeness
- Firmness
- Tenderness
- Texture
- Syrup Density
- Drained Weight

PEACHES

- Flavor
- Symmetry
- Firmness
- General Appearance
- Count
- Tenderness
- Color
- Absence of Defects
- Clearness of Syrup
- Type
- General Character
- Syrup Density
- Style
- Maturity
- Drained Weight
- Uniformity of Size
- Texture

PEARS

- Flavor
- Symmetry
- General Appearance
- Color
- Count
- Absence of Defects
- Type
- General Character
- Style
- Maturity
- Uniformity of Size
- Texture

PINEAPPLE

- Flavor
- Count
- General Appearance
- Absence of Defects
- Color
- General Character
- Style
- Maturity
- Uniformity of Size
- Texture
- Symmetry

PLUMS (GREENGAGE)

- Flavor
- Symmetry
- Absence of Defects
- Style

General Appearance
Color
Count
General Character
Uniformity of Size
Maturity

PLUMS (PURPLE OR RED)

Flavor
Symmetry
General Appearance
Color
Count
Absence of Defects
Style
General Character
Uniformity of Size
Maturity
Firmness
Tenderness
Clearness of Syrup
Syrup Density

PLUMS (YELLOW EGG)

Flavor
Symmetry
General Appearance
Color
Absence of Defects
Style
General Character
Uniformity of Size
Maturity

PRUNES (CANNED DRIED)

Flavor
Absence of Defects
Texture
General Appearance
General Character
Tenderness
Color
Maturity
Syrup Density
Uniformity of Size
Firmness
Drained Weight

RASPBERRIES (BLACK)

Flavor
Absence of Defects
Texture
General Appearance
General Character
Tenderness
Color
Maturity
Syrup Density
Uniformity of Size
Firmness
Drained Weight

RASPBERRIES (RED)

Flavor
Absence of Defects
Texture
General Appearance
General Character
Tenderness
Color
Maturity
Syrup Density
Uniformity of Size
Firmness
Drained Weight

STRAWBERRIES

Flavor
Absence of Defects
Texture
General Appearance
General Character
Tenderness
Color
Maturity
Syrup Density
Uniformity of Size
Firmness
Drained Weight

YOUNGBERRIES

Flavor
Absence of Defects
Maturity
Syrup Density
Texture
General Appearance
General Character
Tenderness
Color
Uniformity of Size
Firmness
Drained Weight
Symmetry

CANNED VEGETABLES

ASPARAGUS

Flavor
Uniformity of Size
Firmness
General Appearance
Count
Tenderness
Color
Absence of Defects
Clearness of Liquor
Type
General Character
Cut
Style
Maturity
Drained Weight
Style
Firmness
Consistency

BEANS

Flavor
Absence of Defects
Firmness
General Character
Tenderness
General Appearance
Color
Maturity
Syrup Density
Uniformity of Size
Texture
Drained Weight
Flavor

BEANS (LIMA)

Flavor
Absence of Defects
Texture
General Character
Tenderness
General Appearance
Color
Maturity
Syrup Clarity
Uniformity of Size
Firmness
Drained Weight
Texture
Wholeness
Style

BEANS (WAX)

Flavor
General Appearance
Uniformity of Size
Flavor
Absence of Defects
Texture
General Character
Tenderness
General Appearance
Maturity
Syrup Clarity
Firmness
Drained Weight
Texture
Wholeness
Style
Color

BEETS

Flavor
Absence of Defects
Texture
General Appearance
Syrup Density
Uniformity of Size
Firmness
Drained Weight

Tenderness
General Appearance
Color
Maturity
Texture
Wholeness
Style

BRUSSELS SPROUTS

Flavor
Absence of Defects
Texture
General Character
Tenderness
General Appearance
Color
Maturity
Syrup Clarity
Uniformity of Size
Firmness
Drained Weight
Texture
Wholeness
Style

CABBAGE

Flavor
Crispness Texture
General Character
General Appearance
Color
Maturity
Firmness
Drained Weight
Texture
Wholeness
Style

CARROTS

Flavor
Absence of Defects
Texture
General Character
Tenderness
General Appearance
Color
Maturity
Syrup Clarity
Uniformity of Size
Firmness
Drained Weight
Texture
Wholeness
Style

CORN (CREAM)

Flavor
Crispness Texture
General Character
General Appearance
Color
Style
Maturity
Firmness
Drained Weight
Texture
Wholeness

CORN (WHOLE KERNEL)

Flavor
Absence of Defects
Texture
General Character
Tenderness
General Appearance
Color
Maturity
Syrup Clarity
Uniformity of Size
Firmness
Drained Weight
Texture
Wholeness
Style

HOMINY

Flavor
Crispness
Texture
General Character
General Appearance
Color
Maturity
Firmness
Drained Weight
Texture
Wholeness
Style

OKRA

Flavor
Absence of Defects
Texture
General Character
Tenderness
General Appearance
Color
Maturity
Syrup Clarity
Uniformity of Size
Firmness
Drained Weight
Texture
Wholeness
Style

PEAS

Flavor
Absence of Defects
Texture
General Character
Tenderness
General Appearance
Color
Maturity
Syrup Clarity
Uniformity of Size
Firmness
Drained Weight
Texture
Wholeness
Style

PEAS (BLACK-EYED)

Flavor
Absence of Defects
Texture
General Character
Tenderness
General Appearance
Color
Maturity
Syrup Clarity
Uniformity of Size
Firmness
Drained Weight
Texture
Wholeness
Style

PUMPKIN

Flavor
Absence of Defects
Texture
General Character
General Appearance
Color
Drained Weight
Texture

RHUBARB

Flavor
Absence of Defects
Texture
General Appearance
General Character
Tenderness
Color
Maturity
Firmness
Drained Weight

SAUERKRAUT

- Flavor
- Absence of Defects
- Texture
- General Appearance
- General Character
- Tenderness
- Color
- Maturity
- Firmness
- Drained Weight

SPINACH

- Flavor
- Absence of Defects
- Texture
- General Appearance
- General Character
- Tenderness
- Color
- Maturity
- Firmness
- Drained Weight

SQUASH

- Flavor
- Absence of Defects
- Texture
- General Appearance
- General Character
- Tenderness
- Color
- Maturity
- Firmness
- Drained Weight

SUCCOTASH

- Flavor
- Absence of Defects
- Texture
- General Appearance
- General Character
- Tenderness
- Color
- Maturity
- Firmness
- Drained Weight

SWEET POTATOES

- Flavor
- Absence of Defects
- Texture
- General Appearance
- General Character
- Tenderness
- Color
- Maturity
- Firmness
- Drained Weight

TOMATOES

- Flavor
- Absence of Defects
- Texture
- General Appearance
- General Character
- Canned Juices and Nectars
- Tenderness
- Color
- Maturity
- Firmness
- Drained Weight

APPLE JUICE

- Flavor
- Color
- Style
- Brix Test for Sweetness
- General Appearance
- Absence of Defects
- Percentage of Free and Suspended Pulp

APRICOT NECTAR

- Flavor
- Style
- Brix Test for Sweetness
- General Appearance
- Absence of Defects

CRANBERRY JUICE

- Flavor
- Color
- Style
- Brix Test for Sweetness
- General Appearance
- Absence of Defects
- Percentage of Free and Suspended Pulp

GRAPEFRUIT JUICE

- Flavor
- Color
- Style
- Brix Test for Sweetness
- General Appearance
- Absence of Defects
- Percentage of Free and Suspended Pulp

LEMON JUICE

- Flavor
- Color
- Style
- Brix Test for Sweetness
- General Appearance
- Absence of Defects
- Percentage of Free and Suspended Pulp

ORANGE JUICE

- Flavor
- Color
- Style
- Brix Test for Sweetness
- General Appearance
- Absence of Defects
- Percentage of Free and Suspended Pulp

PEACH NECTAR

- Flavor
- Color
- Style
- Brix Test for Sweetness
- General Appearance
- Absence of Defects
- Percentage of Free and Suspended Pulp

PEAR NECTAR

- Flavor
- Style
- Brix Test for Sweetness
- General Appearance
- Absence of Defects

PINEAPPLE JUICE

- Flavor
- Color
- Style
- Brix Test for Sweetness
- General Appearance
- Absence of Defects
- Percentage of Free and Suspended Pulp

PLUM NECTAR

- Flavor
- Style
- Brix Test for Sweetness
- General Appearance
- Absence of Defects

PRUNE JUICE

Flavor
Color
Style
Brix Test for Sweetness
General Appearance
Absence of Defects
Percentage of Free and Suspended Pulp

TANGERINE JUICE

Flavor
Style
Brix Test for Sweetness
General Appearance
Absence of Defects

TOMATO JUICE

Flavor
Color
Consistency
General Appearance
Absence of Defects

Dried Fruits

APPLES

Flavor
Type
General Character
General Appearance
Style
Wholeness
Color
Uniformity
Cut
Absence of Defects

CURRANTS (ZANTE)

Flavor
Maturity
General Appearance
Absence of Defects
Texture
Type
Color
General Character

DATES

Flavor
Type
General Character
General Appearance
Size
Maturity
Color
Uniformity
Texture
Style
Absence of Defects

FIGS

Flavor
Style
General Character
General Appearance
Size
Maturity
Color
Uniformity
Texture
Type
Absence of Defects
Wholeness

PEACHES

Flavor
Size
General Character
General Appearance
Uniformity
Maturity
Color
Absence of Defects
Texture
Type

PEARS

Flavor
Type
General Character
General Appearance
Size
Maturity
Color
Uniformity
Texture
Absence of Defects

PRUNES

Flavor
Size
General Appearance
Count
Color
Uniformity
Type
Absence of Defects

RAISINS

Flavor
Type
General Appearance
Style
Color
Size

Frozen Foods

APPLES

Flavor
Uniformity of Size
General Appearance
Absence of Defects
Color
General Character
Type
Maturity
Style

APRICOTS

Flavor
Uniformity of Size
General Appearance
Symmetry
Color
Absence of Defects
Type
General Character
Style

ASPARAGUS

Flavor
Style
General Appearance
Size
Color
Uniformity
Type
Absence of Defects

BEANS (LIMA)

Flavor
Type
General Appearance
Size
Color
Absence of Defects

BEANS (SNAP)

Flavor
Style
General Appearance
Size
Color
Absence of Defects
Type
General Character
Maturity
Texture

BERRIES

Flavor
Size
Maturity
General Appearance
Uniformity
Texture
Color
Absence of Defects
Tenderness
Variety
General Character
Sugar or Syrup Type Ratio

BLUEBERRIES OR HUCKLEBERRIES

Flavor
Uniformity
Texture
General Appearance
Absence of Defects
Firmness
Color
General Character
Sugar or Syrup Ratio

BROCCOLI

Flavor
Uniformity of Size
Texture
General Appearance
Absence of Defects
Firmness
Color
General Character

BRUSSELS SPROUTS

Flavor
Texture
Size
General Appearance
Absence of Defects
Firmness
Color
General Character
Tenderness
Count
Maturity

CAULIFLOWER

Flavor
Absence of Defects
Texture
General Appearance
General Character
Firmness
Color
Maturity
Tenderness
Style

CHERRIES (RED SOUR PITTED)

Flavor
Absence of Defects
Texture
General Appearance
General Character
Firmness
Color
Maturity
Tenderness
Style

CHERRIES (SWEET)

Flavor
Count
General Character
General Appearance
Size
Maturity
Color
Uniformity
Texture
Type
Symmetry
Firmness
Style
Absence of Defects
Sugar or Syrup Ratio

CORN (WHOLE KERNEL)

Flavor
Type
General Appearance
Absence of Defects
Color
General Character

PEACHES

Flavor
Size
General Appearance
Symmetry
Color
Count
Type
Absence of Defects
Style
General Character

PEAS

Flavor
Sieve Size
General Appearance
Absence of Defects
Color
General Character
Type

RASPBERRIES

Flavor
Symmetry
General Appearance
Absence of Defects
Color
General
Character Type
Maturity
Size

RHUBARB

Flavor
Absence of Defects
General Appearance
General Character
Color
Maturity
Type

SPINACH

Flavor
Absence of Defects
General Appearance
General Character
Color

STRAWBERRIES

Flavor
Symmetry
General Appearance
Absence of Defects
Color
Miscellaneous Canned Foods
General Character
Style
Maturity
Size
Firmness

CHILI SAUCE

Flavor
Color
General Character
General Appearance
Absence of Defects
Consistency

FRUIT BUTTERS

Flavor
Type
Finish
Color

Consistency
General Appearance
Absence of Defects
General Character
Percent Soluble Solids

JAMS

Flavor
Type
Consistency
General Appearance
Absence of Defects
Finish
Color
General Character
Percent Soluble Solids

JELLIES

Flavor
Type
Consistency
General Appearance
Absence of Defects
Finish
General Character
Color

MARMALADE

Flavor
Type
Consistency
General Appearance
Absence of Defects
Finish
Color
General Character

MUSHROOMS

Flavor
Absence of Defects
General Appearance
Count
Color
Size
Style
Uniformity of Size

OLIVES (GREEN)

Flavor
Absence of Defects
General Appearance
Count
Color
Size
Style
Uniformity of Size
Type
Symmetry

OLIVES (RIPE)

Flavor
Count
General Appearance
Size
Color
Uniformity of Size
Type
Symmetry
Absence of Defects
General Character

PEANUT BUTTER

Flavor
Color
General Appearance
Type

PICKLES

Flavor
Absence of Defects
General Appearance
Count
Color
Size
Type
Uniformity of Size
Style
Symmetry

PIMENTOS

Flavor
Absence of Defects
General Appearance
Uniformity of Size
Color
Wholeness
Type

PRESERVES

Flavor
Type
General Appearance
Absence of Defects
Color
General Character

SALMON

Flavor
Color
General Appearance
Absence of Defects
General Character
Tenderness
Texture
Symmetry

SHRIMP

Flavor
Absence of Defects
General Character
General Appearance
Count
Texture
Color
Size
Tenderness
Uniformity of Size

TUNA FISH

Flavor
Color
General Character
General Appearance
Absence of Defects
Texture
Tenderness

USDA FOOD STANDARDS DIVISIONS

The U.S. Department of Agriculture (USDA) regulates food standards under the following divisions:

Dairy Products Division
Agricultural Marketing Service
U.S. Department of Agriculture
Washington, DC 20250

Fresh Products Standardization and Inspection Branch
Fruit and Vegetable Division
Agricultural Marketing Service
U.S. Department of Agriculture
Washington, DC 20250

Processed Products Standardization and Inspection Branch
Fruit and Vegetable Division
Agricultural Marketing Service
U.S. Department of Agriculture
Washington, DC 20250

Grain Division (includes soybeans, beans, peas, and rice)
Agricultural Marketing Service
U.S. Department of Agriculture
Washington, D.C. 20250

Livestock, Meat (and Wool) Division
Agricultural Marketing Service
U.S. Department of Agriculture
Washington, DC 20250

Poultry and Poultry Products Division (includes rabbits)
Agricultural Marketing Service
U.S. Department of Agriculture
Washington, DC 20250

Requests for Military Specifications used by the U.S. Department of Defense should be addressed to:

Commanding Officer Naval Supply Depot Attention: Code D.C.I.
5801 Tabor Avenue
Philadelphia, PA 19120

Information on food standards of the Food and Drug Administration is divided into the following parts:

21 CFR Part 14—Cacao Products
21 CFR Part 15—Cereal Flours and Related Products
21 CFR Part 16—Alimentary Pastes
21 CFR Part 17—Bakery Products
21 CFR Part 18—Milk and Cream
21 CFR Part 19—Cheeses, Processed Cheeses, Cheese Foods, and Related Foods
21 CFR Part 20—Frozen Desserts
21 CFR Part 25—Dressings for Food
21 CFR Part 27—Canned Fruits and Canned Fruit Juices
21 CFR Part 29—Fruit Butters, Fruit Jellies, Fruit Preserves, and Related Products
21 CFR Part 36—Shellfish
21 CFR Part 37—Fish
21 CFR Part 42—Eggs and Egg Products
21 CFR Part 45—Oleomargarine, Margarine
21 CFR Parts 51 and 53—Canned Vegetables
21 CFR Part 281—Tea

Information may be obtained by writing:

Food and Drug Administration
5600 Fishers Lane
Rockville, MD 20852

The Washington, DC, FDA headquarters is assisted by the following Field District Offices:

NORTHEAST REGION

New York Office
158-15 Liberty Avenue
Jamaica, NY 11433

New England Regional Office
One Montvale Avenue
Stoneham, MA 02180

CENTRAL REGION

Philadelphia Regional Office
900 U.S. Customhouse
200 Chestnut Street
Philadelphia, PA 19106

Chicago Regional Office
20 N. Michigan Avenue Room 510
Chicago, IL 60602

SOUTHEAST REGION

Atlanta Regional Office
60 Eighth Street NE
Atlanta, GA 30309

SOUTHWEST REGION

Dallas Regional Office
4040 North Central Expressway, Suite 900
Dallas, TX 75204

Kansas City Regional Office
11630 West 80th Street
Lenexa, KS 66214-3338

PACIFIC REGION

San Francisco Regional Office
1301 Clay Street
Oakland, CA 94512-5217

Seattle Regional Office
22201 23rd Drive SE
Bothell, WA 98021-4421

APPENDIX

B

Food Purchasing Guide

This appendix contains information useful in estimating the number of purchase units of foods to buy to serve a specific number of portions. For each purchase unit specified, these data need to be determined:

Weight of the unit of purchase
Yield from weight "as purchased" to weight "as served"
Size and description of portion
Number of portions from each purchase unit
The approximate number of purchase units needed for 25 and 100 portions

EXPLANATION OF INDUSTRY TERMS

Purchasing Definitions

Food as purchased. Foods are described in the forms as purchased—-fresh, frozen, canned, dried. Further descriptive information that would affect the yield is also given, such as: for meats—bone in, bone out; for carrots—with tops, without tops; and for potatoes—to be pared, ready to cook.

Unit of purchase. Sizes of cans, packages, or other containers and weights of units in common use in the wholesale and retail markets are given. Usually, data for the 1 lb unit are given; from these the yield of a purchase unit of any weight can be determined.

Weight per unit. Weights given for purchase units refer to weights as purchased on the market. Legal weights for the contents of such units as bushels, lugs, crates, and boxes may vary in different states. The lowest of these weights is used to ensure that the specified portions can be obtained. Weights for canned goods are the same as those given as net weights on the labels.

Yield as served. "Yield as served" refers to the weight of food "as served" as a percentage of weight "as purchased." Absence of information in this column means a weight yield was not used to determine the number of portions per purchase unit. The same item "as purchased" may have more than one yield, depending on the way it is served. For example, 1 lb of fresh carrots without tops will yield 0.75 lb cooked and 0.82 lb grated raw. The yield does not always refer to a serving that is all edible. As pork chops are usually served with bone, the yield given is the percentage of the "as purchased" weight represented by the cooked chop with the bone in. However, the yield of a chuck roast, usually served without bone, is the percentage of the chuck "as purchased"—that is, cooked and served without the bone.

The amount of ready-to-serve food obtained from a given amount of food "as purchased" may vary widely, depending on the size, grade, and general condition of the food, discards in preparation, and the method and time of cooking. For the yields given, it is assumed that the food used is in good condition (free of rot, insect infestation, bruising), that only usual amounts are discarded in preparation, and that usual cooking methods and time of cooking for the food specified are used.

The yields of meats can be found in the Agriculture Handbook No. 102, *Food Yields Summarized by Different Stages of Preparation*, and unpublished data from the Agricultural Research Service of the U.S. Department of Agriculture and the Fish and Wildlife Service of the U.S. Department of the Interior. Yields for commercially prepared meat combinations are based on minimum meat requirements for meat food products packed for interstate shipment under federal inspection. The contents of cooked poultry meat in poultry products are based on regulations governing the inspection of poultry and poultry products.

Size and description of portion. This portion, for most foods, refers to the amount served. For foods often used in combination with other foods, such as canned milk, nonfat dry milk, eggs, nuts, flour, and uncooked cereals, a portion refers to a measure commonly called for in institution recipes. For meat, poultry, and fish, data for two sizes of portions are given. The larger portion may be appropriate for an adult serving; the smaller may be sufficient for a child's meal or a luncheon menu. The sizes of meat portions are given:

1. In ounces of cooked meat as served, including fat and, in some cases, bone, and
2. In ounces of cooked lean only.

For vegetables and fruits, the portion size is generally the weight of ½ cup rounded to the nearest ounce. Canned, cooked fresh, and cooked dried fruits are served in the liquid, and heated canned vegetables and cooked fresh and frozen vegetables are served drained. It is assumed that solid-pack fruits are served without liquid. Institutions that use different portion sizes than those given may adjust the amount to purchase.

Portions per purchase unit. In obtaining the portions per purchase unit or the yield of a purchase unit in portions, average quantities of refuse and usual weight losses in cooking are assumed. The number of portions may vary from that shown if the condition of the food purchased is poor or unusual waste occurs in preparation and cooking. To obtain the number of units to purchase, the number of portions per purchase unit can be divided into the number of persons to be served. This may be the most practical method of computing the amount to purchase if an institution does not serve 25 or 100 persons or easy multiples of these numbers of persons. For example, if 1 lb of meat will serve 2.57 portions of the size desired, the food manager of an institution serving 580 persons may divide 580 by 2.57 to get the amount of meat needed—226 lb. It is because of such use of these figures, and not because the figures represent this degree of accuracy, that they have been carried to the nearest one-hundredth of a portion. Had the portions per pound of meat been rounded to the nearest number of portions per pound (3), the food manager would purchase only 193 lb of meat, which is 33 lb short of the best estimate of his or her needs.

Approximate purchase units for 25 and 100 portions. See chart. These columns represent 25 and 100 divided by the number of portions per purchase unit in the preceding column. The resulting number of purchase units is always carried to the next even ¼ unit. Thus, the number of purchase units specified is always sufficient if the yield and portion size are as shown in the table. If the purchase units serve many portions, such as a bushel of apples or a No. 10 can of vegetables, this ¼ purchase unit may be very significant. If the purchase unit serves few portions, as in the case of a pound of meat, the ¼ purchase unit is relatively unimportant.

Table B.1 Quantities of Meat to Purchase for 25 and 100 Portions

Meat as purchased	Unit of purchase	Yield, cooked: As served	Yield, cooked: Lean only	Description of portion as served	Size of portion: As served	Size of portion: Lean only	Portions per purchase unit	Approximate purchase units for— 25 portions	Approximate purchase units for— 100 portions
Beef, Fresh or Frozen									
	Pound	*Percent*	*Percent*		*Ounces*	*Ounces*	*Number*	*Number*	*Number*
Brisket:									
Corned, bone out	″	60	41	Simmered	4	2.8	2.40	10½	41¾
					3	2.1	3.20	8	31¼
Fresh:									
Bone in	″	52	36	Simmered, bone out	4	2.8	2.08	12¼	48¼
					3	2.1	2.77	9¼	36¼
Bone out	″	67	46	Simmered	4	2.8	2.68	9½	37½
					3	2.1	3.57	7¼	28¼
Ground Beef:									
Lean	″	75	75	Broiled	3	3.0	4.00	6¼	25
					2	2.0	6.00	4¼	16¾
Regular	″	72	72	Pan-fried	3	3.0	3.84	6¾	26¼
					2	2.0	5.76	4½	17½
Heart	″	39	39		2	2.0	3.12	8¼	32¼
Kidney	″	39	39		2	2.0	3.12	8¼	32¼
Liver	″	69	69	Braised	3	3.0	3.68	7	27¼
					2	2.0	5.52	4¾	18¼
Oxtails	″	29	29						
Roasts:									
Chuck:									
Bone in	″	52	42	Roasted, moist heat, bone out	4	3.2	2.08	12¼	48¼
					3	2.4	2.77	9¼	36¼
Bone out	″	67	54	Roasted, moist heat	4	3.2	2.68	9½	37½
					3	2.4	3.57	7¼	28¼
7-Rib (shortribs removed):									
Bone in	″	65	42	Roasted, dry heat, bone out	4	2.6	2.60	9¾	38½
					3	1.9	3.47	7¼	29
Bone out	″	73	47	Roasted, dry heat	4	2.6	2.92	8¾	34¼
					3	1.9	3.89	6½	25¾
Round:									
Bone in	″	69	56	Roasted, dry heat, medium, bone out	4	3.3	2.76	9¼	36¼
					3	2.5	3.68	7	27¼
Bone out	″	73	60	Roasted, dry heat, medium	4	3.3	2.92	8¾	34¼
					3	2.5	3.89	6½	25¾

(continues)

Table B.1 Quantities of Meat to Purchase for 25 and 100 Portions

Meat as purchased	Unit of purchase	Yield, cooked: As served	Yield, cooked: Lean only	Description of portion as served	Size of portion: As served	Size of portion: Lean only	Portions per purchase unit	Approximate purchase units for— 25 portions	Approximate purchase units for— 100 portions
Rump:									
Bone in	″	58	43	Roasted, dry heat, bone out	4	3.0	2.32	11	43¼
					3	2.2	3.09	8¼	32½
Bone out	″	73	55	Roasted, dry heat	4	3.0	2.92	8¾	34¼
					3	2.2	3.89	6½	25¾
Shortribs	″	67	32	Braised, bone in	6	2.9	1.79	14	56
					4	1.9	2.68	9½	37½
Steaks:									
Club:									
Bone in	Pound	73	33	Broiled, bone in	6	2.7	1.95	13	51½
					4	1.8	2.92	8¾	34¼
Bone out	″	73	42	Broiled	4	2.3	2.92	8¾	34¼
					3	1.7	3.89	6½	25¾
Flank	″	67	67	Braised	3	3.0	3.57	7¼	28¼
					2	2.0	5.36	4¾	18¾
Hip:									
Bone in	″	73	32	Broiled, bone in	6	2.6	1.95	13	51½
					4	1.8	2.92	8¾	34¼
Bone out	″	73	40	Broiled	4	2.2	2.92	8¾	34¼
					3	1.6	3.89	6½	25¾
Minute, cubed	″	75	75	Pan-fried	3	3.0	4.00	6¼	25
					2	2.0	6.00	4¼	16¾
Porterhouse:									
Bone in	″	73	36	Broiled, bone in	8	4.0	1.46	17¼	68½
					6	3.0	1.95	13	51½
					4	2.0	2.92	8¾	34¼
Bone out	″	73	42	Broiled	4	2.3	2.92	8¾	34¼
					3	1.7	3.89	6½	25¾
Round:									
Bone in	″	73	56	Broiled, bone in	4	3.1	2.92	8¾	34¼
					3	2.3	3.89	6½	25¾
Bone out	″	73	60	Broiled	4	3.3	2.92	8¾	34¼
					3	2.5	3.89	6½	25¾
Sirloin (wedge and round):									
Bone in	″	73	44	Broiled, bone in	6	3.6	1.95	13	51½
					4	2.4	2.92	8¾	34¼
Bone out	″	73	48	Broiled	4	2.6	2.92	8¾	34¼

Table B.1 Quantities of Meat to Purchase for 25 and 100 Portions

Meat as purchased	Unit of purchase	Yield, cooked		Description of portion as served	Size of portion		Portions per purchase unit	Approximate purchase units for—	
		As served	Lean only		As served	Lean only		25 portions	100 portions
					3	2.0	3.89	6½	25¾
T-bone:									
Bone in	"	73	34	Broiled, bone in	8	3.8	1.46	17¼	68½
					6	2.8	1.95	13	51½
					4	1.9	2.92	8¾	34¼
Bone out	"	73	41	Broiled	4	2.2	2.92	8¾	34¼
					3	1.7	3.89	6½	25¾
Stew meat (chuck), bone out	"	67	54	Cooked, moist heat	3	2.4	3.57	7¼	28¼
					2	1.6	5.36	4¾	18¾
Tongue:									
Fresh	"	59	59	"	3	3.0	3.15	8	31¾
					2	2.0	4.72	5½	21¼
Smoked	"	51	51	"	3	3.0	2.72	9¼	37
					2	2.0	4.08	6¼	24¾
				Beef, Canned					
Beef, corned	"	100	100	Heated	3	3.0	5.33	4¾	19
					2	2.0	8.00	3¼	12½
	6-pound can	100	100	"	3	3.0	32.00	1	3¼
Beef, chipped	"	125	125	Cooked, moist heat	3	3.0	6.67	3¾	15
					2	2.0	10.00	2½	10
				Lamb, Fresh or Frozen					
Chops:									
Loin	"	76	41	Broiled, bone in	5.0	2.7	2.43	10½	41¼
Rib	"	76	34	"	5.0	2.2	2.43	10½	41¼
Shoulder	"	70	41	"	5.0	2.9	2.24	11¼	44¾
Ground lamb	"	68	68	Broiled patties	3.0	3.0	3.63	7	27¾
					2.0	2.0	5.44	4¾	18½
Roasts:									
Leg:									
Bone in	"	54	45	Roasted, bone out	4.0	3.3	2.16	11¾	46½
					3.0	2.5	2.88	8¾	34¾
Bone out	"	70	58	Roasted	4.0	3.3	2.80	9	35¾

(continues)

Table B.1 Quantities of Meat to Purchase for 25 and 100 Portions

Meat as purchased	Unit of purchase	Yield, cooked: As served	Yield, cooked: Lean only	Description of portion as served	Size of portion: As served	Size of portion: Lean only	Portions per purchase unit	Approximate purchase units for— 25 portions	Approximate purchase units for— 100 portions
					3.0	2.5	3.73	6¾	27
Shoulder:									
Bone in	"	55	41	Roasted, bone out	4.0	3.0	2.20	11½	45½
					3.0	2.2	2.93	8¾	34¼
Bone out	"	70	52	Roasted	4.0	3.0	2.80	9	35¾
					3.0	2.2	3.73	6¾	27
Stew meat,[c] bone out	"	66		Simmered	3.0		3.52	7¼	28½
					2.0		5.28	4¾	19
Pork, Cured (Mild)									
Bacon (24 slices per pound)	"	32		Fried or broiled	2 slices		12.00	2¼	8½
Canadian bacon	"	63	63	Broiled, sliced	2	2.0	5.04	5	20
					1	1.0	10.08	2½	10
Ham:									
Bone in	"	67	54	Roasted, slices and pieces	4	2.9	2.68	9½	37½
					3	2.2	3.57	7¼	28¼
	"	56	44	Roasted, slices	4	2.4	2.24	11¼	44¾
					3	1.8	2.99	8½	33½
Bone out	"	77	72	Roasted, slices and pieces	4	2.9	3.08	8¼	32½
					3	2.2	4.11	6¼	24½
	"	64	60	Roasted, slices	4	2.4	2.56	10	39¼
					3	1.8	3.41	7½	29½
Ground	"	77	77	Patties	3	3.0	4.11	6¼	24½
					2	2.0	6.16	4¼	16¼
Shoulder, Boston butt:									
Bone in	"	67	52	Roasted, bone out	4	3.1	2.68	9½	37½
					3	2.3	3.57	7¼	28¼
Bone out	"	74	58	Roasted	4	3.1	2.96	8½	34
					3	2.3	3.95	6½	25½
Shoulder, picnic:									
Bone in	"	56	41	Roasted, bone out	4	3.9	2.24	11¼	44¾
					3	2.2	2.99	8½	33½
Bone out	"	74	53	Roasted	4	3.9	2.96	8½	34
					3	2.2	3.95	6½	25½

APPENDIX

C

Purchase Specifications

Table C.1 Purchase Specifications for Baby and Junior Foods

Product	Pack	USDA Grade Min. Score	Min. Net Min. Drained Wt.	Size of Count	Remarks
APPLESAUCE	24/4½ oz.				Baby
APRICOT Puree	12/1#				
BANANAS	24/4½ oz.				Baby
BARLEY Cereal	12/8 oz.				Baby
BEEF	24/4½ oz.				Kosher Baby
BEEF NOODLE Dinner	24/7½ oz.				Junior Foods
CARROTS	24/4½ oz.				Baby
CHICKEN	24/4½ oz.				Kosher Baby
COTTAGE CHEESE and Pineapple	24/7½ oz.				Junior Foods
HIGH PROTEIN	12/8 oz.				Baby
LAMB	24/4½ oz.				Kosher Baby
OATMEAL	12/8 oz.				Baby
PEACHES	24/4½ oz.				Baby
PEACH PUREE	12/1#				
PEARS	24/4½ oz.				Baby
PEAR PUREE	12/1#				
PEAS	24/4½ oz.				Baby
PRUNES	24/4½ oz.				Baby
RICE CEREAL	12/8 oz.				Baby
SPINACH	24/4½ oz.				Baby
STRING BEANS	24/4½ oz.				Baby
TURKEY & RICE Dinner	24/7½ oz.				Junior Foods
VEAL	24/4½ oz.				Kosher Baby

Table C.1 Purchase Specifications for Beef—Wholesale & Fabricated Cuts

Product	Weight Range	Requirements
BRISKET, BONELESS, CURED	9–12#	NO. 1, DECKEL OFF, CLOSELY TRIMMED, BRIGHT APPEARANCE, NO INDICATION OF RUST OR OTHER DISCOLORATION. (SEE USDA IMPS. SERIES 600.)
BRISKET, BONELESS, FRESH	10–12#	CLOSELY TRIMMED: DECKEL OFF.
CHUCK, SQUARE CUT, BONE-IN	93–106#	TO BE CUT BETWEEN 5TH AND 6TH RIBS. FORESHANK, BRISKET, SHORT PLATE, AND RIB OFF. ½″ MAXIMUM COVERING SUET.
CHUCK, ARMBONE	103–118#	TO BE CUT BETWEEN 5TH AND 6TH RIBS. FORESHANK, BRISKET, SHORT PLATE, AND RIB OFF. ½″ MAXIMUM COVERING SUET.
CROSSCUT CHUCK	120–138#	TO INCLUDE 5 RIBS, SQUARE CUT CHUCK, FORESHANK AND BRISKET. ½″ MAXIMUM COVERING SUET.
FOREQUARTER	183–210#	TO INCLUDE 12 RIBS, CUT INTO MATCHED WHOLESALE MARKET CUTS. ½″ MAXIMUM COVERING SUET.
GROUND BEEF, REGULAR	10# CONTAINER	PACKED COMPACTLY, NOT MORE THAN 25% FAT. SHALL BE WELL MIXED, SO THAT FAT AND LEAN MEAT ARE EVENLY DISTRIBUTED.
HINDQUARTER	167–190#	TO INCLUDE 1 RIB, CUT INTO MATCHED WHOLESALE MARKET CUTS. ½″ MAXIMUM COVERING SUET.
INSIDES, DRIED, SLICED	10# CONTAINERS	BRIGHT APPEARANCE, SLICES TO BE A MINIMUM OF 3″ LONG AND 2″ WIDE.
LIVER, WHOLE	8–10#	SELECTION NO. 1, UNBLEMISHED, FRESH CHILLED OR FROZEN; LIGHT BROWN COLOR.
OXTAILS	1½–2#	STEER, NO. 1, COLOR OF FAT: WHITE TO SLIGHTLY YELLOW.
PASTRAMI, PLATE; RED	3½–4#	NO CORNER PIECES—ALL BONES AND FAT TO BE REMOVED. MILDLY CURED WITH SPICES ADDED. SMOKED 6–8 HOURS.
RIB, FULL	33–38#	SEVEN RIBS, 8″ CHUCK END, 10″ RIB END. ½″ MAXIMUM COVERING SUET.
RIB, BONELESS, TIED, ROAST-READY	12–16#	SAME AS ABOVE, WITH RIBS, FEATHER BONES AND INTERCOSTAL MEAT REMOVED. STRING-TIED GIRTHWISE EVERY 2″, ALSO TIED LENGTHWISE. ½″ MAXIMUM COVERING SUET.
ROUND, BONE-IN, FULL CUT	83–95#	RUMP AND SHANK ON; FLANK OFF; CUT CHICAGO STYLE.
ROUND, INSIDE, BONELESS	18–20#	INSIDE PORTION OF ABOVE, WITH ALL BONES REMOVED.
ROUND, OUTSIDE, BONELESS	11–13#	OUTSIDE PORTION OF BONE-IN ROUND, WITH ALL BONES REMOVED.
SHOULDER, CLOD, ROAST-READY	16–18#	MUST BE ONE PIECE WITHOUT UNDUE SCORING. NOT LESS THAN 1″ THICK AT ANY ONE POINT. STRING-TIED GIRTHWISE.
SHORTLOIN, TRIMMED	26–30#	10″ RIB END, DIAMOND-BONE CUT.
SHORTRIBS, TRIMMED	4–5#	BONE-IN, TO INCLUDE 4 RIBS, CLOSELY TRIMMED, APPROXIMATELY 3″ WIDE. DECKEL AND ALL COVERING FAT REMOVED.
SIDE OF BEEF	350–400#	CUT INTO 5 MATCHED PRIMAL CUTS. LOIN TO BE WHOLE.
STEWING BEEF (CUBES)	AMOUNT AS SPECIFIED	UNIFORM CUBES, APPROXIMATELY 1″. SURFACE FAT NOT TO EXCEED ¼″ AT ANY ONE POINT. MAXIMUM FAT CONTENT: 20%.
STRIP-LOIN, BONE-IN	16–19#	10″ WIDE.
STRIP-LOIN, BONELESS	12–14#	10″ WIDE.
TONGUE, FRESH	4–5#	SHORT CUT; SELECTION NO. 1. FAT COVERING NOT TO EXCEED ½″ AT ANY ONE POINT.
TONGUE, SMOKED	3–5#	SHORT CUT, NO. 1. (SEE USDA IMPS. SERIES 600.)
TONGUE, CURED	3–5#	DRY-PACK, NO. 1, SHORT-CUT. (SEE USDA IMPS. SERIES 600.)
TRIPE, HONEYCOMB		FRESH FROZEN OR FRESH, NO. 1, SCALDED AND THOROUGHLY SCRAPED; RESILIENT.
TRIPE, PLAIN		FRESH FROZEN OR FRESH, NO. 1, SCALDED AND THOROUGHLY SCRAPED; RESILIENT.
TENDERLOIN, FULL	6–7#	NOT MORE THAN ½″ COVERING SUET.
TENDERLOIN, BUTT	4# UP	NOT MORE THAN ½″ COVERING SUET.

Index

Grades for Fish and Shellfish
A—Highest quality (devoid of blemishes)

B—Good quality(more blemishes)

C—Good quality but less attractive appearance (lacking in appearance, suitable for soups, casseroles)
PUFI Seal
"**P**acked **U**nder **F**ederal **I**nspection"
Ensures that fish items are produced under continuous government inspection

Traditional Weight Ranges for Lobsters

Chickens	approximately 1 pound
Quarters	1 to 1 1/4 pounds
Selects	1 1/2 to 2 1/2 pounds
Jumbos	2 1/2 to 5 pounds
Monsters	over 5 pounds

Lobster
Sized by Count means
Lobster tails 10/12 = 10 lb lot of to -12 tails
Product Yield
State on specification the expected yield:
Example: No more than 2% dead oysters for each bushel purchased

Example: no more than 2% broken fish sticks per 20lb case
Packing
Crushed ice
Self draining
Packed in seaweed
Live in shell- moisture proof
IQF = Individually Quick Flash frozen
Slab packed, layered, ice packed, cello packed, in marinade, MAP (3 wks)
SURIMI
Fish based paste used to shape and form products

Sealegs are flaked and formed fish juice and scraps with red dye added
Point of Origin
Maine- lobsters
Alaskan salmon
Chilean Sea bass

Remember truth in Menu laws…….. (mother in-law shrimp)
Variables
Check market reports Weather conditions
Seasons

Breading
Lightly breaded = 65% shrimp (35%breading)

Breaded = 50% breading and 50% shrimp

Market Forms of Fish
Whole or round = intact as caught
Drawn = viscera removed
Dressed = viscera, scales, head, tail and fins removed
Steaks= cross-section slices containing backbone
Fillet= boneless sides of fish, skin on or off
Butterfly fillets with bones removed
Sticks= cross section of fillets
Preservation Methods for Fish Products
Frozen
Dried
Smoked
Refrigerated
Ice packed

Cello packed
Chill packed
Live

Live-in-shell
Canned
Questions to Consider Before Taking Advantage of Seafood "Bargains"
Can employees handle the item if it is new to them?
Does the operation have the proper equipment to prepare and serve the item?
If the item is currently on the menu, should the menu price be dropped?
Should the operation use the bargain as a loss leader on the menu?
SHRIMP
Head on
Headless
PUDS
Shell on
Deveined

Stock tag
90 days requirement for <u>bulk raw shellfish</u>

<u>Red tide=</u> *mites in water; bacteria which is poisonous* oysters

Oysters; the destruction of Vibrio bacteria in raw oysters without destroying the raw feel and taste of the oyster.

Fresh (raw) oysters can be made safe from Vibrio bacteria by Fresher Under Pressure· A pressure of 200 to 300 MPa for 5 to 15 minutes at 25C inactivated: CDC
Reported that foodborne illness caused by
Fish and shellfish was slightly greater than
The combined cases from beef and poultry!

Parasitic worms , round or nematode worms, flatworm, tapeworm are a few of these parasites tha can be hosted by mankind. Most common are scombroid and ciguartera
Number of cases per billion pounds consumed
Trust your supplier

Train your staff to clean thoroughly and cook fish and shellfish properly

Chapter 12
Main Topics

Quality Loss and Spoilage
Receiving Control
Dry Storage Control

Why do we store food?
Quality Loss and Storage
Chemical Changes

Spoilage Organisms

Chemical Changes
Loss of or change in color
Loss of firmness, texture
Loss of nutrients
Loss of flavor

Spoilage Organisms
Molds
Are they all bad?

What do they need to live?
Yeast
When is this okay?
What does Yeast produce?

Bacteria
Signs
Smells
Grows best on/in…..
Dry Storage
Controlled Temp and Humidity
Venting

Special storage
Storage rules??????
Receiving Control
Personnel
Equipment
Returns
Discrepancies
Storage

Types of Receiving
Invoice
Standing Order
Blind order

Odd Hours Receiving
Drop Shipment Receiving
Mailed Deliveries
COD Deliveries
Receiving Practices
Good?
Bad?
Are you kidding me?

Poultry Chapter
Chickens, Turkeys, Ducks, Geese, Guineas, Pigeons, Rabbits, Swans, Quail, Wild ducks, Pheasants
Poultry Facts
Today; we can produce a heavier chicken in about half the time it once took
Consumption of poultry is steadily rising
A hen lays an average of 260 eggs a year

Nearly 1/3 of the meat we eat today is poultry
Poultry Regulation
Must be inspected for **wholesomeness** before and after slaughter

Approved poultry will bear the "Inspected and Passed" round circle stamp with a number identifying the establishment where the product was processed.

Grading however is voluntary

Specification Guidelines
Kind (correlates to Item Name)
Class (either age or sex)
Quality (brand name) or Grade
Style (whole, halves, qtrd, or in parts, RTC)
Size or weight
Type (fresh-killed, fresh-frozen, frozen storage…)
Container or Packaging (count per container)
Temperature required
Kinds
Chickens;
Rock Cornish or Cornish Game Hen
Broiler or Fryer
Roaster-Chicken
Capon
Hen, Fowl, or Baking or Stewing Chicken
Cock or Rooster
Turkeys:
Mature Turkey or Old Turkey (HEN or TOM)
Ducks;
Broiler duckling or Fryer duckling
Roaster Duckling
Mature Duck or Old Duck

(Continued next slide)
Kinds (continued)
Geese:
Young Goose
Mature Goose or Old Goose
Guineas:
Young guinea
Mature or Old Guine
Pigeons:
Squab
Pigeon
KIND and Class : Chickens
Rock Cornish or Game Hen:
Young immature chicken 5-6 weeks old
Weighs no more than 2lb RTC weight
Rock Cornish Fryer, Roaster or Hen:
Weighs over 2lbs
Kind and Class : Chickens
Broiler or Fryer:

Kind and Class : Chickens
Capon:
Surgically Unsexed male chicken under 8 months of age

Hen, Fowl or Baking or Stewing Chicken:
Mature female chicken older than 10 months
Meat is less tender than that of a roaster
Kind and Class : Chickens
Cock or Rooster:
Male chicken with coarse skin, toughened and darkened meat and hardened breastbone

Poultry Grades
Grade A
Grade B
Grade C
Best
Least marketable
Grading (continued)
Conformation
Fleshing
Fat coverage
Freedom from pinfeathers
Vestigial feathers (hair and down)
Exposed flesh
Disjointed or broken bones
Missing parts
Discoloration and bruises
Freezing defects
STYLE
Style (continued)
TYPE
Specific term used to indicate the time that has elapsed since time of slaughter
and
Whether the bird is fresh, storage, or frozen

Included in type would be the specification that poultry is to be raw, raw-breaded, (maximum breading specified as 30% to total weight), cooked, or cooked-breaded.
TYPE
Conformation ratio of meat to bone
Container or Packaging
Chickens:
12 to 24 per container
Turkeys, Ducks or Geese: Individually wrapped and depending on size, come packed:
2 to the box
4 to the box
6 to the box
Containers/Packaging (continued)
Containers should be plainly marked to show:
Net weight
Kind
Class
Style
Grade
Certificate of inspection number
Containers/Packaging (continued)
Case sizes include:
25, 30 and 50 lb of fresh killed poultry, RTC or parts unwrapped
Shipment Temperature
Frozen 0 F
Fresh 36 F
NO POULTRY SHOULD BE ACCEPTED IF ABOVE 40 F
Convenience Poultry products
60/40 pulled chicken
Diced
Tenders
Buffalo wings
Breaded
Nuggets (% fill)
Bnls, sknls, breast (24 cs)

Turkey:
Deli breast meat
Hotel style -breast only

What Kind of chicken would you recommend to purchase for chicken a la king?
Why?
What Kind of Turkey would you recommend to purchase for Turkey pot pie?

Kind and Class : TURKEY
Fryer-Roaster Turkey:
Young immature turkey,
Under 16 weeks of age
Either sex
Tender meated
Flexible breastbone cartilage
Young Turkey:
Under 8 months
Tender meated
Either sex

Kind and Class : TURKEY
Yearling Turkey:
Fully matured turkey
Under 15 months of age
Reasonably tender-meated
Sex is optional
Mature Turkey or Old Turkey (HEN or TOM):
Old turkey usually in **excess of 15 months**
either sex
Coarse skin, toughened flesh
Hen is female
Tom is male
Kind and Class : Duck
(Duck is 100% dark meat)
Broiler Duckling or Fryer Duckling:
Young duck under 8 weeks of age
Either sex
Tender - meated
Roaster Duckling:
Young duck under 16 weeks of age
Either sex
Tender-meated
Kind and Class : Duck
Mature Duck or Old Duck:
Over 6 months of age
Either sex
Toughened flesh

Kind and Class: Geese
Young Goose:
Either sex
Tender – meated

Mature Goose or Old Goose:
Either sex
Toughened flesh
Kind and Class: Guineas
Young Guineas:
Either sex
Tender-meated
Mature Guinea or Old Guinea:
Either sex
Toughened flesh
Kind and Class: Pigeons
Squab:
Young, immature pigeon
Either sex
Extra tender-meated
Pigeon:
Mature pigeon
Either sex
Coarse skin and toughened flesh
Kinds: **Game Birds**

Peafowl

Peafowl are originally from India. They are of the order Galliforms and related to pheasants. In their native land they eat snakes, plant seeds, fruits, worms and insects. Peafowl are often allowed to roam at large and can become pests. A gathering of peafowl is called a muster.
Peafowl come in several colors, ours are India Blues. The male, called a Peacock, is a beautiful iridescent blue on his neck and head. The females, referred to as peahens, are a greenish color. The tail of a peacock is more correctly called a train, the train starts in the back with 100 to 150 coverts, the fan like feathers. The true tail is underneath with about 20 retrices or true tail feathers. The train color will change when struck by light at different angles causing an array of colors to sparkle, blues, violets, greens, and oranges. The peacock will display his tail vibrating the coverts and shaking his wings to attract the peahen.
The Peahen will nest on the ground laying a clutch of 3 to 12 eggs. The eggs, a little larger then ducks eggs, will be incubated for 28 days, before hatching out peachicks. Peachicks are ready to fly within one week, so they can start roosting with their mothers at night. Peafowl rarely fly preferring to run under cover when startled, but at twilight they will fly to roost in a tree.
Peafowl do not like to be alone, it is usual to keep a trio of two peahens and one peacock. Peacocks will lose their train after breeding season about august, and a new one will begin to grow that winter. Each year the train will be larger until the bird reaches an overall length of approximately 8 feet.

Common Snipe
26 - 29 cm long with a 44 - 50 cm wingspread. Sexes similar, but females are heavier with a longer bill.

Bill is straight, averaging 6 - 7 cm long.

Stocky body with short legs.

(continued)
A 5 lb *roasting chicken* yields about **30%** AS meat, not including giblets or picked neck meat.
Figure Yield
The chef needs chicken for a barbeque serving to HS seniors. What do you recommend to order? Figure the qty to order.
Recommend a purchase for:
Children's lunch menu
Barbequed chicken
Sliced roast chicken dinner
Pot pie
Chicken soup

Size or Weight and Portions obtained
Geese, Ducks and Turkeys:
Specify weight
Chicken, squab, and other small birds:
Specify individual or group weight such as a dozen with an individual allowed weight variation
Quail:4.5 lb per dozen with no single bird varying more than a ½ oz over 6 oz
Important Facts
Turkeys over 18 lb yield more meat per pound than those under 18 lb, due to the fact that the skeleton size for a 30 lb turkey is about the same size as one for an 18 lb bird.
Normally, a roasted turkey from the RTC to AS state yields around 50% AS meat.
Figure yield
Banquet needs to serve 350 (6 ozportions)
of Roast Turkey. You want to buy 30 lb average turkeys. How many should you purchase?
BREED
Yield and Flavor
Peking duck
Long Island Duck
Homer or Carneaux pigeons for squab meat
Leading Brand Names:
Con-Agra
Tyson
P